FRANKLIN D. ROOSEVELT
AND THE NEW DEAL

FRANKLIN D. ROOSEVELT AND THE NEW DEAL 1932 ★ 1940

By WILLIAM E. LEUCHTENBURG

HARPER PERENNIAL

NEW YORK • LONDON • TORONTO • SYDNEY • NEW DELHI • AUCKLAND

HARPER ● PERENNIAL

A hardcover edition of this book was published in 1963 by Harper & Row, Publishers, Incorporated, in the New American Nation Series, edited by Henry Steele Commager and Richard B. Morris.

FRANKLIN D. ROOSEVELT AND THE NEW DEAL. Copyright © 1963 by William E. Leuchtenburg. All rights reserved. Printed in the United States of America. No part of this book may be used or reproduced in any manner whatsoever without written permission except in the case of brief quotations embodied in critical articles and reviews. For information, address HarperCollins Publishers, 195 Broadway, New York, NY 10007.

HarperCollins books may be purchased for educational, business, or sales promotional use. For information, please e-mail the Special Markets Department at SPsales@harpercollins.com.

FIRST HARPER PAPERBACK PUBLISHED 1963.
FIRST HARPER PERENNIAL EDITION PUBLISHED 2009.

Library of Congress Cataloging-in-Publication Data is available upon request.

ISBN 978-0-06-183696-1

19 20 21 LSC 14 13 12 11 10 9 8

For Jean

Contents

Illustrations

Editors' Introduction

THE crisis and the failure of the twenties set the stage in America for the thirties, set the stage and almost wrote the plot. And once the inherited crisis had been set on the road to solution, another, and even more extraneous, crisis emerged to stamp its character on the latter part of the decade: the crisis of totalitarianism and war abroad. How could the New Deal develop a character of its own, one independent of past and future?

Who can doubt that it did, and who can doubt that the explanation is to be found very largely in the character and personality of Franklin D. Roosevelt? Not since Washington had any President dominated his administration, and his time, as did F.D.R. If it is true that much of his progress, domestic and foreign, was imposed on him by circumstances beyond his control, it is no less true that to an extraordinary degree he imposed his own will and purpose upon that program.

The character of the Republican ascendancy of the twenties had been pervasively negative; the character of the New Deal was overwhelmingly positive. "This nation asks for action, and action now," Roosevelt said in his first inaugural address, and asked for "power to wage war against the emergency." "Let us move forward in strong and active faith" were the last words that he wrote. Apathy, resignation, defeat, despair—these were the foes that Roosevelt routed from the scene; action, advance, confidence, hope—these were the sentiments that he inspired in his followers, the vast majority of the American people.

It is the stuff of good history, this—a leadership that was buoyant and dynamic; a large program designed to enable the government to catch up with a generation of lag and solve the problems that crowded upon it; a people quickened into resolution and self-confidence; a nation brought to realize its responsibilities and its potentialities. How it lends itself to drama! The sun rises on a stricken field; the new leader raises the banner and waves it defiantly at the foe; his followers crowd about him, armies of recruits emerge from the shadows and throng into the ranks; the bands play, the flags wave, the army moves forward, and soon the sound of battle and the shouts of victory are heard in the distance. In perspective we can see that it was not quite like that, but that was the way it seemed at the time.

Not to all Americans, to be sure. Many of Roosevelt's contemporaries reacted to "That Man"—and to the New Deal—the way the Federalists had reacted to Jefferson and the Whigs to Jackson. They saw dictatorship and revolution where the majority of Americans saw leadership and a democratic resurgence. They were, of course, mistaken. The New Deal was in fact pretty much what the phrase implied: not a new game with new rules, but a reshuffle of cards that had too long been stacked against the workingman and the farmer and the small shopkeeper. To be sure, there was no guarantee that these players would now hold the winning cards, but at least they had as good a chance as any of the others. In the perspective of almost a generation it is the conservative character of the New Deal and of its leader that is most impressive. The "Roosevelt Revolution" was in fact the culmination of half a century of historical development. It was deeply rooted in American experience; it relied on familiar instruments of politics and law; even its style was characteristically American.

Even before recovery had been achieved (and some indeed would question whether it would not have been achieved in any event; others, whether it could have been achieved short of war), and while the reform program was still incomplete, the shadow of totalitarianism fell across the land and lengthened with every passing day. Roosevelt, like Wilson, was almost instinctively isolationist; like Wilson he came to understand that isolation was no longer a meaningful concept and that what happened in Italy, Germany, and Japan powerfully affected not only American security but American freedom and democracy. Reluctantly, during his second administration, Roosevelt began to move toward some kind of intervention in the Old World scene and toward preparation of the American people for participation in the still obscure destinies in store for them.

This is the story that Mr. Leuchtenburg tells in what is a remarkable feat of condensation and interpretation. He belongs to the generation that came to maturity after the New Deal, and he is able to look at it and analyze it dispassionately, even critically; he writes at a time when most of the important evidence is in and available—the great masses of documents, the formidable collections of manuscripts, the mounting collections of journals and diaries and memoirs. He has drawn on all of these, on the vast arsenal of newspapers and journals, and on personal interviews with key actors in the great drama of that time, to give us a judicious, penetrating, and luminous analysis. He does not allow President Roosevelt to steal the show—not all of the time. He does not accept at face value the interpretations which the New Dealers have written into history; he is too familiar with the historical past to believe that all the enterprises and techniques called "new" were really more than variations on past experience. To the survey of this passionately controversial chapter of our history he brings a mind cool and skeptical, but a pen that can be ardent enough.

This volume is one of the New American Nation Series, a comprehensive, co-operative survey of the history of the area now embraced in the United States from the days of discovery to our own time. Each volume of this series is part of a carefully planned whole, and fitted as well as is possible into other volumes in the series; each is designed to be complete in itself. Other volumes in the series will deal with the coming of World War II, and with constitutional and cultural developments during the thirties. If there is overlapping, it has seemed to the editors that this is less regrettable than omission and that something is to be gained from looking at the same period from independent points of view.

HENRY STEELE COMMAGER
RICHARD BRANDON MORRIS

Preface

THE Great Depression was one of the turning points of American history. Even at the time, men were aware of a dividing line. In ordinary conversation in the thirties, people would begin by saying "Since '29 . . . ," or "In the days before the crash. . . ." What the Lynds have written about "Middletown" applies to all of America in the 1930's: "The great knife of the depression had cut down impartially through the entire population cleaving open lives and hopes of rich as well as poor. The experience has been more nearly universal than any prolonged recent emotional experience in the city's history; it has approached in its elemental shock the primary experiences of birth and death."

In recent years, historians, writing from the perspective of a quarter of a century, have tended to minimize the significance of the changes wrought by the thirties. They have stressed, quite properly, the continuity between the New Deal reforms and those of other periods, and especially the many debts the New Dealers owed the progressives. They have noted, too, that if much was altered in the 1930's, much too remained the same, and the basic values of the nation stayed fairly constant. Yet, in recognizing the strands of continuity, they have too often obscured the extraordinary developments of the decade. The six years from 1933 through 1938 marked a greater upheaval in American institutions than in any similar period in our history, save perhaps for the impact on the South of the Civil War. It is this upheaval—the "Roosevelt Revolution"—that is the subject of this book: the devastation wrought by the depression and the remarkable improvisation of new political and social institutions to cope with it.

In writing this book, I have had the help of a great many people, and I am happy to acknowledge the many kindnesses they have shown me. Every student of the New Deal owes a special debt to the staff of the Franklin D. Roosevelt Library at Hyde Park, expertly directed for many years by Herman Kahn and more recently by Elizabeth Drewry. On numerous visits to the Library, I have had the skillful help of Jerry Deyo, Robert Jacoby, Joseph Marshall, William Nichols, Edgar Nixon, and George Roach. Librarians in every part of the country have aided me in my research; I received especially courteous assistance from Mrs. Irene Bray of the Industrial Relations Library, U.C.L.A.; Marjorie Edwards, Swarthmore College; Howard Gotlieb, Yale University; Gene Gressley, University of Wyoming; Jack Haley, University of Oklahoma; Martin Schmitt, University of Oregon; Louis Starr and the staff of the Columbia Oral History Collection; Ermine Stone, Sarah Lawrence College; and Pearl von Allmen, University of Louisville. The staff of the Center for Advanced Study in the Behavioral Sciences, Stanford, California, where I was a Fellow in 1961–62, helped me in countless ways.

The manuscript was read in its entirety by John Blum of Yale, Robert Burke of the University of Washington, Robert Cross of Columbia, Frank Freidel of Harvard, Richard Hofstadter of Columbia, Barry Karl of Washington University, and Christopher Lasch of Iowa, and I am deeply obligated to them for criticisms and suggestions which markedly improved the book. I have had the encouragement and assistance of the editors, Richard B. Morris and Henry Steele Commager, from the very beginning. Students in my graduate seminars at Columbia in the past decade have, both by their questioning and by their own research in the sources, added greatly to my understanding of this period.

I am indebted to George Biddle, Clifford Hope, Alfred Landon, Mrs. Eugene Meyer, Frances Perkins, Norman Thomas and Lindsay Warren for granting me access to their papers and to Mrs. Albert Levitt for permission to quote from the papers of Jay Pierrepont Moffat. Yale University granted me permission to quote from the papers of Edward House and Henry Stimson, and Columbia University permission to quote from its numerous collections. I am indebted to a number of people for granting me access to their memoirs at the Columbia Oral History Collection. J. Joseph Huthmacher kindly allowed me to examine materials in the Robert Wagner Papers at Georgetown University. My colleagues James Shenton and Robert Dallek lent me microfilms of the Monsignor Ryan manuscripts and the William Dodd–Franklin Roosevelt correspondence. Morton Borden of Montana State University gained me access to the James Murray Papers. Richard Abrams of the University of California secured help for me in footnote checking, a chore that was ably performed by Mrs. Karen Swanson. I am greatly indebted to Thomas Harding, Theressa Gay, Katherine Holbrook, Carol Moodie, and Harry P. Jeffrey, Jr., who worked with me in the final weeks of research and writing. Beulah Hagen and the staff of Harper & Row gave me the benefit of their professional experience. Mrs. Sigmund Diamond assisted me in preparing the index. The manuscript was typed by three proficient typists: Mrs. Helen Aplin, Mrs. Kathryn Harrison, and Mrs. Leonore Russell of Palo Alto.

My wife, Jean McIntire Leuchtenburg, improved the manuscript by her intelligent reading and my spirits by her patience and good humor.

W.E.L.

Dobbs Ferry, N.Y.
December 6, 1962

FRANKLIN D. ROOSEVELT
AND THE NEW DEAL

CHAPTER 1

The Politics of Hard Times

THE Democratic party opened the 1932 campaign confident of victory. The crash of 1929 had made a mockery of Republican claims to being "the party of prosperity." In the three years of Herbert Hoover's Presidency, the bottom had dropped out of the stock market and industrial production had been cut more than half. At the beginning of the summer, *Iron Age* reported that steel plants were operating at a sickening 12 per cent of capacity with "an almost complete lack" of signs of a turn for the better. In three years, industrial construction had slumped from $949 million to an unbelievable $74 million. In no year since the Civil War were so few miles of new railroad track laid.[1]

By 1932, the unemployed numbered upward of thirteen million. Many lived in the primitive conditions of a preindustrial society stricken by famine. In the coal fields of West Virginia and Kentucky, evicted families shivered in tents in midwinter; children went barefoot. In Los Angeles, people whose gas and electricity had been turned off were reduced to cooking over wood fires in back lots. Visiting nurses in New York found children famished; one episode, reported

[1] *Federal Reserve Bulletin*, XVIII (1932), 616; *Iron Age*, CXXX (1932), 31; *Historical Statistics of the United States* (Washington, 1960), p. 379; George Boyd, "Railway Construction at Low Ebb," *Railway Age*, XCIV (1933), 153. See too William Howard Shaw, *Value of Commodity Output since 1869* (New York, 1947), p. 68; Harold Barger and Sam Schnurr, *The Mining Industries, 1899–1939* (New York, 1944), p. 280; Spurgeon Bell, *Productivity, Wages and National Income* (Washington, 1940), pp. 96, 110, 136; Simon Kuznets, *Gross Capital Formation, 1919–1933*, National Bureau of Economic Research, Bulletin 52 (New York, 1934), p. 12.

Lillian Wald, "might have come out of the tales of old Russia." A Philadelphia storekeeper told a reporter of one family he was keeping going on credit: "Eleven children in that house. They've got no shoes, no pants. In the house, no chairs. My God, you go in there, you cry, that's all."[2]

At least a million, perhaps as many as two millions were wandering the country in a fruitless quest for work or adventure or just a sense of movement. They roved the waterfronts of both oceans, rode in cattle cars and gondolas of the Rock Island and the Southern Pacific, slept on benches in Boston Common and Lafayette Square, in Chicago's Grant Park and El Paso's Plaza. From Klamath Falls to Sparks to Yuma, they shared the hobo's quarters in oak thickets strewn with blackened cans along the railroad tracks. On snowy days, as many as two hundred men huddled over fires in the jungle at the north end of the railway yards in Belen, New Mexico. Unlike the traditional hobo, they sought not to evade work but to find it. But it was a dispirited search. They knew they were not headed toward the Big Rock Candy Mountain; they were not, in fact, headed anywhere, only fleeing from where they had been.[3]

On the outskirts of town or in empty lots in the big cities, homeless men threw together makeshift shacks of boxes and scrap metal. St. Louis had the largest "Hooverville," a settlement of more than a thousand souls, but there was scarcely a city that did not harbor at least one. Portland, Oregon, quartered one colony under the Ross Island bridge and a second of more than three hundred men in Sullivan's Gulch. Below Riverside Drive in New York City, an encampment of squatters lined the shore of the Hudson from 72nd Street to

---

[2] James Myers to Olive Van Horn, December 16, 1931, Dorothy Wysor to Edward Costigan, December 28, 1931, Costigan MSS., V.F. 1; Lillian Wald to James J. Walker, January 22, 1932, Wald MSS., Correspondence; Mauritz Hallgren, "Mass Misery in Philadelphia," *The Nation*, CXXXIV (1932), 275–276. The location of manuscripts and abbreviations used are indicated in the Bibliography. In the case of manuscripts located in more than one place, footnote citations of the William Allen White MSS. refer to the collection at the Library of Congress and citations of the Lewis Schwellenbach MSS. to the collection at the University of Washington, Seattle, unless otherwise indicated.

[3] Edward Anderson, *Hungry Men* (New York, 1935); A. Wayne McMillen, "Migrant Boys: Some Data from Salt Lake City," *Social Service Review*, VII (1933), 64–83; Thomas Minehan, *Boy and Girl Tramps of America* (New York, 1934); U.S. Department of Labor, *Twentieth Annual Report of the Chief of the Children's Bureau* (Washington, 1932), pp. 5–9.

110th Street. In Brooklyn's Red Hook section, jobless men bivouacked in the city dump in sheds made of junked Fords and old barrels. Along the banks of the Tennessee in Knoxville, in the mudflats under the Pulaski Skyway in New Jersey, in abandoned coke ovens in Pennsylvania's coal counties, in the huge dumps off Blue Island Avenue in Chicago, the dispossessed took their last stand.[4]

"We are like the drounding man, grabbing at every thing that flotes by, trying to save what little we have," reported a North Carolinian. In Chicago, a crowd of some fifty hungry men fought over a barrel of garbage set outside the back door of a restaurant; in Stockton, California, men scoured the city dump near the San Joaquin River to retrieve half-rotted vegetables. The Commissioner of Charity in Salt Lake City disclosed that scores of people were slowly starving, because neither county nor private relief funds were adequate, and hundreds of children were kept out of school because they had nothing to wear. "We have been eating wild greens," wrote a coal miner from Kentucky's Harlan County. "Such as Polk salad. Violet tops, wild onions. forget me not wild lettuce and such weeds as cows eat as a cow wont eat a poison weeds."[5]

As the party in power during hard times, the Republicans faced almost certain defeat in the 1932 elections. President Herbert Hoover could escape repudiation only if the Democrats permitted internal divisions to destroy them. There was some prospect that the Democrats might do just that. National Democratic party leaders criticized Hoover not because he had done too little but because he had done too much. The main criticism they leveled at Hoover was that he was a profligate spender. In seeking to defeat progressive measures, Republicans in Congress could count on the votes of a majority of Democrats on almost every roll call.[6] But when, in their determination to

[4] Charles Walker, "Relief and Revolution," *Forum*, LXXXVIII (1932), 73–74; Boris Israel, "Shantytown, U.S.A.," *New Republic*, LXXV (1933), 39–41; Portland *Morning Oregonian*, February 1, 27, 1933.

[5] H.P.F. to Robert Doughton, February 10, 1932, Doughton MSS., Drawer 2; Louise Armstrong, *We Too Are the People* (Boston, 1938), p. 10; Harold Ware and Lement Harris, "In the 'Jungle,'" *The Nation*, CXXXV (1932), 55–56; George Yager to George Dern, March 26, 1932, Edward Costigan MSS., V.F. 1; *The Nation*, CXXXIV (1932), 651.

[6] The Democratic party, Hiram Johnson lamented, "can be described as the weak and timid echo of the Republican party in this Congress." Johnson to Charles K. McClatchy, February 29, 1932, Johnson MSS. Cf. T.R.B., "Washing-

balance the budget, Democratic leaders reached the point of advocating a federal sales tax, many of the congressional Democrats balked.[7] Under the leadership of Representative Robert "Muley" Doughton of North Carolina, rebellious Democrats joined with Fiorello La Guardia's insurgent Republicans to vote down the sales tax and adopt income and estate taxes instead.[8] The sales tax fight fixed the lines of combat at the forthcoming Democratic convention. Progressive Democrats were determined to overturn the national party leadership at Chicago in June and choose a liberal presidential nominee.

By the spring of 1932, almost every prominent Democratic progressive had become committed to the candidacy of New York's Governor Franklin Delano Roosevelt. Liberal Democrats were somewhat uneasy about Roosevelt's reputation as a trimmer, and disturbed by the vagueness of his formulas for recovery, but no other serious candidate had such good claims on progressive support. As governor of New York, he had created the first comprehensive system of unemployment relief, sponsored an extensive program for industrial welfare, and won western progressives by expanding the work Al Smith had begun in conservation and public power. In his campaign for the presidential nomination, he frequently voiced the aspirations of the reformers. In Albany, he pleaded the cause of "the forgotten man at the bottom of the economic pyramid"; in St. Paul, he stressed the need for "imaginative and purposeful planning"; at Oglethorpe University, he declared:

ton Notes," *New Republic*, LXX (1932), 95–96; Henry Morrow Hyde MS. Diary, January 18, 1932; Jouett Shouse to Claude Bowers, November 1, 1932, Shouse MSS., Correspondence, 1932–1959.

[7] Doughton protested: "It looks like our National Democratic leaders have betrayed us. Mr. Garner and Mr. Rainey, Speaker and Floor Majority Leader, stand on the Republican platform of the Sales tax. . . . It looks like we are headed for destruction. . . ." Doughton to Josephus Daniels, March 12, 1932, Doughton MSS., Drawer 2. See too "Soak the Poor," Wilburn Cartwright MSS.

[8] See Fiorello La Guardia to North American Newspaper Alliance, March 25, 1932, La Guardia MSS.; Lewis Kimmel, *Federal Budget and Fiscal Policy 1789–1958* (Washington, 1959), pp. 148–150. Irked by this rebuff, Speaker John Garner staged an old-time revival meeting on the floor of the House and asked every member who wanted to balance the budget to rise. Aided by this performance, Garner drove through a series of excises to make up for the rejected sales tax. The Revenue Act of 1932 fixed the tax structure for the rest of the decade. Since New Deal Congresses retained most of the 1932 excises, they sapped the effectiveness of the more progressive fiscal policy pursued in the Roosevelt years. E. Cary Brown, "Fiscal Policy in the 'Thirties: A Reappraisal," *American Economic Review*, XLVI (1956), 869.

"The country needs and, unless I mistake its temper, the country demands bold, persistent experimentation."[9]

If progressives admired Roosevelt for his record, they loved him even more for the enemies he had made. In the 1928 campaign, Smith had chosen John J. Raskob, a Republican business executive in the Du Pont hierarchy, to serve as the party's national chairman. After the 1928 debacle, Raskob had secured an unusually powerful position within the party, chiefly by using his own funds to keep it solvent. An arch reactionary and a leader of the wets, Raskob muted the economic issues of the depression by insisting that prohibition was the greatest question before the country. He mobilized all the resources of the national party organization against Roosevelt's bid for the nomination.

After his decisive defeat in 1928, Smith seemed resigned to retirement from politics. "There's no chance for a Catholic to be President," Smith confided to a western Democrat in 1930. "Not in my lifetime or in yours."[10] But as convention time neared, Smith came to feel that he deserved a second chance. Moreover, he had become embittered by what he felt was Roosevelt's ingratitude in not consulting him on questions in Albany, and piqued by his sudden loss of power to a man he regarded as his own protégé. When Roosevelt made his appeal for the "forgotten man," Smith, red-faced and raw-voiced, retorted: "I will take off my coat and fight to the end against any candidate who persists in any demagogic appeal to the masses of the working people of this country to destroy themselves by setting class against class and rich against poor."[11]

Although Smith had formidable support in 1932, few men seriously thought he could win the nomination, so bitter had been the feeling aroused against him in 1928. Democratic leaders, who foresaw victory over Hoover in November, did not want to jeopardize their chances by creating a new Donnybrook over rum and Romanism.[12] The real sig-

[9] Samuel Rosenman (ed.), *The Public Papers and Addresses of Franklin D. Roosevelt* (13 vols., New York, 1938–50), I, 625, 631, 646. See T.R.B., "Washington Notes," *New Republic,* LXX (1932), 205–206; Norman Davis to Cordell Hull, April 7, 1932, Davis MSS., Box 27; Arthur P. Mullen, *Western Democrat* (New York, 1940), p. 252.

[10] Mullen, *Western Democrat,* p. 260.

[11] *The New York Times,* April 14, 1932; Arthur Schlesinger, Jr., *The Crisis of the Old Order, 1919–1933* (Boston, 1957), pp. 273–277, 282, 291–292; Raymond Clapper MS. Diary, February 11, 1932.

[12] Joseph E. Davies denounced Smith's attack on Roosevelt's "forgotten man"

nificance of the Smith candidacy lay in the possibility that, since the convention was still governed by the two-thirds rule, Smith could use his power to kill off Roosevelt's candidacy and force a compromise choice. Hence, a whole army of would-be compromise candidates vied for the nomination. Marylanders like Dean Acheson grouped around the conservative, highly popular Governor Albert Ritchie, and the Chicago machine tied up Illinois' fifty-eight votes by advancing first its bewigged, pink-whiskered Senator J. Hamilton Lewis, then the conservative Chicago banker, Melvin J. Traylor, who could not appear personally at the convention because of a run on his bank.[13] More serious was the candidacy of Texas's favorite son, the Speaker of the House of Representatives, John Nance Garner. An erratic leader of his party, the Speaker had advocated both retrenchment and heavy spending. He had the endorsement of the powerful publisher William Randolph Hearst, who had called him, with unwitting precision, "a sound and sincere Democrat; in fact, another Champ Clark."[14]

Roosevelt seemed unstoppable. He had the shrewd counsel of his devoted aide Louis Howe, whose faith in his leader had not wavered even in the bleakest hours of Roosevelt's affliction with paralysis. Roosevelt's manager, the strapping James A. Farley, ran the Governor's campaign adroitly. A New York Irish Catholic Democratic politician, he fitted none of the stereotypes of the hinterland bigot; he came not from Tammany Hall but from Rockland County, neither smoked nor drank, was well-mannered, intelligent, and talked to delegates with a disarming modesty and directness. The Governor himself had carefully devised a farm program which spoke to the dissatisfaction of leaders in the hinterland, and he won to his cause most of the South, and much of the agrarian West. By the spring of 1932, Roosevelt was far out in front of the field, and aiming for nomination on the first ballot.

Then came near disaster. When Roosevelt, badly advised, entered

speech as an "invitation to again engage in a Kilkenny cat fight, which will completely convince the country that the Democratic Party is incapable of ruling itself. . . ." Davies to F.D.R., April 14, 1932, Roosevelt Family Papers (Halsted), Box 1.

[13] Roy Peel and Thomas C. Donnelly, *The 1932 Campaign* (New York, 1935), p. 42; Richard Byrd to Eleanor Roosevelt, September 17, 1931, Roosevelt Family Papers (Halsted), Box 1.

[14] San Francisco *Examiner*, January 3, 1932.

the Massachusetts primary, Smith trounced him 3–1, and put him to rout elsewhere in New England.[15] Two days after his smashing triumph in the Bay State, Smith showed unexpected strength in Pennsylvania. A man who would in a few years forge a new coalition around urban liberalism, Roosevelt in 1932 had little following in the industrial Northeast. Early in May, Speaker Garner, aided by Hearst, William McAdoo, and the Texas Society of California, ran well ahead of Roosevelt in the California primary. Suddenly Roosevelt's nomination was in jeopardy.

At the Democratic national convention late in June, the anti-Roosevelt coalition came within an ace of success. Roosevelt took a decided lead on the first ballot with 666¼ votes to Smith's 201¾, Garner's 90¼, and scattered support for favorite sons. Yet he was still more than a hundred votes short of the needed two-thirds, and the next two roll calls revealed only modest gains. Moreover, the Governor's lines had begun to crack. Only the cajoling of Louisiana's Huey Long prevented a break in the Mississippi delegation that might have started a rout. When the convention recessed after the third ballot, the Roosevelt leaders had a margin of only a few hours in which to try to save the day.

Late that afternoon, when the fourth roll call reached California, the rangy, straight-backed William McAdoo strode to the platform. California, he announced, had come to Chicago not to deadlock a convention but to nominate a President. Quickly, the meaning of his words became clear. California had abandoned Garner and was casting its forty-four votes for Roosevelt. Before McAdoo finished, the Smith galleries jeered him mercilessly, but he had heard those jeers before. His mission at Chicago did something to even scores for 1924, when Smith and the two-thirds rule had denied him victory.[16] With the switch of California, it was all over. The other states, all save the diehard Smith forces, climbed on the band wagon, and within minutes Franklin Roosevelt had been named as the Democratic nominee.

Roosevelt had won a victory that seemed all but lost a few hours

[15] J. Joseph Huthmacher, *Massachusetts People and Politics 1919–1933* (Cambridge, 1959), pp. 232–237; Worcester *Gazette,* April 27, 1932, clipping in James Michael Curley Scrapbooks.

[16] Russell Posner, "California's Role in the Nomination of Franklin D. Roosevelt," *California Historical Society Quarterly,* XXXIX (1960), 121–140; McAdoo to E. M. House, August 8, 1932, House MSS.

before. The crucial change was the defection of the Garner forces from the anti-Roosevelt front. Accounts of how and why this occurred vary widely.[17] One simple answer is that Garner bolted to Roosevelt in return for the vice-presidential nomination. However, Garner, who preferred the much more powerful post of Speaker, did not want to be Vice-President. Some have stressed the influence of William Randolph Hearst, whose agent cautioned Garner that if Roosevelt crumbled, the nomination might go either to Hearst's old enemy Smith or to the internationalist Newton Baker. Yet it was Garner himself who made the fateful decision, not for personal gain, but because he was a good party man who feared a deadlocked convention and a compromise choice would hurt the party as it had before.[18]

Upsetting all precedents, Roosevelt flew to Chicago—in a tri-motored plane that was buffeted by squalls and twice had to land to refuel—to deliver the first acceptance speech ever made to a nominating convention, incidentally helping dispel the notion at the outset that he was too frail to undertake a strenuous campaign. The Governor pointed out: "I have started out on the tasks that lie ahead by breaking the absurd traditions that the candidate should remain in professed ignorance of what has happened for weeks until he is formally notified of that event many weeks later. . . . You have nominated me and I know it, and I am here to thank you for the honor. Let it . . . be symbolic that in so doing I broke traditions. Let it be from now on the task of our Party to break foolish traditions. . . . I pledge you, I pledge myself," he concluded, "to a new deal for the American people."[19] The very next day, an alert cartoonist plucked from Roosevelt's speech the words "new deal"—a phrase to which the Governor had attached no special significance—and henceforth they were to be the hallmark of the Roosevelt program.

The Democratic convention displayed little more interest than the

[17] "Of the 55,000 Democrats allegedly to have been in Chicago for the recent Convention," Roosevelt's law partner remarked a few days later, "unquestionably 62,000 of them arranged the . . . shift." Frank Freidel, *Franklin D. Roosevelt, The Triumph* (Boston, 1956), p. 309.

[18] Bascom Timmons, *Garner of Texas* (New York, 1948), pp. 159–168; Tom Connally, as told to Alfred Steinberg, *My Name Is Tom Connally* (New York, 1954), pp. 141–145; Breckinridge Long to Key Pittman, July 5, 1932, Pittman MSS., Box 149; Arthur Krock, COHC; Newton Baker to Jouett Shouse, July 14, 1932, Shouse MSS., Correspondence, 1932–1959.

[19] *Public Papers,* I, 647–659.

Republicans in the crucial problems of the depression. Roosevelt's platform, drafted at the Governor's request by A. Mitchell Palmer, was a conservative document. Senator Key Pittman protested: "The platform has the merit of being short, and the demerit of being cold. There is not a word in it with regard to the 'forgotten man.' "[20] The convention made only modest changes. In its final form, the Democratic platform differed little from that of the Republican on economic questions, although the Democrats espoused somewhat more ambitious welfare programs. The two platforms, observed one commentator, were "mere rehashes of old proposals, political croquettes concocted from the leftovers of former years and dressed up with a little fresh verbal parsley."[21]

At both conventions the delegates showed far more concern over prohibition than over unemployment. "Here we are," wrote John Dewey, "in the midst of the greatest crisis since the Civil War and the only thing the two national parties seem to want to debate is booze."[22] The Democrats, who had been torn apart by the liquor issue in the twenties, determined this would not happen again. It was ridiculous, observed a Missourian, for a jobless wet Democrat to wrangle with a jobless dry Democrat over liquor when neither could afford the price of a drink.[23] By convention time, the wets held a decided edge. While the G.O.P. took an equivocal stand on prohibition, the Democrats voted a wringing wet plank. In the debate on the plank favoring repeal of the eighteenth amendment, the wets paraded the aisles chanting "How Dry I Am," and booed and hissed the leader of the drys, Senator Cordell Hull of Tennessee. Everything else in the program proved anticlimactic.

While the platform aroused little enthusiasm, the nominee kindled little more. In the contest for the nomination, Roosevelt's foes had pictured him as a timeserver, a man lacking in convictions and incapable of decisive action, and this view found wide acceptance, es-

[20] Key Pittman to F.D.R., June 16, 1932, Pittman MSS., Box 16.

[21] A. R. Hatton, cited in Peel and Donnelly, *The 1932 Campaign*, p. 124. Financial journals were pleased by the conservatism of the Democratic platform. *Barron's*, XII (July 4, 1932), 4; *Commercial and Financial Chronicle*, CXXXV (July 2, 1932), 1–2.

[22] Leo Gurko, *The Angry Decade* (New York, 1947), p. 42. Cf. Mauritz Hallgren, "Beer, Bums, and Republicans," *The Nation*, CXXXIV (1932), 717; Cleveland Trust Company, *Business Bulletin*, XIII (July 15, 1932).

[23] Charles Hay to James A. Farley, August 27, 1931, Hay MSS.

pecially in the eastern press. A man who had first won national attention as the paladin of the anti-Tammany reformers, Roosevelt had temporized during an inquiry into Tammany corruption. A man who had won allegiance as a Wilsonian idealist, he had capitulated to William Randolph Hearst's demand that he repudiate the movement to bring the United States into the League of Nations. These two episodes, coupled with Roosevelt's attempts to straddle issues such as the liquor question, led some of the keenest political observers to serious misjudgments. Elmer Davis believed the Democrats had nominated "the man who would probably make the weakest President of the dozen aspirants."[24] Walter Lippmann thought him the master of the "balanced antithesis," "a pleasant man who, without any important qualifications for the office, would very much like to be President."[25]

Roosevelt's campaign did little to reassure critics who thought him a vacillating politician. His speeches sounded painfully discordant themes. He assailed the Hoover administration because it was "committed to the idea that we ought to center control of everything in Washington as rapidly as possible," but he advanced policies which would greatly extend the power of the national government. He struck out at the disastrous high-tariff policies of the Republicans, but by the end of the campaign, taunted by Hoover, he had eaten so many of his words that no real difference separated the two candidates on the tariff issue. At times he talked the language of his New Nationalist advisers, at other times, as at Columbus, he spoke the idiom of the New Freedom.[26] His Topeka speech left farm leaders, as it was to leave historians, arguing over precisely what he intended. He would initiate a far-reaching plan to help the farmer; but he would do it in such a fashion that it would not "cost the Government any money,"

[24] Davis, "The Collapse of Politics," *Harper's*, CLXV (1932), 388.

[25] New York *Herald-Tribune*, April 28, 1932. See too Lippmann to Newton Baker, July 29, 1932, Baker MSS., Box 149. "It is useless to blame this very unrooseveltian Roosevelt for having no ideas; he really tries his best to have them," snapped the veteran Washington correspondent Charles Willis Thompson. "The Democrats have nominated nobody quite like him since Franklin Pierce. . . ." Thompson, "Wanted: Political Courage," *Harper's*, CLXV (1932), 726–727.

[26] *Public Papers*, I, 808, 835–836; *The New York Times*, November 5, 1932; Raymond Moley, *After Seven Years* (New York, 1939), pp. 45–52; Rexford Tugwell, "The Progressive Orthodoxy of Franklin D. Roosevelt," *Ethics*, LXIV (1953), 1–22; Felix Frankfurter to Louis Brandeis, August 7, 1932, Brandeis MSS., G9.

nor, unlike Hoover's farm program, would it "keep the Government in business."[27]

Yet all this was as nothing compared to his oscillations on fiscal policy. He would increase aid to the unemployed, but he would slash federal spending. On this one point he was specific; he would cut government spending 25 per cent. At Sioux City, Iowa, in September, Governor Roosevelt stated: "I accuse the present Administration of being the greatest spending Administration in peace times in all our history. It is an Administration that has piled bureau on bureau, commission on commission, and has failed to anticipate the dire needs and the reduced earning power of the people." In Pittsburgh the next month, he declared: "I regard reduction in Federal spending as one of the most important issues of this campaign. In my opinion, it is the most direct and effective contribution that Government can make to business."[28] One of his New Deal administrators reflected subsequently: "Given later developments, the campaign speeches often read like a giant misprint, in which Roosevelt and Hoover speak each other's lines."[29]

In retrospect, it is clear that a fair amount of the New Deal had been foreshadowed during the campaign. In his acceptance speech, Roosevelt advocated a reforestation program which would employ the nation's youth, an obvious anticipation of the Civilian Conservation Corps. In Portland, Oregon, he came out for public power development and government regulation of utilities. Roosevelt in his Columbus speech, and the Democratic party in its platform, both promised regulation of financial dealings on Wall Street, a pledge which was later fulfilled by a series of regulatory acts. On several occasions, the Governor stressed the need for planning, and in his Commonwealth Club address in San Francisco he adumbrated govern-

[27] *Public Papers*, I, 655. For the contrasting views of historians, see Daniel Fusfeld, *The Economic Thought of Franklin D. Roosevelt and the Origins of the New Deal* (New York, 1956), p. 302, n. 26; Freidel, *Roosevelt, The Triumph*, pp. 347–349; Gilbert Fite, *George N. Peek and the Fight for Farm Parity* (Norman, Okla., 1954), p. 239; Gertrude Almy Slichter, "Franklin D. Roosevelt and the Farm Problem, 1929–1932," *Mississippi Valley Historical Review*, XLIII (1956), 255, n. 51.

[28] *Public Papers*, I, 761, 809. By June 30, 1933, Roosevelt asserted, the "true deficit" might rise above $1,600,000,000—"a deficit so great that it makes us catch our breath." *Ibid.*, p. 805.

[29] Marriner Eccles, *Beckoning Frontiers* (New York, 1951), p. 95.

ment-business co-operation to achieve a new economic order. Some of his remarks foretold the subsequent development of welfare measures. In the very Pittsburgh speech in which he advocated budget-balancing, Roosevelt added: "If starvation and dire need on the part of any of our citizens make necessary the appropriation of additional funds which would keep the budget out of balance, I shall not hesitate to tell the American people the full truth and ask them to authorize the expenditure of that additional amount."[30]

Some writers have even insisted that Roosevelt foretold virtually all of the New Deal during the campaign.[31] Yet what is striking in the light of later developments is what he did not mention in 1932—deficit spending, a gigantic federal works program, federal housing and slum clearance, the NRA, the TVA, sharply increased income taxes on the wealthy, massive and imaginative relief programs, a national labor relations board with federal sanctions to enforce collective bargaining —and this does not begin to exhaust the list. Roosevelt did not advocate such proposals in part because they had not yet crystallized, but even more because his primary task was to get himself elected, and, as the front runner, he saw little point in jeopardizing his chances by engaging in controversies that might cost him more than they would win.

Yet in contrast to Hoover, Roosevelt seemed attractive enough to induce a whole regiment of Republican progressives to bolt the party. By Election Day, Roosevelt had won the support of an impressive number of G.O.P. senators—Robert La Follette, Jr., of Wisconsin, Hiram Johnson of California, George Norris of Nebraska, Smith Wildman Brookhart of Iowa, and Bronson Cutting of New Mexico. Some came out of deep dislike of the President. "I think Hoover is the most pitiful failure we have ever had in the White House," observed the former Oregon Senator Jonathan Bourne, Jr.[32] Others, disturbed at first by reports that Roosevelt was a vacillator, came to dismiss such

[30] *Public Papers,* I, 810.

[31] Schlesinger, *Crisis of Old Order,* p. 538, n. 21; Samuel Rosenman, *Working with Roosevelt* (New York, 1952), p. 85.

[32] Bourne to Ida Arneson, July 9, 1932, Bourne MSS. "As an American," explained Hiram Johnson, "I have really shuddered at what four more years of Hoover may mean to this country." Johnson to M. C. Zumwalt, October 24, 1932, Johnson MSS. For Cutting's rejection of Hoover, see Santa Fe *New Mexican,* October 26, 1932, extra edition.

talk as the propaganda of the "Power Trust."[33] "Frankly I had been under the impression until I met him that he was of the 'Little Boy Blue–Little Lord Fauntleroy' type," the public power advocate Morris Cooke confessed. Now he felt certain Roosevelt was a fighter.[34]

If Roosevelt's program lacked substance, his blithe spirit—his infectious smile, his warm, mellow voice, his obvious ease with crowds—contrasted sharply with Hoover's glumness. While Roosevelt reflected the joy of a campaigner winging to victory, Hoover projected defeat. From the onset of the depression, he had approached problems with a relentless pessimism. After one gloomy White House session, Secretary of State Henry Stimson noted, "It was like sitting in a bath of ink to sit in his room." The President seemed wedded to failure. Gutzon Borglum, the sculptor, remarked: "If you put a rose in Hoover's hand it would wilt."[35]

Even in his earliest days in office, Hoover had sought to escape from the world of politics Roosevelt relished and to insulate himself from the demands of popular government. A man of impressive accomplishments, he had little understanding of the nuances of the art of governing. Hoover, observed Willmott Lewis, Washington correspondent of the London *Times,* "can calculate wave lengths, but cannot see color. . . . He can understand vibrations but cannot hear tone."[36] No President ever worked harder in the White House than Herbert Hoover, but he was never able to convince the nation that he cared deeply how people were suffering and that he shared with them the sorrows and the blighted prospects the depression had brought.

Nothing revealed so sharply Hoover's lack of rapport with the people nor caused him such a loss of favor as his inept handling of the Bonus Army. Through the month of June, 1932, singing the old war songs

[33] National Popular Government League, *Bulletin No. 156* (Washington, 1932). In the 1932 campaign, the utilities spread the rumor that Roosevelt was an untrustworthy eccentric. J. David Stern, COHC.

[34] Cooke to George Norris, June 23, 1932, Cooke MSS., Box 90. Clarence Darrow wrote: "It seems to me as if Roosevelt is the most available democrat in sight, and still from all sources I am told he sadly lacks 'guts'—but of course most men do. The interests are conspiring to beat him, which commends him." Darrow to Frank Walsh, February 10, 1932, Walsh MSS., Correspondence.

[35] Henry Stimson MS. Diary, June 18, 1931; Grace Tully, *F.D.R. My Boss* (New York, 1949), p. 60.

[36] Walter Lippmann, "The Peculiar Weakness of Mr. Hoover," *Harper's,* CLXI (1930), 1–7; Irving Bernstein, *The Lean Years* (Boston, 1960), p. 249.

and bearing placards "Cheered in '17, Jeered in '32," veterans of World War I massed on Washington to petition Congress to pay them at once war service bonuses Congress had agreed to pay in 1945. They set up a large shanty town on the Anacostia Flats outside the capital, and when they overran the flats, they took over some land and unused government buildings on Pennsylvania Avenue. At its height, the Bonus Army numbered over twenty thousand. When Congress rejected this petition in boots reminiscent of Coxey's march of forty years before, some of the veterans went home, but many others, homeless, jobless, and aimless, stayed on in Washington.[37]

The encampment of the Bonus Army presented the head of the District police, Pelham Glassford, who had been the youngest brigadier general in the A.E.F., with a difficult problem. He opposed the bonus and had tried to discourage the men from coming; when they refused to leave, he tried to lure them home by having the Marine Band play "Carry Me Back to Old Virginny" and "My Old Kentucky Home." Yet he reasoned that, although the men were encamped illegally, their only offense was poverty. If they could have afforded it, they would have booked rooms in downtown hotels like other petitioners. His motorcycle cops furnished iodine and pills to the veterans, played baseball with them, gave them advice on digging latrines, even furnished some tents and bedding. For a time General Glassford handled the finances of the Bonus Army. He believed the greatest mistake the government could make would be a show of violence.[38]

Glassford's composure was not matched by that of Hoover and his aides. As the Bonus Army squatted down for what might be a protracted stay, Hoover lost his sense of proportion and some of his more volatile lieutenants panicked. The White House was put under guard, the gates chained, the streets about it cleared of people—all of which added to the impression that the President had cut himself off from the country. Then, precipitately, the administration decided it must have immediate possession of the unused buildings which the veterans

[37] W. W. Waters, as told to William C. White, *B.E.F.* (New York, 1933), pp. 1–144; Jack Douglas, *Veterans on the March* (New York, 1934), pp. 25–175; *B.E.F. News,* July 2, 1932.

[38] Glassford to the Commissioners, D.C., July 22, 1932, Glassford MSS., Box 4; John Dos Passos, "The Veterans Come Home to Roost," *New Republic,* LXXI (1932), 177; Fleta Campbell Springer, "Glassford and the Siege of Washington," *Harper's,* CXLV (1932), 641–655.

had occupied. On July 28, Glassford reluctantly undertook the assignment of dislodging the veterans. Two brief skirmishes ensued, in one of which two veterans were killed. The Commissioners now had the pretext for requesting federal troops, and Hoover and Secretary of War Patrick Hurley, who was convinced he faced a Communist uprising of menacing proportions, quickly obliged. The Army, under General Douglas MacArthur, carried out its mission with few casualties, but with a thoroughness that the situation scarcely required. Four troops of cavalry with drawn sabers, six tanks, and a column of steel-helmeted infantry with fixed bayonets entered downtown Washington. After clearing the buildings on Pennsylvania Avenue, they crossed the Anacostia Bridge, thousands of veterans and their wives and children fleeing before them, routed the bonusers from their crude homes, hurled tear gas bombs into the colony, and set the shacks afire with their torches. That night, Washington was lit by the burning camps of Anacostia Flats.[39]

Hoover's use of armed force against American veterans raised a storm of protest. "What a pitiful spectacle is that of the great American Government, mightiest in the world, chasing unarmed men, women and children with army tanks," observed the Washington News.[40] The administration made matters still worse by vilifying the Bonus Army as a rabble of Communists and criminals. General MacArthur described the marchers as "a mob . . . animated by the essence of revolution." If the President had let the "insurrectionists" carry on for one more week, MacArthur stated, "the institutions of our Government would have been very severely threatened."[41] Far from being a menacing band of revolutionaries, the Bonus Army was a whipped, melancholy group of men trying to hold themselves together when their spirit was gone. The Communist faction had to be protected from other bonus marchers who threatened physical violence. There were some criminals in the group, but as one commentator re-

[39] Ferry K. Heath to the Commissioners of the District of Columbia, July 26, 27, 1932, Glassford MSS., Box 4; Agnes Meyer MS. Diary, June 6, 1932; Bernstein, *Lean Years,* pp. 448–454; Schlesinger, *Coming of New Deal,* pp. 262–263; Thomas Stokes, *Chip Off My Shoulder* (Princeton, 1940), pp. 301–304. Some of the veterans added to the blaze by setting their shacks afire.

[40] Washington *News,* July 30, 1932.

[41] *The Memoirs of Herbert Hoover* (3 vols., New York, 1951–52), III, 228–229.

marked, the record for serious crime in President Harding's cabinet was proportionately higher than in the B.E.F. Washington experienced more crime in August after the Army had departed than in July when it was there. To cap the episode, the administration claimed the U.S. Army had been guilty of no violence, and had not fired the camp, two statements which the newsreels belied.[42]

Months before the Bonus Army incident, Hoover seemed destined for defeat in November. "There is not a single soul thus far I have met, stand-pat, Progressive or otherwise, who believes Hoover can be elected," Hiram Johnson wrote early in February.[43] After the rout of the Bonus Army, Hoover encountered the bitter animosity of men and women who held him personally responsible for their plight. In Detroit, he was greeted with the cry: "Down with Hoover, slayer of veterans." In the auto city, he drove through miles of streets lined with sullen, silent men. When he got up to speak, his face was ashen, his hands trembled.[44] Toward the end, Hoover was a pathetic figure, a weary, beaten man, often jeered by crowds as a President had never been jeered before.

Hoover had, by his own *hubris*, placed himself in an impossible position. Nothing he said could free him from the inconsistency of claiming credit for prosperity but shirking blame for depression. When he accepted the Republican nomination at the Stanford stadium in August, 1928, he had declared: "We in America today are nearer to the final triumph over poverty than ever before in the history of any land." In his inaugural address in March, 1929, he had asserted: "In no nation are the fruits of accomplishment more secure." "The test of the rightfulness of our decisions," he told Congress in December, 1929, "must be whether we have sustained and advanced . . . pros-

[42] Bernstein, *Lean Years,* pp. 454–456; Schlesinger, *Coming of New Deal,* pp. 518–519, n. 15; Glassford, "Circumstances Leading up to the Use of Federal Forces against the Bonus Army," Glassford MSS., Box 4; Glassford to John Chamberlain, January 25, 1952, Glassford MSS., Box 1. Cf. Benjamin Gitlow, *The Whole of Their Lives* (New York, 1948), p. 228; Don Lohbeck, *Patrick J. Hurley* (Chicago, 1956), pp. 102–118, 486–488.

[43] Hiram Johnson to Charles K. McClatchy, February 2, 1932, Johnson MSS. See Miguel Otero to Bronson Cutting, January 25, 1932, Otero MSS.; "40 Years a Legislator," p. 196, Elmer Thomas MSS.

[44] Stokes, *Chip Off My Shoulder,* pp. 304–305. Cf. Theodore Joslin, *Hoover Off the Record* (Garden City, N.Y., 1934), p. 324.

perity."[45] Good times, he insisted, were not the result of good fortune, but of the wisdom of the Republican party and business leaders. His promises, observed Elmer Davis, "ran to that excess which above all things offended the Greek temperament, which seemed above all things to invite the correcting interposition of Nemesis. . . . Never in American history did a candidate so recklessly walk out on a limb and challenge Nemesis to saw it off."[46]

On Election Day, Nemesis responded to the challenge. Franklin Roosevelt swept to victory with 22,800,000 votes to Hoover's 15,750,000. With a 472–59 margin in the Electoral College, he captured every state south and west of Pennsylvania. Roosevelt carried more counties than a presidential candidate had ever won before, including 282 that had never gone Democratic. Of the forty states in Hoover's victory coalition four years before, the President held but six. Save for 1912, when the party was divided, no Republican candidate had ever been defeated so badly.

[45] *The New York Times,* August 12, 1928; William Starr Myers (ed.), *The State Papers and Other Public Writings of Herbert Hoover* (2 vols., Garden City, N.Y., 1934), I, 12, 166.

[46] Elmer Davis, "Hoover and Hubris," *New Republic,* LXXIII (1932), 7–8. Cf. *The New York Times,* November 9, 1932, editorial on the *"hubris"* of the Republicans. One writer commented: "What a pity that we cannot send Mr. Hoover to Delphi to learn some act of atonement to placate the implacable gods." "The United States Before the Election," *Round Table,* XXII (1932), 794.

CHAPTER 2

Winter of Despair

"YES, we could smell the depression in the air," one writer remembered of "that historically cruel winter of 1932–3, which chilled so many of us like a world's end. . . . It was like a raw wind; the very houses we lived in seemed to be shrinking, hopeless of real comfort."[1] The interval between Roosevelt's election in November, 1932, and his inauguration in March, 1933, proved the most harrowing four months of the depression. Three years of hard times had cut national income more than half; the crash of five thousand banks had wiped out nine million savings accounts. In the summer of 1932, the economy had rallied once more, but by the end of the year recovery seemed as elusive as ever.[2]

[1] Harold Clurman, *Fervent Years* (New York, 1957), pp. 107, 112.
[2] The most perspicacious business forecast anticipated that 1933 would be another year of depression, with exports, dividends, and industrial wage rates even lower than in 1932, and business failures greater. Cleveland Trust Company, *Business Bulletin,* XIII (December 15, 1932). Mr. Hoover and his supporters have contended that recovery began in the summer of 1932 and that only anxiety about the policies Roosevelt would pursue when he took office produced the crisis of March, 1933. *The Memoirs of Herbert Hoover* (3 vols., New York, 1951–52), III, 39–40, 176–216; William Starr Myers and Walter Newton, *The Hoover Administration* (New York, 1936). Such a contention is not finally subject to proof or disproof; suffice it to say that while some indices showed improvement in 1932, others did not; that the disruption was far greater than this argument suggests; and that it seems only commonsensical to suppose that people withdrew money from banks, not because they feared Roosevelt, but because they had a well-founded distrust of the banks. Furthermore, the belief that the depression persisted only because Roosevelt would not endorse the

"I come home from the hill every night filled with gloom," one Washington correspondent noted. "I see on streets filthy, ragged, desperate-looking men, such as I have never seen before."[3] By the end of Hoover's reign, more than fifteen million workers had lost their jobs. U.S. Steel's payroll of full-time workers fell from 225,000 in 1929 to zero on April 1, 1933; even the hands employed part-time in 1933 numbered only half as many as the full-time force of 1929. In Seattle, jobless families whose lights had been cut off spent every evening in darkness, some even without candles to light the blackened room. In December, 1932, a New York couple moved to a cave in Central Park, where they lived for the next year. That same month, Rexford Tugwell wrote in his diary: "No one can live and work in New York this winter without a profound sense of uneasiness. Never, in modern times, I should think, has there been so widespread unemployment and such moving distress from sheer hunger and cold."[4]

As the depression deepened, amorphous resentment finally took form in one overwhelming question: Who was to blame? The answer came readily enough. Throughout the 1920's, publicists had trumpeted one never-ending refrain: that the prosperity of the decade had been produced by the genius of businessmen. If businessmen had caused prosperity, who but they must be responsible for the depression? To the deep displeasure of the financial world, President Hoover started a hunt for personal devils. Frustrated by the failure of the economy to respond to his nostrums, Hoover encouraged a Senate investigation into Wall Street in the spring of 1932 to make an example of the "bears," who he believed had been organizing "raids" on the stock market.[5]

---

policies Hoover had pursued for the previous three years is not without its comical aspects. Cf. George Soule, "This Recovery: What Brought It? And Will It Last?" *Harper's*, CLXXIV (1937), 337–345; Thomas Wilson, *Fluctuations in Income and Employment* (London, 1942), p. 161.

[3] Henry Morrow Hyde MS. Diary, January 5, 1933.

[4] Irving Bernstein, *The Lean Years* (Boston, 1960), pp. 506–507; "Interoffice to Ross," February 17, 1933, J. D. Ross MSS.; Joseph Mitchell, *McSorley's Wonderful Saloon* (Harmondsworth, Eng., 1954), pp. 147–155; Tugwell, "Notes from a New Deal Diary," December 24, 1932.

[5] For Hoover's views, an important source is a memorandum by Morris Tracy of a conversation with Hoover on February 29, 1932, in the files of the Washington Bureau of the United Press. For Wall Street resentment, see Richard Whitney to Ogden Mills, August 26, 1932; Mills to Whitney, August 27, 1932, Whitney to Mills, September 2, 1932, Mills MSS., Box 11.

Under the persistent probing of Ferdinand Pecora, the committee shattered the image of the investment banker as a man of probity whose first concern was the welfare of his customers and who operated in an institution that was a model of fair play. Pecora revealed that the most respected men on Wall Street had rigged pools, had profited by pegging bond prices artificially high, and had lined their pockets with fantastic bonuses.[6] The leading financial houses, the committee learned, invited insiders to purchase securities at a price much below that paid by the public. When officers of Charles Mitchell's National City Bank faced ruin because they could not cover their investments, the bank gave them interest-free loans while ruthlessly selling out their own customers. The bankers seemed bereft of a sense of obligation even to their own institutions. Albert Wiggin, president of the Chase National Bank, had even sold short the stock of his own bank.

Pecora's inquiry was only one of a series of spectacular episodes that destroyed the financier as a folk hero. In March, 1932, the Swedish match king, Ivar Kreuger, shot himself in a Paris apartment. Mourned at the time of his death as a financial titan, he was revealed a month later to have been a swindler who had forged $100,000,000 in bonds. Kreuger's ability to deceive the investment house of Lee, Higginson and, in turn, American investors suggested that the morality of international finance and the business acumen of Lee, Higginson left much to be desired.[7] While auditors attempted to untangle the Kreuger empire, a new bombshell exploded. The firm of Samuel Insull, the leading utility magnate in the country, chairman of the boards of sixty-five companies and emperor of a utility holding company domain that reached into thirty-two states, folded. Insull, an idol

[6] The Stock Exchange, Pecora later wrote, "was in reality neither more nor less than a glorified gambling casino where the odds were weighted heavily against the eager outsiders." Pecora, *Wall Street under Oath* (New York, 1939), p. 263. Pecora was hired as counsel after the 1932 elections. Cf. Arthur Capper to Joe H. Mercer, January 5, 1932, Capper MSS.

[7] Robert Shaplen, *Kreuger* (New York, 1960); "Ivar Kreuger," *Fortune*, VII (May, 1933), 51–57, 78–84; (June, 1933), 59–63, 78–95; VIII (July, 1933), 68–76. The author of the anonymous *Fortune* series was Archibald MacLeish. By the end of 1932, newspapers were advertising "The Match King" with Warren William and Lily Damita: "SIZZLING HOT from the headlines by Warners . . . this story of a man who made suckers of millions, and who was himself made a sucker by an actress you all know!" Cincinnati *Enquirer*, December 31, 1932.

of the twenties, was now pilloried as an unscrupulous charlatan.[8]

Yet it was less their financial transgressions than their social irresponsibility which caused the loss of faith in America's business leaders. At a time when millions lived close to starvation, and some even had to scavenge for food, bankers like Wiggin and corporation executives like George Washington Hill of American Tobacco drew astronomical salaries and bonuses. Yet many of these men, including Wiggin, manipulated their investments so that they paid no income tax at all. In Chicago, where teachers, unpaid for months, fainted in classrooms for want of food, wealthy citizens of national reputation brazenly refused to pay taxes or submitted falsified statements.[9] In Detroit, the hardest hit of any large city, Henry Ford set the standard for businessmen by shrugging off all responsibility for the welfare of the jobless. Detroit bankers, in fact, insisted that before the city would be granted a loan to maintain relief, it would have to cut relief pittances still further.

The real responsibility for their poverty, insisted John Edgerton in his presidential address to the National Association of Manufacturers in October, 1930, lay with the jobless themselves. If, he asked, "they do not . . . practice the habits of thrift and conservation, or if they gamble away their savings in the stock market or elsewhere, is our economic system, or government, or industry to blame?" The tone of lament in Edgerton's remarks was common. One of the most popular themes of business literature of the period was that the wealthy businessman had suffered more than the worker.[10]

[8] James Bonbright and Gardiner Means, *The Holding Company* (New York, 1932). Cf. Forrest McDonald, *Insull* (Chicago, 1962).

[9] Teachers lost their homes, their insurance policies, nearly every possession of value, but kept at their desks although they did not know if they would ever be paid. During the 1932–33 school year, 81 per cent of the children in white rural schools in Alabama were schoolless; Georgia closed more than a thousand schools with an enrollment of over 170,000 pupils; Dayton schools in early 1933 opened only three days a week. Avis Carlson, "Deflating the Schools," *Harper's*, CLXVII (1933), 705–714.

[10] *Proceedings of the Thirty-fifth Annual Meeting of the National Association of Manufacturers,* New York City, October 6–9, 1930, pp. 14–15. "As it now stands," complained *Barron's* "labor has reaped all the reward, and the new capital supplied is with little or no return." *Barron's*, XIII (January 16, 1933), 7. Cf. *Commercial and Financial Chronicle*, CXXXVI (January 7, 1933), 18; *Sales Management*, XI (1932), 361.

By the time Roosevelt took office, the country had been whipped to a fury at the performance of bankers and businessmen. The New York bankers especially were the target of popular wrath. "The best way to restore confidence in the banks," Senator Burton Wheeler of Montana stormed at Mitchell, "would be to take these crooked presidents out of the banks and treat them the same as we treated Al Capone when he failed to pay his income tax."[11] Senator Carter Glass of Virginia told his colleagues: "One banker in my state attempted to marry a white woman and they lynched him."[12]

In primitive societies, homage was paid to the witch doctors who could control those forces—sun, rain, and wind—which could spell the difference between plenty and privation in a primitive economy and before which most men stood helpless. The crisis of the Great Depression revealed that the American people had come to view their financial "wizards" in the same light; the businessman had been thought of as a magic-maker who could master the forces of a complex industrial society which the common man viewed with awe, and which were as much out of his control as the wind or tides. By the winter of 1932, the businessman had lost his magic and was as discredited as a Hopi rainmaker in a prolonged drought. "It will be many a long day before Americans of the middle class will listen with anything approaching the reverence they felt in 1928 whenever a magnate of business speaks," observed Gerald Johnson. "We now know they are not magicians. When it comes to a real crisis they are as helpless as the rest of us, and as bewildered."[13]

The persistence of the depression raised questions not merely about business leadership but about capitalism itself. When so many knew want amidst so much plenty, something seemed to be fundamentally wrong with the way the system distributed goods. While the jobless wore threadbare clothing, farmers could not market thirteen million bales of cotton in 1932. While children trudged to school in shoes

[11] Portland *Morning Oregonian,* February 23, 1933. "Certainly the New York bankers have proved that they are no heroes," Agnes Meyer, the wife of Hoover's chief of the Reconstruction Finance Corporation, noted in her diary. "The wealthy classes as I have learned to know them through the depression are not much to be admired. They are overcome by fear and selfishness. There is very little social outlook." Agnes Meyer MS. Diary, April 29, 1932.

[12] Henry Morrow Hyde MS. Diary, March 11, 1932.

[13] "The Average American and the Depression," *Current History,* XXXV (1932), 672–673.

soled with cardboard, shoe factories in Lynn and Brockton, Massachusetts, had to close down six months of the year. With ten billion dollars in bank vaults, hundred of cities felt compelled to issue scrip because there was not enough currency in the town. Some groups even resorted to barter.[14] While people went without food, crops rotted in the fields. One observer reported in 1932: "Beginning in the Carolinas and extending clear into New Mexico are fields of unpicked cotton that tell a mute story of more cotton than could be sold for enough, even, to pay the cost of picking. Vineyards with grapes still unpicked, orchards of olive trees hanging full of rotting fruits and oranges being sold at less than the cost of production."[15] Western ranchers, unable either to market their sheep or feed them, slit their throats and hurled their carcasses into canyons.[16] In the plains states, breadlines marched under grain elevators heaped high with wheat.[17]

The savage irony of want amidst plenty drove farmers to violent action. As farm income dipped sharply while taxes and mortgage obligations remained constant, thousands of farmers lost their land for failure to pay taxes or meet payments. One account reported that on a single day in April, 1932, one-fourth of the entire area of the state of Mississippi went under the hammer of auctioneers.[18] "Right here

[14] Henry Morton Robinson, *Fantastic Interim* (New York, 1943), pp. 217–218; "Cooperative Self-Help Activities Among the Unemployed," *Monthly Labor Review*, XXXVI (1933), 979–1038, 1229–1240; Wayne Weishaar and Wayne Parrish, *Men Without Money* (New York, 1933). Since the barter groups reflected the tradition of self-reliance, they won an inordinate amount of attention. Yet, in a pecuniary economy, a man needed money, and self-help groups solved few of his problems. The radical Louis Budenz commented on the barter groups: "Potatoes became the opium of the people." John Chamberlain, "Crusoe Economics," *New Republic*, LXXVII (1933), 302.
[15] *New Republic* XX (1932), 211.
[16] U.S. Congress, House of Representatives, *Unemployment in the United States*, Hearings before a Subcommittee of the Committee on Labor, U.S. House of Representatives, 72d Cong., 1st Sess., on H.R. 206, H.R. 6011, H.R. 8088 (Washington, 1932), pp. 98–100.
[17] "One February in the early thirties, when bread-lines were decorating our cities and towns, I had to go out to Salt Lake City," a writer recalled. "As I passed through the high plains I noticed piles of sand beside the tracks and I asked the conductor what the stuff was doing there. His answer was, 'Mister, that isn't sand, that's wheat.'" Paul Sears, "O, Bury Me Not," *New Republic*, XCI (1937), 9.
[18] *Literary Digest*, CXIII (May 7, 1932), 10. Cf. Lawrence Jones and David Durand, *Mortgage Lending Experience in Agriculture* (Princeton, 1954), pp. 95–110.

in Mississippi some people are about ready to lead a mob," remarked Governor Theodore Bilbo. "In fact, I'm getting a little pink myself."[19] In the fall of 1931, Iowa farmers had broken out into an open rebellion that spread to other parts of the country. In Logan, Iowa, five hundred farmers, milling around the courthouse, prevented the sale of Ernest Ganzhorn's farm; in Wilmar, Minnesota, farmers saved the land of a man who had tilled it for fifty-seven years; in Bucks County, Pennsylvania, three hundred farmers forced the sale of John Hanzel's farm for only $1.18, then leased it back to him.

In the summer of 1932, Milo Reno, the mercurial leader of the Iowa Farmers' Union, formed a Farm Holiday Association to organize farmers to refuse to ship food into Sioux City for thirty days or until such time as they got the "cost of production." Farmers blocked highways with logs and spiked telegraph poles, smashed windshields and headlights, and punctured tires with their pitchforks. When authorities in Council Bluffs arrested fifty-five pickets, one thousand angry farmers threatened to storm the jail, and the pickets were released on bail. In Nebraska, where farmers carried placards reading "Be Pickets or Peasants," strikers halted a freight train and took off a carload of cattle. Wisconsin dairy farmers, led by the magnetic Walter Singler, a rangy, goateed organizer who sported a red vest and a ten-gallon hat, dumped milk on the roadsides and fought pitched battles with deputy sheriffs. The attempt to boost farm prices failed, but the farmers used still more violent means to halt foreclosures. At Storm Lake, Iowa, rope-swinging farmers came close to hanging a lawyer conducting a foreclosure; in the LeMars area, five hundred farmers marched on the courthouse steps and mauled the sheriff and the agent of a New York mortgage company.[20] "If we don't get beneficial service from the Legislature," warned a leader of the Nebraska holiday movement, "200,000 of us are coming to Lincoln and we'll tear that new State Capitol Building to pieces."[21]

The farm strikes were only one of a series of spectacular incidents

[19] Hilton Butler, "Bilbo, The Two-Edged Sword," *North American Review*, CCXXXII (1931), 496.

[20] Theodore Saloutos and John D. Hicks, *Agricultural Discontent in the Middle West: 1900–1939* (Madison, 1951), pp. 435–448; A. William Hoglund, "Wisconsin Dairy Farmers on Strike," *Agricultural History*, XXXV (1961), 24–34. The "holiday" movement was paralleled by strikes of dairy farmers.

[21] *The New York Times*, January 22, 1933.

—the bonus march episode, Communist-led demonstrations of the unemployed, the forty-eight-mile-long motor car "Coal Caravan" of ten thousand striking miners in southern Illinois—that led men to speculate whether the country faced imminent revolution.[22] In February, 1933, thousands of former members of barter groups seized the county-city building in Seattle. In the Blue Ridge, miners smashed company store windows and storekeepers were given the choice of handing out food or having it seized. Unemployed workers in Detroit invaded self-service groceries in groups, filled their baskets, and left without paying. Iowa leagues of the unemployed enlisted jobless gas and light workers to tap gas and electric lines. In Des Moines, workers boarded streetcars in groups of ten or twenty and told the cowed conductors to "charge the fares to the mayor."[23] In Chicago, a group of fifty-five was charged with dismantling an entire four-story building and carrying it away brick by brick.[24]

No longer could property holders count on the protection of the law. From the anthracite country of Pennsylvania to Sweetwater County in Wyoming, jobless miners worked coal seams on company property, sank their own shafts, and marketed the ore. At the height of operations, they "bootlegged" some $100,000 worth of coal each day. When the owners went to court, juries in mining towns would not convict.[25] In Idaho, Governor C. Ben Ross stopped a sheriff's sale of a ranch in Canon County and announced he would halt all such transactions in the future. Minnesota's Governor Floyd Olson proclaimed a moratorium on all mortgage sales until the legislature could

[22] "I am alarmed at the increasing undercurrent of hate directed against our bankers and big industrial leaders," noted one financier. "The mention of revolution is becoming quite common." Rudolph Spreckels to Franklin Hichborn, June 14, 1932, Hichborn MSS. "The word revolution is heard at every hand," observed one writer. George Leighton, "And if the Revolution Comes . . . ?" *Harper's*, CLXIV (1932), 466.

[23] Malcolm Ross, *Machine Age in the Hills* (New York, 1933), p. 60; Boston *Transcript*, February 21, 1933, clipping in James Michael Curley Scrapbook 88; Nathaniel Weyl, "Organized Hunger," *New Republic*, LXXIII (1932), 120; Clayton Fountain, *Union Guy* (New York, 1949), pp. 35–37.

[24] Denver *Post*, February 3, 1933. The owner had sold the building, covering more than an acre, only to discover that it had vanished.

[25] Bernstein, *Lean Years*, pp. 424–425; Louis Adamic, *My America* (New York, 1938), pp. 316–324; Denver *Post*, February 5, 1933; Sylvan Lebow, "Bootleg Coal," *Review of Reviews*, XCV (March, 1937), 36; John Janney, "When Public Opinion Runs Out on the Law," *American*, CXXIII (February, 1937), 16–17, 83–89.

act, and Governor William Langer of North Dakota mobilized the militia to halt foreclosures. Despairing of help from lawmen, New Mexico cattlemen threatened to form a vigilante committee to apprehend rustlers who had stolen more than a thousand head of cattle in two months.[26]

It was frequently remarked in later years that Roosevelt saved the country from revolution. Yet the mood of the country during the winter of 1932–33 was not revolutionary. There was less an active demand for change than a disillusionment with parliamentary politics, so often the prelude to totalitarianism in Europe. The country was less rebellious than drifting, although there were clearly limits to how long it would be willing to drift. As early as the summer of 1931, Gerald Johnson had noted the "fathomless pessimism" which the depression had induced. "The energy of the country has suffered a strange paralysis," he observed. "We are in the doldrums, waiting not even hopefully for the wind which never comes."[27]

Both American and European observers were astonished by the strangely phlegmatic response of Americans to the depression.[28] In England, thousands of British workers marched on Buckingham Palace and Downing Street, and in Newfoundland a mob broke into the cabinet chamber in St. John's and roughed up the Prime Minister; but in the United States, a country thought of as the land of lawlessness, most of the unemployed meekly accepted their lot. After a trip through New England mill towns in 1931, one commentator noted of the workers: "There was something dead in them, as from exhaustion or perhaps too much idleness, without any personal winsomeness or any power of demand. . . . The situation is infinitely pathetic, not to say appalling." More than a year later, Elmer Davis observed that people were asking "Do you think we are going to have a revolution?"

[26] Denver *Post*, February 12, 1933; George Mayer, *The Political Career of Floyd B. Olson* (Minneapolis, 1951), p. 125; Portland *Sunday Oregonian*, February 5, 1933.
[27] "Bryan, Thou Shouldst Be Living: A Plea for Demagogues," *Harper's*, CLXIII (1931), 388. Cf. Paul Lazarsfeld, *Die Arbeitslosen von Marienthal* (Leipzig, 1933).
[28] Philippe Soupault, "La grande inquiétude des paysans et les calmes des cités industrielles américaines," *L'Europe Nouvelle*, XIV (1931), 1148–1151; Marquis Childs, "America Faces the Winter," *The Spectator*, CXLIX (1932), 574–575; Jay Franklin, "Why This Political Apathy?" *Current History*, XXXVI (1932), 265–269; Reno *Evening Gazette*, November 5, 1932.

—but asking it "apathetically, as if nothing they might do could either help it or hinder it."[29]

Certainly the 1932 elections represented no mandate for radicalism. The Communists, with the greatest opportunity they would ever have, polled only 120,000 votes, and the Socialists won proportionately far less than in 1920 or in 1912. The voters turned out progressives like Senator Smith Brookhart in Iowa and Governor Philip La Follette in Wisconsin, as well as conservative Senators like Reed Smoot of Utah and James Watson in Indiana. They revealed the same churlish resentment at the party in power during hard times which upended the Scullin Labor government in Australia, which put the Conservative Baldwin government in power in Britain, which, in short, concerned itself less with ideology and a program for recovery than with finding a target for its frustrated anger.

Many Americans came to despair of the whole political process, a contempt for Congress, for parties, for democratic institutions, which was caught by the relentless cynicism of *Of Thee I Sing*.[30] "There is no doubt in the world," wrote William Dodd, "that both political parties have been bankrupted."[31] The "lame duck" session from December, 1932, through February, 1933, further damaged the prestige of Congress. At a time of mounting crisis, Congress failed to produce a single important piece of economic legislation. Senator Borah said

[29] Louis Adamic, "Tragic Towns of New England," *Harper's*, CLXII (1931), 755; Davis, "The Collapse of Politics," *Harper's*, CLXV (1932), 386. When the banks in one small town failed, noted another writer, people expressed not indignation but the feeling that the community was the victim of "conditions," which had spread like some terrible malaise. At the same time, he remarked that the people no longer had the same faith in the old leaders. "There is a strong feeling of suspense in the air, an overcast sultriness. Anything may happen." George Clark, "Beckerstown: 1932," *Harper's*, CLXV (1932), 590–591.

[30] In his review of *Of Thee I Sing*, Joseph Wood Krutch wrote: "But it is doubtful if there has ever been a time before when the general public would have found pleasing so raucously contemptuous a treatment of the whole spectacle of our government from the President on up (or down) to the Supreme Court itself." *The Nation*, CXXXIV (1932), 56. "When you think of the political satire in the theater of the pre-Roosevelt days," another critic noted two years later, "you find that it was all cynical, mostly bitter, and sometimes too bitter to be funny." Hiram Motherwell, "Political Satire Scores on the Stage," *Today*, II (July 28, 1934), 24.

[31] William Dodd to E. Edgar Colby, June 8, 1932, Dodd MSS., Box 40. Professor of history at the University of Chicago, Dodd would be Roosevelt's first ambassador to Germany.

he had never seen Congress spend so much time on trivialities during a crisis, and Hiram Johnson wrote: "We're milling around here utterly unable to accomplish anything of real consequence."[32] Hoover hoped Congress would adjourn before it did any harm, business wished that Congress would not meet at all, and some critics echoed the demand of antiparliamentary leaders on the Continent that the "talking shops" be closed. Bickering and quarreling, the Senate watched helplessly while Louisiana's Huey Long paralyzed the legislature with a filibuster.[33]

The longer the depression persisted, the more people began to conjecture whether they were not witnessing the death of an era.[34] No longer did America seem a land bright with promise. In the spring of 1931, West African natives in the Cameroons sent New York $3.77 for relief for the "starving"; that fall Amtorg's New York office received 100,000 applications for jobs in Soviet Russia. On a single weekend in April, 1932, the *Ile de France* and other transatlantic liners carried nearly 4,000 workingmen back to Europe; in June, 500 Rhode Island aliens departed for Mediterranean ports.[35] Four days before his death, Calvin Coolidge declared: "In other periods of depression it has always been possible to see some things which were solid and upon which you could base hope, but as I look about, I now see nothing to give ground for hope—nothing of man."[36]

Society seemed to be disintegrating, a disjointedness caught in the fragmented "Newsreel" of Dos Passos' *U.S.A.*, in the theater scene in

[32] Henry Morrow Hyde MS. Diary, December 27, 1932; Hiram Johnson to Charles K. McClatchy, January 16, 1933, Johnson MSS.

[33] Charles A. Beard, "Congress Under Fire," *Yale Review*, XXII (1932), 35–51; Theodore Joslin, *Hoover Off the Record* (Garden City, N.Y., 1935), pp. 338–339. "In the meantime democratic gov. becomes a farce and nobody can see the outcome," noted Agnes Meyer. "Our economic life ebbs lower every day." Agnes Meyer MS. Diary, January 13, 1933.

[34] "One wonders if this paralysis of the power to produce and consume will drive us all to a life on the level of the Middle Ages," speculated a welfare worker. Amy Maher to Edward Costigan, February 16, 1932, Costigan MSS. V.F. 1. A writer asked: "Will future scholars date from 1929 the decline and fall of western civilization—as we now hear on every hand . . . ?" "The United States in the New Year," *Round Table*, XXII (1932), 308.

[35] Bernstein, *Lean Years*, p. 294; *Business Week*, October 7, 1931, pp. 32–33; *The New York Times*, May 1, 26, 1932.

[36] Charles Andrews, "I'm all burned out," in Edward Connery Lathem (ed.), *Meet Calvin Coolidge* (Brattleboro, 1960), p. 216. Cf. W. Cameron Forbes MS. Diary, December 31, 1931.

Nathanael West's *A Cool Million,* in the final anguished writing of Hart Crane, who took his life in 1932. The very word "depression" came to have a special meaning. Like a manic-depressive patient, noted one writer, the nation had moved from the unstable euphoria of 1929 to the depression of 1932 which was characterized by panic, utter hopelessness about the future, an inability to initiate or sustain any activity, and a mood of apprehension. "Corresponding to the fear of the patient that he will never recover," he observed, "is the current forecast of permanent depression."[37]

Many believed that the long era of economic growth in the western world had come to an end. The depression seemed to demonstrate with painful precision the corollary some had drawn from Frederick Jackson Turner's frontier thesis—that the consequence of the passing of the frontier was a "closed system." Less than two years after he took office, Roosevelt closed the last of the public domain.[38] Even more portentous were the statistics on population decline. For the first time in three centuries, the curve of American population growth was leveling off; an economy accustomed to the demand of a swiftly expanding population would have to adapt itself painfully to a stationary population. In the next decade, popular writers warned of the "baby crop shortage"; economists worried over surpluses and excess plant capacity; demographers fretted about the impending "deficit in our vital account"; and Stuart Chase predicted that, if present trends continued, the country in 1960 would face the crisis of "10,000,000 empty desks in schools and colleges."[39] New Deal theorists and business executives came to a like conclusion: that the stationary population pattern and the absence of new industries meant that the economy had reached a

[37] Elisha Friedman, "Business Psychosis," *The New York Times,* May 15, 1932. Cf. Arnold Toynbee, *Survey of International Affairs 1931* (London, 1932), pp. 1–5.

[38] Curtis Nettels, "Frederick Jackson Turner and the New Deal," *Wisconsin Magazine of History,* XVII (1934), 257–265; James Malin, "Mobility and History," *Agricultural History,* XVII (1943), 177–178; E. Louise Peffer, *The Closing of the Public Domain* (Stanford, 1951), pp. 223–224.

[39] *Rocky Mountain News,* March 1, 1933; Chicago *Daily Times,* September 10, 1934; George Wehrwein, "The Second Conservation Crusade," *Journal of Land and Public Utility Economics,* XII (1936), 422; Philip Hansen, "Population," *American Journal of Sociology,* XLVII (1942), 816–817; Alfred Lotka, "Modern Trends in the Birth Rate," *Annals of the American Academy of Political and Social Science,* CLXXXVIII (1936), 6; Stuart Chase, "Population Going Down," *Atlantic Monthly,* CLXIII (1939), 183.

point of saturation for many years to come. The idea of a "mature" economy was less an economic datum than an index of the despondency that overtook western societies in the 1930's.

Many argued that the country could get out of the morass of indecision only by finding a leader and vesting in him dictatorial powers.[40] Some favored an economic supercouncil which would ignore Congress and issue edicts; Henry Hazlitt proposed abandoning Congress for a directorate of twelve men.[41] Others wished to confer on the new president the same arbitrary war powers Woodrow Wilson had been granted.[42] Even businessmen favored granting Roosevelt dictatorial powers when he took office. Distressed by the chaotic competition in industries such as oil and textiles, alarmed by the outbursts of violence, convinced of the need for drastic budget slashing, they despaired of any leadership from Congress. "Of course we all realize that dictatorships and even semi-dictatorships in peace time are quite contrary to the spirit of American institutions and all that," remarked Barron's. "And yet—well, a genial and lighthearted dictator might be a relief from the pompous futility of such a Congress as we have recently had. . . . So we return repeatedly to the thought that a mild species of dictatorship will help us over the roughest spots in the road ahead."[43]

Meanwhile, Roosevelt and Hoover were playing cat and mouse. Immediately after the election, Hoover confronted a joint Anglo-French request for a postponement of the war debt installments which would fall due on December 15. It was a nettlesome problem, and the President hit on the bold scheme of inviting Roosevelt to confer with him to work out a common position on the debts. Hoover's canny move came to naught. Both men agreed that the debts should be paid, but beyond this they divided sharply. Hoover, who believed that the main

[40] "The time has come," declared Representative E. W. Pou of North Carolina, the dean of the House, "when some autocratic power should be put in the hands of somebody who can be trusted." Portland *Morning Oregonian*, February 11, 1933.

[41] Henry Hazlitt, "Without Benefit of Congress," *Scribner's*, XCII (1932), 13–18.

[42] Key Pittman to F.D.R., June 16, 1932, Pittman MSS., Box 16; Daniel Roper to William Dodd, November 16, 1932, Dodd MSS., Box 41; Henry Morrow Hyde MS. Diary, January 4, 1933.

[43] *Barron's*, XIII (February 13, 1933), 12. Cf. James Jackson to F.D.R., Mar. 28, 1933, FDRL PPF 279. For the Mussolini vogue, see Seward Collins, "The Revival of Monarchy," *American Review*, I (1933), 253, and Paul Y. Anderson, "Wanted: A Mussolini," *The Nation*, CXXXV (1932), 9.

cause of the depression lay abroad, thought the best hope of recovery lay in re-establishing the international gold standard. The New Dealers, on the other hand, gave domestic reforms highest priority, distrusted the international bankers who clamored for cancellation, and discounted the war debts issue as politically explosive but economically insignificant. A strained meeting between Hoover and Roosevelt at the White House on November 22 led to nothing.

Despite this rebuff, Hoover persisted in a bootless campaign to win Roosevelt's agreement to join him in appointing a single body of delegates to the debt negotiations. Two months of maneuvering only bred more distrust. Hoover's aides could barely conceal their contempt for the President-elect. Secretary of the Treasury Ogden Mills dismissed Roosevelt as a "Pollyanna," and Stimson thought Hoover had made the Governor "look like a peanut."[44] Roosevelt in turn suspected that Hoover was trying to trap him into endorsing his policies. Moreover, the President-elect had no intention of letting any of Hoover's unpopularity rub off on him.[45]

On February 17, 1933, Hoover once again appealed for co-operation from Roosevelt, but a peculiar brand of co-operation it was. He requested the President-elect to help "restore confidence" by promising to balance the budget and to maintain a sound currency when he took office. In fact, Hoover asked nothing less than that the man who had defeated him at the polls repudiate his own program and accept Hoover's view of the depression. "I realize that if these declarations be made by the President-elect," Hoover wrote Senator Reed of Pennsylvania a few days later, "he will have ratified the whole major program of the Republican Administration; that is, it means the abandonment of 90% of the so-called 'new deal.' "[46] Roosevelt all but

[44] "Confidential Files, Governor Harrison," November 11, 1932, "Telephone Conversation with Secretary Mills re formulating program for nonpartisan action in Congress during short session," George Leslie Harrison MSS., Conversations, II; Henry Stimson MS. Diary, December 21, 1932.

[45] Arthur Schlesinger, Jr., *The Crisis of the Old Order* (Boston, 1957), pp. 441–448; Herbert Hoover, "Memorandum of Conversation between President Hoover and Governor Roosevelt, January 20, 1933," in Henry Stimson MS. Diary, January 20, 1933; Rexford Tugwell, "Notes from a New Deal Diary," December 20, January 22, 1933; Raymond Moley, *After Seven Years* (New York, 1939), pp. 68–79, 86–90, 94–98; Robert Ferrell, *American Diplomacy in the Great Depression* (New Haven, 1957), pp. 232–246.

[46] W. S. Myers and W. H. Newton, *The Hoover Administration: A Documented Narrative* (New York, 1936), pp. 338–341, 351.

ignored Hoover's "cheeky" letter. Roosevelt's refusal to acquiesce, Hoover told Stimson, was the act of a madman.[47]

Roosevelt was everywhere—grinning, joking, waving his cigarette holder, giving the impression that he had not a care in the world, and leaving sober observers with the feeling that he was an immature politician with little sense of the seriousness of the tasks that lay ahead. His failure to co-operate with Hoover, Secretary Stimson noted, "showed a most laughable, if it were not so lamentable, an ignorance of the situation in which Roosevelt is going to find himself when he gets in on March 4th."[48] In fact, the President-elect was quietly going about the business of building a legislative program—sifting memoranda, conferring with advisers, wearing out whole teams of experts with his amazing energy, and impressing skeptics with his ability to grasp the crux of an argument or to retain the most minute details of economic data.

During his governorship, Roosevelt had consulted frequently with college professors on legislative policy. In March, 1932, he gave the green light to a proposal by Samuel Rosenman, his counsel in Albany, that he once more go to the universities to canvass ideas for the campaign. Rosenman turned to Raymond Moley, a shrewd, knowledgeable Barnard College professor, and Basil O'Connor, the Governor's law partner. Moley, who quickly established himself as the leader of the group, recruited two Columbia colleagues: the handsome Rexford Guy Tugwell, an authority on agriculture, and the precocious Adolf Berle, Jr., whose brilliant study, *The Modern Corporation and Private Property*, written with Gardiner Means, was published that summer. Roosevelt referred privately to this quintet as his "privy council," but in September, James Kieran of *The New York Times* called them the "brains trust" and it was this name, later shortened to "brain trust," which caught on. During the campaign, especially before the Chicago convention, they actually functioned as a Brain Trust. They argued economic doctrine through long spring evenings at the Governor's fireside in Albany, held audiences for economists in a hotel suite in New York City, and wrangled over drafts of campaign

[47] Henry Stimson MS. Diary, February 22, 1933; Schlesinger, *Coming of New Deal,* pp. 476–477; Moley, *After Seven Years,* pp. 139–144.

[48] Henry Stimson MS. Diary, December 20, 1932. The people wanted leadership, noted William Bullitt, but "they feel they are not going to get it from Roosevelt." William Bullitt to Edward House, House MSS., September 3, 1932. Cf. Clarence Cannon to William Hirth, November 25, 1932, Hirth MSS.

speeches. After the election, the Brain Trust disbanded. Moley, however, continued to serve as minister without portfolio in the months
before the inauguration; he interviewed experts, assigned men to draft
bills, and hammered out the legislation of the Hundred Days.[49]

Roosevelt welcomed all the advice he could get, but he had no
intention of being confined by rigid adherence to any particular doctrine. When Moley presented him with two sharply conflicting statements of tariff policy during the 1932 campaign and asked the
Governor to choose between them, Roosevelt rendered his adviser
speechless by telling him to "weave the two together." Both before
and after Roosevelt took office, many deplored the absence of a coherent New Deal ideology. Moley later commented: "But to look
upon these policies as the result of a unified plan was to believe that
the accumulation of stuffed snakes, baseball pictures, school flags, old
tennis shoes, carpenter's tools, geometry books, and chemistry sets in a
boy's bedroom could have been put there by an interior decorator."[50]

Yet too much can be made of the "ideological innocence" of the
New Deal.[51] By the time Roosevelt took office, he and his advisers had
developed, or familiarized themselves with, a body of theory a good
deal more coherent than is commonly suggested. The New Dealers
drew on a thousand books, on a storehouse of past political experiences.
From the Populists came a suspicion of Wall Street and a bevy of ideas
for regulating agriculture; from the mobilization of World War I
they derived instrumentalities for central direction of the economy;
from urban social reformers of the Jane Addams tradition arose a concern for the aged and the indigent. Frances Perkins, who would be the
first woman ever to serve in the cabinet, had been schooled at Hull-
House, Harry Hopkins at Christadora House; Adolf Berle, Henry
Morgenthau, Jr., and Herbert Lehman had all worked at Henry
Street. The New Dealers shared John Dewey's conviction that organized social intelligence could shape society, and some, like Berle, reflected the hope of the Social Gospel of creating a Kingdom of God
on earth.

[49] Moley, *After Seven Years*, pp. 5–148; Samuel Rosenman, *Working with Roosevelt* (New York, 1952), pp. 56–92; Moley, "There Are Three Brains Trusts," *Today*, I (April 14, 1934), 3 ff. After the Chicago convention, they added to their ranks General Hugh Johnson, a flamboyant, hard-driving associate of Bernard Baruch's.
[50] Moley, *After Seven Years*, pp. 48, 369–370.
[51] The phrase "ideological innocence" is Alistair Cooke's in *A Generation on Trial: U. S. A. v. Alger Hiss* (New York, 1950), p. 18.

Most influential were the theorists of the New Nationalism: Theodore Roosevelt, Charles Van Hise, Simon Patten, Walter Weyl, Herbert Croly, and the Walter Lippmann of *Drift and Mastery*.[52] Occasionally, there was an apostolic succession; both Tugwell and Miss Perkins had studied under Patten. Roosevelt's advisers of the early New Deal scoffed at the nineteenth-century faith in natural law and free competition; argued for a frank acceptance of the large corporation; and dismissed the New Freedom's emphasis on trust-busting as a reactionary dogma that would prevent an organic approach to directing the economy. Every member of the Brain Trust agreed, as Moley later explained, in a "rejection of the traditional Wilson-Brandeis philosophy that if America could once more become a nation of small proprietors, of corner grocers and smithies under spreading chestnut trees, we should have solved the problems of American life."[53]

The free market of Adam Smith, the New Dealers argued, had vanished forever. Tugwell pressed on Roosevelt his conviction that the competitive economy had never functioned well, that it could not be restored in any event, and that reforms which aimed at reconstituting the old order were futile.[54] The enormously influential studies of Berle and Gardiner C. Means reinforced these views. They concluded that monopoly and oligopoly had become not the exception but the rule; that the market no longer performed its classic function of maintaining an equilibrium between supply and demand; and that, in the new "administered market," the two thousand men who controlled American economic life manipulated prices and production. The new administration, the Brain Trust reasoned, had to strip itself of the illusion that it might recreate the classical society of small competitors and proceed with the structural reforms needed to stabilize the economy.[55]

[52] Thorstein Veblen had a special place: his emphasis on the hostility between technology and finance, his skepticism about an apocalyptic struggle ending in a dictatorship of the proletariat, and his advocacy of an elite of social engineers attracted men like Tugwell, Jerome Frank, and Isador Lubin.

[53] Moley, *After Seven Years*, p. 24.

[54] Tugwell, *The Democratic Roosevelt* ·(Garden City, N.Y., 1957), p. 213. "The cat is out of the bag. There is no invisible hand. There never was," Tugwell wrote. "We must now supply a real and visible guiding hand to do the task which that mythical, nonexistent, invisible agency was supposed to perform, but never did." Tugwell, *The Battle for Democracy* (New York, 1935), p. 213.

[55] Berle and Means, *The Modern Corporation and Private Property* (New

Each member of the Brain Trust believed in some kind of business-government co-operation, yet in their common aversion to other alternatives—laissez-faire, trust-busting, or socialism—they managed to conceal how sharply they differed on what they meant by co-operation. Moley wanted government to work with the businessman, but it was business rather than government on which he relied. Berle, a clergyman's son, hoped to evangelize the businessman rather than resort to the naked power of the state. Tugwell distrusted business and placed his confidence in government planners who would administer the economy. These conflicting conceptions in the midst of apparent harmony have made it possible for different writers to see in the "First New Deal" either a surrender to large business interests which the reforms of 1935 and after remedied, or an ambitious program of national planning which later developments cut off.

The word that appears most frequently in the writings of New Deal theorists is "balance." Like eighteenth-century savants captivated by Montesquieu, they believed the best society was one in which no important element held preponderant power. They thought the depression had been precipitated by a number of imbalances, but above all they were disturbed by the disproportionate power wielded by business in comparison to that possessed by agriculture. Although critics stamped the Roosevelt administration a "labor government," the New Dealers had far less interest in labor than in the farmer. "Since my graduate-school days," Tugwell confessed, "I have always been able to excite myself more about the wrongs of farmers than those of urban workers."[56] To be sure, they hoped to see the adoption of a series of welfare measures such as the abolition of child labor and the creation of social insurance systems; but when they talked of lack of purchasing power, it was farmers more than workers they had in mind. They gave highest priority to raising farm prices in order to restore the balance between industry and agriculture and to provide business with a vast home market.[57]

York, 1932); Schlesinger, *Coming of New Deal,* pp. 190–198; Richard Kirkendall, "A. A. Berle, Jr., Student of the Corporation, 1917–1932," *Business History Review,* XXXV (1961), 43–58.

[56] Tugwell, "Notes from a New Deal Diary," December 31, 1932.

[57] William McAdoo to Francis Heney, December 3, 1932, Heney MSS.; Tugwell, *Roosevelt,* p. 215; Louis Howe, "Balanced Government, The Next Step," *American Magazine,* CXV (June, 1933), 11–12, 84–86.

The New Dealers wished to co-operate with business, but they distrusted financiers. They determined to eliminate the abuses of the financial system by subjecting it to federal regulation. Many of the New Nationalists agreed this should be done, but they thought such regulation was not central. The main thrust for disciplining Wall Street came from the old Wilsonians. While Roosevelt in 1933 relied mostly on the New Nationalists, he lent an ear to such Brandeisians as Felix Frankfurter too. Roosevelt, born into the stable Knickerbocker society of Edith Wharton and schooled at Groton by Rector Endicott Peabody in the duty of service to the less fortunate, had the squire's sense of *noblesse oblige* and contempt for the grasping speculator. More than most of his advisers, he believed the government should curb "the lone wolf, the unethical competitor, the reckless promoter, the Ishmael or Insull."[58]

Roosevelt, who instinctively accepted the underconsumptionist explanation of the cause of the depression, was still groping for a lever to boost wages and prices. Some economists, influenced by the American William Trufant Foster and the English theorists J. A. Hobson and John Maynard Keynes, recommended government spending as the way to revive purchasing power. Although they had won some converts —even the orthodox Edwin R. A. Seligman had put forward a cautious defense of deficit spending—their influence on the Roosevelt circle was still slight. Tugwell and Berle, who drew on the institutional school of Veblen, Patten, and John Commons, agreed that the crisis centered in a failure of purchasing power but espoused structural reform rather than deficit spending.[59]

On the eve of Roosevelt's inauguration there was at least as much popular demand for economy as for mammoth government spending. Iowa farmers who denounced eastern bankers felt just as irate at officeholders or teachers whose salaries were blamed for the intolerable tax burden. In Routt County, Colorado, where more than one-third of the taxable area had already been sold for taxes, two hundred taxpayers threatened to go on strike unless county officials accepted a 10 per cent salary cut. Minnesota farmers compelled the radical Governor Floyd Olson to trim his budget to the lowest point since 1921. Early

[58] *Public Papers*, I, 755. Cf. Frank Freidel, *Franklin D. Roosevelt: The Apprenticeship* (Boston, 1952), pp. 14–51.
[59] Joseph Dorfman, *The Economic Mind in American Civilization* (5 vols., New York, 1959), V, 672–673, 722; Schlesinger, *Coming of New Deal*, Ch. 23.

in 1933, the president of the United States Chamber of Commerce, Henry Harriman, regarded as the spokesman for enlightened business, advocated a slash of more than $1 billion in government spending by returning federal activities to the 1925 basis of operations.[60]

Roosevelt gave respectful attention to such counsel, because he had always believed in budget balancing, and because the advice came from such a wide spectrum of opinion within the party. Bernard Baruch, who exerted uncommon influence over southern Democratic senators, exhorted: "With the monotony and persistence of old Cato, we should make one single and invariable dictum the theme of every discourse: 'Balance budgets. . . . Cut governmental spending—cut it as rations are cut in a siege. Tax—tax everybody for everything.' " Old Wilsonians like Breckinridge Long proposed to eradicate government agencies "by the simple expedient of failing to appropriate for bureaus."[61] Progressives such as Robert Wagner, Robert La Follette, and Fiorello La Guardia all opposed deficit spending.[62]

In shaping their legislative program, the New Dealers had to take account of the rising clamor for inflation. In the lame-duck session, congressmen introduced more than fifty inflation proposals, and in the recent campaign one minor party had even nominated "Coin" Harvey for President on a program of coining free silver at 16 to 1. Influential

[60] Denver *Post*, February 14, 1933; Mayer, *Floyd Olson*, p. 137; Jackson (Miss.) *Daily Clarion Ledger*, January 5, 1933. Cf. Lewis Kimmel, *Federal Budget and Fiscal Policy* (Washington, 1959), pp. 167–174.

[61] U.S. Congress, Senate, *Investigation of Economic Problems*, Hearings before Committee on Finance, U.S. Senate, 72d Cong., 2d Sess., pursuant to S. Res. 315 (Washington, 1933), p. 18; Baruch to F.D.R., April 1, 1932, Roosevelt Family Papers (Halsted), Box 1; Baruch to Elmer Thomas, April 2, 1932, in "40 Years a Legislator," pp. 210–216, Elmer Thomas MSS.; Long to House, November 11, 1932, House MSS.

[62] Howard Zinn, *La Guardia in Congress* (Ithaca, 1958), p. 195. Cf. Arthur Mann, *La Guardia* (Philadelphia, 1959), pp. 309, 322. Charles A. Beard cautioned that elaborate public works might "withdraw millions and billions from fruitful capital investments." Beard, "A 'Five-Year Plan' for America," *Forum*, LXXXVI (1931), 4. In the spring of 1932, *The Nation* commented: "Senator Borah and the other Senators who are planning to cut expenses by a straight 25 per cent are on the wisest, the inevitable track. Mr. Hoover again revealed his feebleness when he admitted that he had cut only $365,000,000 out of the budget." *The Nation*, CXXXIV (1932), 384. Alvin Hansen, who in a few years would be John Maynard Keynes's leading American disciple, opposed massive public works spending. Hansen, *Economic Stabilization in an Unbalanced World* (New York, 1932); Hansen to Edward Costigan, August 5, 1932, Costigan MSS., V.F. 1.

leaders, among them the Iowa editor Henry Wallace, espoused Irving
Fisher's doctrine of "reflation," a euphemism for boosting prices to the
point where they could be stabilized. "The word inflation grates
harshly on the ears of many bankers," Wallace explained. "Therefore,
we speak of reflation and save their feelings."[63]

The Brain Trust, although it did not shudder at inflation as sac-
rilegious, preferred to deal with debts by less crude devices than tam-
pering with the currency. When Wallace went to New York, he was
disturbed to find that Roosevelt was being advised by Tugwell, a hard-
money man.[64] For the most part, Roosevelt shared the orthodox views
of his advisers, but he studiously avoided a firm commitment to the
gold standard. Distrustful of the President-elect's monetary views,
Carter Glass spurned an invitation to become Secretary of the Treasury
and James P. Warburg turned down a bid to be Under Secretary. Both
men sensed that while Roosevelt had not yet committed himself, he
would resort to inflation if commodity prices continued to fall.[65]

While the New Dealers churned ideas and blue-penciled drafts of
legislation, the long financial illness which had gripped the country
reached a crisis. As early as October, 1932, the governor of Nevada
had proclaimed a bank holiday, and in the next two months an epidemic of bank failures in the western states created growing alarm.[66]
The real panic began with the collapse of Detroit's banks in February,
1933. Early in the morning of St. Valentine's Day, February 14, Governor William Comstock of Michigan declared an eight-day bank
holiday which tied up the funds of 900,000 depositors and froze $1.5
billion in bank deposits.[67] The next week brought frightful tidings of

[63] Wallace to William Hirth, February 17, 1932, Hirth MSS.: Theodore
Christianson, "The Mood of the Mid-West," *Current History,* XXXVI (1932),
141–142; Cincinnati *Enquirer,* December 31, 1932; John Simpson to F.D.R.,
August 27, 1932, Simpson MSS.

[64] Wallace to William Hirth, August 20, 1932, Hirth MSS.

[65] Glass's decision was motivated in part by reasons of health. For the symp-
tomatic Glass episode, see Glass to F.D.R., February 7, 1933, Glass MSS., Box
6; Jay Pierrepont Moffat MS. Diary, January 29, February 18, 1933; Henry
Morrow Hyde MS. Diary, February 5, 1933; *The New York Times,* February
22, 1933; Rixey Smith and Norman Beasley, *Carter Glass* (New York, 1939),
pp. 329–338.

[66] Reno *Evening Gazette,* November 1, 1932; Hiram Johnson to Charles K.
McClatchy, January 29, 1933, Johnson MSS.; Jackson Reynolds, COHC, pp.
164–165.

[67] Detroit *Free Press,* February 15, 1933; Harry Barnard, *Independent Man:
The Life of Senator James Couzens* (New York, 1958), pp. 213–247.

a fall in security prices, a flight of gold to Europe, and the precarious state of some of the country's leading banks. On February 24, following a run on Baltimore banks, Governor Ritchie of Maryland declared a three-day bank holiday. By the end of the month, banks in every section of the country were in trouble—a Wichita bank closed, Little Rock limited withdrawals, the fourth largest bank in Washington, D.C., closed its doors. People stood in long queues with satchels and paper bags to take gold and currency away from the banks to store in mattresses and old shoe boxes. It seemed safer to put your life's savings in the attic than to trust the greatest financial institutions in the country.[68]

In Hoover's last days in office, the old order tottered on the brink of disaster. "World literally rocking beneath our feet," noted Agnes Meyer. "Hard on H to go out of office to the sound of crashing banks. Like the tragic end of a tragic story. . . . The history of H's administration is Greek in its fatality."[69] As the debacle in Maryland followed the disaster in Michigan, Franklin Roosevelt wondered aloud whether the whole house of cards might not have fallen by Inauguration Day.[70] On March 1, the governors of Kentucky and Tennessee proclaimed bank holidays; that night the governors of California, Louisiana, Alabama, and Oklahoma pursued the same course. By March 4, the day of Roosevelt's inauguration, thirty-eight states had closed their banks, and banks operated on a restricted basis in the rest. Shortly before dawn, Governor Herbert Lehman of New York and Governor Henry Horner of Illinois suspended the banks in the two great states that dominated the financial life of the nation. On the morning of March 4, as brokers crowded forward to hear the signal to begin trading, Richard Whitney climbed the rostrum to announce the closing of the Stock Exchange. Kansas City's Board of Trade suspended operation; Savannah ended trading in naval stores; New York closed its markets in cocoa, metal and rubber; and, for the first time since 1848, the Chicago Board of Trade shut its doors. "I do not know how it may have been in other places, but in Chicago, as we saw it, the city seemed to have died," wrote one woman. "There was something awful—ab-

---

[68] C. C. Colt and N. S. Keith, *28 Days: A History of the Banking Crisis* (New York, 1933).

[69] Agnes Meyer MS. Diary, February 25, 1933. See Jay Pierrepont Moffat MS. Diary, February 25, 1933.

[70] Hiram Johnson to Charles K. McClatchy, February 26, 1933, Johnson MSS.

normal—in the very stillness of those streets. I recall being startled by the clatter of a horse's hooves on the pavement as a mounted policeman rode past."[71] In the once-busy grain pits of Chicago, in the canyons of Wall Street, all was silent.

[71] Louise Armstrong, *We Too Are the People* (Boston, 1938), p. 50; F. Cyril James, *The Growth of Chicago Banks* (2 vols., New York, 1938), II, 1059–1065, 1078.

CHAPTER 3

The Hundred Days' War

"F̲IRST of all," declared the new President, "let me assert my firm belief that the only thing we have to fear is fear itself—nameless, unreasoning, unjustified terror. . . ." Grim, unsmiling, chin uplifted, his voice firm, almost angry, he lashed out at the bankers. "We are stricken by no plague of locusts. . . . Plenty is at our doorstep, but a generous use of it languishes in the very sight of the supply. Primarily this is because rulers of the exchange of mankind's goods have failed through their own stubbornness and their own incompetence, have admitted their failure, and have abdicated. . . . The money changers have fled from their high seats in the temple of our civilization. We may now restore that temple to the ancient truths."

The nation, Roosevelt insisted, must move "as a trained and loyal army willing to sacrifice for the good of a common discipline." He would go to Congress with a plan of action, but if Congress did not act and the emergency persisted, the President announced, "I shall not evade the clear course of duty that will then confront me. I shall ask the Congress for the one remaining instrument to meet the crisis —broad Executive power to wage a war against the emergency, as great as the power that would be given to me if we were in fact invaded by a foreign foe."[1]

In the main part of his Inaugural Address, his program for recovery, he had little new to offer. What he did say was so vague as to be

[1] Samuel Rosenman (ed.), *The Public Papers and Addresses of Franklin D. Roosevelt* (13 vols., New York, 1938–50), II, 11–15.

open to any interpretation; currency, he opined, should be "adequate but sound." "He is for sound currency, but lots of it," one congressman complained.[2] Yet this was a new Roosevelt; the air of casual gaiety, of evasiveness, had vanished—the ring of his voice, the swing of his shoulders, his call for sacrifice, discipline, and action demonstrated he was a man confident in his powers as leader of the nation. In declaring there was nothing to fear but fear, Roosevelt had minted no new platitude; Hoover had said the same thing repeatedly for three years. Yet Roosevelt had nonetheless made his greatest single contribution to the politics of the 1930's: the instillation of hope and courage in the people. He made clear that the time of waiting was over, that he had the people's interests at heart, and that he would mobilize the power of the government to help them. In the next week, nearly half a million Americans wrote their new President. He had made an impression which Hoover had never been able to create—of a man who knew how to lead and had faith in the future.

Roosevelt moved swiftly to deal with the financial illness that paralyzed the nation. On his very first night in office, he directed Secretary of the Treasury William Woodin to draft an emergency banking bill, and gave him less than five days to get it ready. Woodin found the Treasury corridors prey to rumor, the bankers empty of ideas and queasy with fear of new calamities. To buy Woodin time to prepare legislation, and to protect the nation's dwindling gold reserves, Roosevelt assumed the posture of a commander in chief in wartime. On Sunday afternoon, March 5, he approved the issue of two presidential edicts—one called Congress into special session on March 9; the other, resting on the rather doubtful legal authority of the Trading with the Enemy Act of 1917, halted transactions in gold and proclaimed a national bank holiday.[3]

The very totality of the bank holiday helped snap the tension the country had been under all winter. "Holiday" was a delightful euphemism, and the nation, responding in good spirit, devised ingenious ways to make life go on as it always had. In Michigan, Canadian money circulated; in the Southwest, Mexican pesos; the Dow Chemical Company paid its workers in coins made of Dowmetal, a mag-

[2] *Literary Digest*, CXV (March 11, 1933), 5.
[3] Arthur Schlesinger, Jr., *The Coming of the New Deal* (Boston, 1959), pp. 4–5; Agnes Meyer MS. Diary, March 6, 1933.

nesium alloy; the *Princetonian* printed twenty-five-cent scrip for students. In New York City, wealthy ladies invaded automats to get twenty nickels; speakeasies extended credit; Roseland Dance Hall accepted I.O.U.'s from dancers with bank books; the Hotel Commodore sent a bellhop to a nearby church to get silver from the collection plate in exchange for bills. In most of the nation, inconveniences were minor; only in Michigan, in its fourth week of a bank moratorium, was the situation becoming critical. In Detroit, movies played to near-empty theaters, and city laborers, unable to buy food with their pay checks, fainted at work.[4]

On March 9, the special session of Congress convened in an atmosphere of wartime crisis. Shortly before 1 P.M., Roosevelt's banking message was read, while some newly elected congressmen were still trying to find their seats. The House had no copies of the bill; the Speaker recited the text from the one available draft, which bore last-minute corrections scribbled in pencil. Members found the proposal an exceptionally conservative document. Roosevelt's assault on the bankers in his inaugural address had invited speculation that he might advocate radical reforms, even nationalization of the banks. Instead, the emergency banking measure extended government assistance to private bankers to reopen their banks. The bill validated actions the President had already taken, gave him complete control over gold movements, penalized hoarding, authorized the issue of new Federal Reserve bank notes, and arranged for the reopening of banks with liquid assets and the reorganization of the rest. The bill was largely the work of bankers such as George Harrison and of Hoover's Treasury officers. Arthur Ballantine, Hoover's Under Secretary of the Treasury, continued in the same post under Roosevelt, while Ogden Mills, Hoover's Secretary of the Treasury, hovered in the background. The emergency banking bill represented Roosevelt's stamp of approval for decisions made by Hoover's fiscal advisers.[5]

With a unanimous shout, the House passed the bill, sight unseen,

---

[4] *Time*, XXI (March 6, 1933), 18; (March 13, 1933), 45–46; *Business Week* (March 15, 1933), 6–7; William Manchester, "The Great Bank Holiday," *Holiday*, XXVII (February, 1960), 60–61, 137–143; Meyer Levin, *The Old Bunch* (New York, 1937), pp. 835–837. Cf. Walton Hamilton, "When the Banks Closed," in Daniel Aaron (ed.), *America in Crisis* (New York, 1952), pp. 267–284.

[5] *The New York Times*, March 10, 1933; Arthur A. Ballantine, "When All

after only thirty-eight minutes of debate. Speaker Rainey observed that the situation recalled the world war, when "on both sides of this Chamber the great war measures suggested by the administration were supported with practical unanimity. . . . Today we are engaged in another war, more serious even in its character and presenting greater dangers to the Republic." Representative Robert Luce of Massachusetts explained to his Republican colleagues: "The majority leaders have brought us a bill on which I myself am unable to advise my colleagues, except to say that this is a case where judgment must be waived, where argument must be silenced, where we should take matters without criticism lest we may do harm by delay." The Senate, over the objections of a small band of progressives, approved the bill unamended 73–7 at 7:30 that evening and at 8:36 that same night it received the President's signature.[6] One congressman later complained: "The President drove the money-changers out of the Capitol on March 4th—and they were all back on the 9th."[7]

On Sunday night, March 12, an estimated sixty million people sat around radio sets to hear the first of President Roosevelt's "fireside chats." In warmly comforting tones, the President assured the nation it was now safe to return their savings to the banks. The next morning, banks opened their doors in the twelve Federal Reserve Bank cities. Nothing so much indicated the sharp shift in public sentiment as the fact that people were now more eager to deposit cash than to withdraw it. "The people trust this admin. as they distrusted the other," observed Agnes Meyer. "This is the secret of the whole situation." Since the President had said the banks were safe, they were; Roosevelt, observed Gerald Johnson, "had given a better demonstration than Schopenhauer ever did of the world as Will and Idea." Deprived of cash for several days, people had been expected to withdraw funds to meet

the Banks Closed." *Harvard Business Review,* XXVI (1948), 138–139; Ogden Mills to George Dayton, September 15, 1933, Mills MSS., Box 11; Jackson Reynolds, COHC, p. 171.

[6] *Congressional Record,* 73d Cong., 1st Sess., pp. 70, 79; *Rocky Mountain News,* March 10, 1933. Hiram Johnson reported "a distinct, nasty undercurrent of feeling in the Senate . . . that a fast one has been put over." Johnson to John Francis Neylan, March 10, 1933, Johnson MSS. Cf. Pat McCarran to George Thatcher, March 13, 1933, McCarran MSS., File 764.

[7] William Lemke to the Farmers' Union convention in Omaha, quoted in Lorena Hickok to Harry Hopkins, November 23, 1933, Hopkins MSS.

immediate needs, yet in every city deposits far exceeded withdrawals. The crisis was over. "Capitalism," Raymond Moley later concluded, "was saved in eight days."[8]

On March 10, Roosevelt fired his second message at Congress. He requested sweeping powers to slice $400 million from payments to veterans and to slash the pay of federal employees another $100 million. "Too often in recent history," the President warned, "liberal governments have been wrecked on rocks of loose fiscal policy."[9] The economy bill reflected the influence of Director of the Budget Lewis Douglas, who thought that Hoover had indulged in "wild extravagance."[10] Convinced that the proposal echoed the demand of Wall Street for wage slashing and that it was cruel to veterans, the Democratic caucus in the House refused to support the President. One congressman protested that the bill would benefit "big powerful banking racketeers," while another called it "a slaughter of the disabled servicemen of the United States." At any other time, such appeals would have carried both chambers, but they made little headway against the power of the President in a time of crisis. Representative John Young Brown of Kentucky declared: "I had as soon start a mutiny in the face of a foreign foe as start a mutiny today against the program of the President of the United States." "When the *Congressional Record* lies on the desk of Mr. Roosevelt in the morning he will look over the roll call," warned Congressman Clifton Woodrum of Virginia, "and from that he will know whether or not the Members of his own party were willing to go along with him in his great fight to save the country."[11] Although more than ninety Democrats broke with Roosevelt in the House, most of them heeded Woodrum's counsel. After only two days' debate, Congress passed the economy bill. Under the leadership of Franklin Roosevelt, the budget balancers had won a victory for orthodox finance that had not been possible under Hoover.

[8] Agnes Meyer MS. Diary, March 14, 1933; George Fort Milton to William McAdoo, March 12, 1933, Milton MSS.; Gerald Johnson, *Roosevelt: Dictator or Democrat?* (New York, 1941), p. 222; Raymond Moley, *After Seven Years* (New York, 1939), p. 155.

[9] *Public Papers*, II, 50.

[10] Schlesinger, *Coming of New Deal*, p. 9; Turner Catledge, "A Hard-Hitter Strikes at the Budget," *The New York Times Magazine*, March 19, 1933, pp. 3, 15.

[11] *Congressional Record*, 73d Cong., 1st Sess., pp. 206, 209, 211, 214.

The staccato rhythm of the Hundred Days had begun: on Thursday, Congress adopted the bank bill; on Saturday, it passed the economy measure; on Monday, March 13, the President asked Congress to fulfill the Democratic pledge of an early end to prohibition. Roosevelt's victory had speeded a remarkable revolution in sentiment; even some veteran dry congressmen now voted for liquor. In February, 1933, the wets had broken a fainthearted Senate filibuster, and the lame-duck Congress, in a startling reversal of attitude, voted to repeal the Eighteenth Amendment.[12] While the Twenty-first Amendment was making its way through the states, Roosevelt requested quick action to amend the Volstead Act by legalizing beer of 3.2 per cent alcoholic content by weight.

Roosevelt's message touched off a raucous, rollicking debate. The drys, who had succeeded in killing a beer bill only a few weeks before, rehearsed the arguments that had been so convincing for more than a decade, but to no avail. Representative John Boylan of New York protested that this was "the same old sob story you have been telling us for the last 12 years. Why, I almost know your words verbatim—the distressed mother, the wayward son, the unruly daughter, the roadhouse, and so forth, and so forth. . . ."[13] Impatient congressmen chanted: "Vote—vote—we want beer"; within a week both houses had passed the beer bill, and added wine for good measure, although congressmen protested that 3.2 wine was not "interesting." On March 22, Roosevelt signed the bill.

As the day of liberation approached, breweries worked feverishly to meet the anticipated demand. Operating around the clock, a St. Joseph, Missouri, factory turned out ten tons of pretzels a day, but still ran two months behind. Emanuel Koveleski, president of the New York Bartenders' Union, proclaimed: "We're all set. The bartenders will be all fine, clean, upstanding young men." On April 7, 1933, beer was sold legally in America for the first time since the advent of prohibition, and the wets made the most of it. In New York, six stout brewery horses drew a bright red Busch stake wagon to the Empire State Building, where a case of beer was presented to Al Smith, "the

---

[12] Representative Clarence Cannon to Bishop James Cannon, Jr., January 5, 1933, James Cannon, Jr., MSS.; Hiram Johnson to Charles K. McClatchy, February 19, 1933, Johnson MSS.

[13] *Congressional Record*, 73d Cong., 1st Sess., p. 290.

martyr of 1928." In the beer town of St. Louis, steam whistles and sirens sounded at midnight, while Wisconsin Avenue in Milwaukee was blocked by mobs of celebrants standing atop cars and singing "Sweet Adeline." The most important ritual took place at the Rennert Hotel bar in Baltimore: H. L. Mencken downed a glass of the new beer and assured anxious connoisseurs it was a worthy brew.[14]

Two weeks after Roosevelt took office, the country seemed a changed place. Where once there had been apathy and despondency, there was now an immense sense of movement. If the country did not know in what direction it was moving, it had great expectations; the spell of lassitude had been snapped. On the walls of Thomas A. Edison, Inc., in West Orange, New Jersey, President Charles Edison posted a notice:

> President Roosevelt has done his part: now you do something.
>
> Buy something—buy anything, anywhere; paint your kitchen, send a telegram, give a party, get a car, pay a bill, rent a flat, fix your roof, get a haircut, see a show, build a house, take a trip, sing a song, get married.
>
> It does not matter what you do—but get going and keep going. This old world is starting to move.[15]

In a single fortnight, Roosevelt, wrote Walter Lippmann, had achieved a recapture of morale comparable to the "second battle of the Marne in the summer of 1916."[16]

Yet what had Roosevelt done? He had pursued a policy more ruthlessly deflationary than anything Hoover had dared. With $4 billion tied up in closed banks, he had embarked on a retrenchment program which still further eroded purchasing power. In his circumspect treatment of the banks, in his economy message, in his beer bill, Roosevelt had summed up the program of the arch-conservative du Pont wing of the Democratic party, the very men who had fought him so bitterly

[14] *The New York Times*, March 27, 1933; *Time*, XXI (April 17, 1933), 13.
[15] *Time*, XXI (April 3, 1933), 43.
[16] New York *Herald-Tribune*, March 15, 1933. "Only a few weeks ago anybody looking at the situation in the United States could only have done so with feelings of the gravest anxiety," observed Neville Chamberlain, Chancellor of the Exchequer, in the House of Commons on March 22. "To-day, thanks to the initiative, courage, and wisdom of the new President, a change has taken place which might almost be called miraculous. . . . A new sense of hope and anticipation for the future is coming back to the American people, and that confidence is being reflected over here in the City of London. . . ." Charles T. Hallinan, "Roosevelt as Europe Sees Him," *Forum*, LXXXIX (1933), 348.

in Chicago. Under the leadership of John Raskob, they had proposed to bring the budget into balance by cutting federal spending and by creating new sources of revenue from legalized sales of liquor. Now Roosevelt had done for them what they had been unable to achieve themselves.[17]

No one knew this better than Roosevelt. "While things look superficially rosy," he wrote Colonel House, "I realize well that thus far we have actually given more of deflation than of inflation. . . . It is simply inevitable that we must inflate and though my banker friends may be horrified, I still am seeking an inflation which will not wholly be based on additional government debt."[18] Roosevelt had originally planned a quick session of Congress which would adjourn as soon as it had dealt with the banking crisis, but it responded so well to his first proposals that he decided to hold it in session. On March 16, the President charted a new course by sending his farm message to Congress.

In framing a farm bill, Secretary of Agriculture Henry Wallace preferred the "domestic allotment" plan developed by several economists —John Black of Minnesota and Harvard, W. J. Spillman of the Department of Agriculture, and Beardsley Ruml of the Rockefeller Foundation—and advanced most persuasively by Milburn L. Wilson of Montana State College. The domestic allotment proposal aimed to deal with the crucial problem of depressed prices and mounting surpluses.[19] Hoover's Farm Board had tried to raise prices without curbing production, only to see a half-billion dollars go down the drain with no substantial benefit to farmers.[20] Wilson and the other advocates of the domestic allotment plan proposed to restrict acreage; levy a tax

[17] Gerald Johnson, *Roosevelt: Dictator or Democrat?*, p. 217; Marriner Eccles, *Beckoning Frontiers* (New York, 1951), pp. 124–125; George Fort Milton to Daniel Roper, March 27, 1933, Milton MSS., Box 13.

[18] F.D.R. to E. M. House, April 5, 1933, House MSS.

[19] "The Malthusian thesis of a population pressing upon the food supply," observed Tugwell, had changed to a "food supply pressing upon the population." Tugwell, *The Battle of Democracy* (New York, 1935), p. 109.

[20] In the summer of 1931, faced by a glut of millions of bales of cotton, the chairman of the Federal Farm Board had wired cotton-state governors "to induce immediate plowing under of every third row of cotton now growing." *The New York Times*, August 13, 1931; Federal Farm Board, *Third Annual Report* (Washington, 1932); Murray Benedict, *Farm Policies of the United States* (New York, 1953), pp. 257–267; Milton Garber to L. A. Dillingham, February 15, 1932, Garber MSS.

on the processors of agricultural commodities (for example, the miller who converted wheat into flour) ; and pay farmers who agreed to limit production benefits based on "parity," which would give the farmer the same level of purchasing power he had had before the war.[21]

Roosevelt had his own ideas about a satisfactory farm program— he disliked dumping, wanted decentralized administration, and stipulated the plan should obtain the consent of a majority of the farmers —but, above all, he insisted that farm leaders themselves agree on the kind of bill they wanted. In this fashion, he avoided antagonizing farm spokesmen by choosing one device in preference to another, and threw the responsibility for achieving a workable solution on the farm organizations. Not a few farm leaders disputed Roosevelt's conception of the problem. Processors and old McNary-Haugenites opposed reducing acreage and favored instead a combination of marketing agreements with dumping surpluses abroad. But they were willing to bargain and so was the administration. Wallace, who wanted a farm act before planting time, proposed an omnibus bill which would embody different alternatives, and that, after a series of conferences with farm spokesmen, was what he got.[22]

The House quickly passed the farm bill without change, but the Senate balked at speedy action. The measure shocked conservatives and drew the wrath of lobbyists for the processors—millers, packers, canners, and others—who objected to the proposed processing tax.[23] At the same time, the radical wing of the farm movement protested that the farmer deserved nothing less than government guarantee of his "cost of production," a conception even more elusive than "parity." Tugwell shrewdly observed: "For real radicals such as Wheeler, Frazier, etc., it is not enough; for conservatives it is too much; for Jefferson Democrats it is a new control which they distrust. For the economic

[21] Gertrude Almy Slichter, "Franklin D. Roosevelt and the Farm Problem, 1929–1932," *Mississippi Valley Historical Review*, XLIII (1956), 238–258; "The Voluntary Domestic Allotment Plan for Wheat," Food Research Institute, Stanford University, *Wheat Studies*, IX (1932), 23–62.

[22] Rexford Tugwell, *The Democratic Roosevelt* (Garden City, N.Y., 1957), pp. 275–277; "Memorandum on the Origins of AAA," Tugwell MSS., FDRL; George Peek MS. Diary, November, 1932–April, 1933; William Hirth to Louis Howe, December 21, 1932, Hirth MSS.; Mordecai Ezekiel, Rexford Tugwell, Louis Bean, O. C. Stine, COHC.

[23] Schlesinger, *Coming of New Deal*, pp. 40–41; Josiah Bailey to Josephus Daniels, May 5, 17, 1933, Daniels MSS., Box 698.

philosophy which it represents there are no defenders at all. Nevertheless, in spite of everything, it will probably become law."[24]

As the Senate debated the farm bill, inflation sentiment grew like the green bay tree. Despite administration opposition, the Senate on April 17 came within ten votes of adopting an amendment by Senator Burton Wheeler of Montana for the free coinage of silver, and Roosevelt was informed that "well over ten" senators had voted against the Wheeler amendment, or had refrained from voting, who favored some kind of inflation. The situation in the House, where nearly half the members had signed a petition to take up a bill to issue greenbacks, was little better. Speaker Rainey confided: "I am an irreconcilable Bryan 16 to 1 man."[25]

On April 18, Senate leaders warned Roosevelt that a new inflationary amendment to the farm bill, sponsored by Senator Elmer Thomas of Oklahoma, could not be defeated. Recognizing that the situation would soon be out of hand, Roosevelt decided to accept the Thomas proposal if it was rewritten to give the President discretionary powers rather than making any specific course of inflationary action mandatory. In its revised form, the Thomas amendment authorized the President to bring about inflation through remonetizing silver, printing greenbacks, or altering the gold content of the dollar. That night, hearing of Roosevelt's capitulation to the soft-money men, Lewis Douglas cried: "Well, this is the end of Western civilization."[26]

The following day, Roosevelt, confined in bed by a sore throat, announced with a smile to the 125 newspapermen gathered in his bedroom that the United States was off the gold standard.[27] The country had, to be sure, been on a greatly modified gold basis for some weeks. The President wrote later of an encounter with Secretary Woodin on April 20: "His face was wreathed in smiles, but I looked at him and

---

[24] Tugwell, "Notes from a New Deal Diary," March 31, 1933. Cf. Gilbert Fite, "Farmer Opinion and the Agricultural Adjustment Act, 1933," *Mississippi Valley Historical Review*, XLVIII (1962), 656–673.

[25] Moley, *After Seven Years*, pp. 156–158; Elbert Thomas to Delbert Draper, April 22, 1933, Elbert Thomas MSS., Box 14; Jeannette Nichols, "Silver Inflation and the Senate in 1933," *Social Studies*, XXV (1934), 12–18; Henry Morrow Hyde MS. Diary, April 17, 19, 1933.

[26] Moley, *After Seven Years*, p. 160; "40 Years a Legislator," pp. 224–236, Elmer Thomas MSS.

[27] Henry Morrow Hyde MS. Diary, April 19, 1933; Arthur Whipple Crawford, *Monetary Management under the New Deal* (Washington, 1940), pp. 36–40.

said: 'Mr. Secretary, I have some very bad news for you. I have to announce to you the serious fact that the United States has gone off the gold standard.' Mr. Woodin is a good sport. He threw up both hands, opened his eyes wide and exclaimed: 'My heavens! What, again?' "[28] Yet Roosevelt's embargo of the export of gold represented a decisive turn—nothing less than the jettisoning of the international gold standard.[29]

Roosevelt took the country off gold, not simply to forestall the inflationists, but because he was deeply concerned by the deflation of the first six weeks of the New Deal. By going off gold, he sought to free himself to engage in domestic price-raising ventures. The President's action horrified conservatives. Bernard Baruch insisted: "It can't be defended except as mob rule." But Roosevelt's historic decision won applause not only from farm-state senators but, unexpectedly, from the House of Morgan. Morgan himself publicly announced his approval, and Russell Leffingwell wrote the President: "Your action in going off gold saved the country from complete collapse. It was vitally necessary and the most important of all helpful things you have done."[30]

As Congress continued to wrangle over the farm bill, rebellion once more broke out in the Corn Belt. In late April, a mob of farmers, masked in blue bandannas, dragged Judge Charles C. Bradley from his bench in LeMars, Iowa, took him to a crossroads out of town, and nearly lynched him in a vain effort to get him to promise not to sign mortgage foreclosures. A few days later, to force the hand of Congress, the Farmers' Holiday Association, led by Milo Reno and A. C. Townley, called a national farmers' strike for May 13.

On May 12, racing to nip the farm strike in the bud, Congress passed the Agricultural Adjustment Act. Since it provided for alternative systems of subsidizing farm staples, the law simply postponed the quarrel over farm policy; within a few months, it would erupt again. But for the moment farm leaders were content to count their

[28] Franklin D. Roosevelt, *On Our Way* (New York, 1934), p. 6.
[29] Gustav Cassel, *The Downfall of the Gold Standard* (Oxford, 1936), pp. 116–134.
[30] Schlesinger, *Coming of New Deal*, p. 202; Leffingwell to F.D.R., October 2, 1933, FDRL PPF 866. Cf. *Commercial and Financial Chronicle* (April 22, 1933), p. 2653. On June 5, the President signed a joint resolution abrogating the gold clause in public and private contracts.

blessings. The act gave the farmer the price supports he desired, and more besides. The Thomas amendment held out to the debt-ridden farmer the prospect of freshly printed greenbacks; the supplementary Farm Credit Act of June 16 promised to keep the sheriff and the mortgage company away from his door. Within eighteen months, the Farm Credit Administration, a merger of government farm loan agencies under the energetic Henry Morgenthau, Jr., and his deputy, William Myers of Cornell, would refinance a fifth of all farm mortgages. After more than a half century of agitation, the farmer had come into his own.[31]

By the spring of 1933, the needs of more than fifteen million unemployed had quite overwhelmed the resources of local governments. In some counties, as many as 90 per cent of the people were on relief. Roosevelt was not indifferent to the plea of mayors and county commissioners for federal assistance, but the relief proposal closest to his heart had more special aims: the creation of a civilian forest army to put the "wild boys of the road" and the unemployed of the cities to work in the national forests. On March 14, the President asked four of his cabinet to consider the conservation corps idea, a project which united his belief in universal service for youth with his desire to improve the nation's estate. Moreover, Roosevelt thought that the character of city men would benefit from a furlough in the country. The next day, his officials reported back with a recommendation not only for tree-army legislation but for public works and federal grants to the states for relief. The President opposed massive public works spending, since he favored retrenchment and believed there were few worthwhile projects, but Secretary of Labor Frances Perkins; Harry Hopkins, who had headed his relief program in New York; and Senators Robert La Follette, Jr., of Wisconsin and Edward Costigan of Colorado won him over.[32] On March 21, the President sent an unemployment relief message to Congress which embraced all three

[31] Schlesinger, *Coming of New Deal*, pp. 44–45.

[32] As early as December, 1931, Costigan and La Follette had introduced a bill for federal grants to the states for unemployment relief. Such measures had had the warm support of men like Bronson Cutting of New Mexico and George Norris of Nebraska in the Senate and Fiorello La Guardia in the House. New York's Senator Robert Wagner had advanced a panoply of proposals: public works, unemployment insurance, and a system of employment exchanges. President Hoover's opposition had killed or emasculated all of these recommendations.

recommendations. Congress took only eight days to create the Civilian Conservation Corps. In little more than a week, the Senate whipped through a bill authorizing half a billion dollars in direct federal grants to the states for relief, and the House gave its approval three weeks later. Before the session ended, Roosevelt had achieved the third goal of his relief program—public works—in the National Industrial Recovery Act.[33]

According to one account, Herbert Hoover, when queried a short time after he left office if there was anything he should have done while President that he had not done, replied: "Repudiate all debts."[34] This was an almost universal sentiment in the spring of 1933. Creditors understood that if they were to avert outright repudiation, they must work out procedures to ease the burden of debt that threatened to deprive millions of all they owned. Especially critical was the plight of homeowners. In 1932, a quarter of a million families lost their homes. In the first half of 1933, more than a thousand homes were being foreclosed every day; Philadelphia averaged 1,300 sheriff's sales a month.[35]

In June, Congress adopted the Home Owners' Loan Act amidst cries that the law bailed out real-estate interests rather than the homeowner. Without having to scale down the debt he was owed, the mortgagor could turn in defaulted mortgages for guaranteed government bonds. Yet, however much the act was tailored to the interests of financial institutions, it proved a lifesaver for thousands of Americans. When the Home Owners' Loan Corporation opened for business in Akron, a double column stretched for three blocks down Main Street by seven in the morning; when the doors opened, five hundred people pressed into the lobby. In the end, the HOLC would help refinance one out of every five mortgaged urban private dwellings in America.[36]

[33] Edgar Nixon, *Franklin D. Roosevelt and Conservation, 1911–1945* (2 vols., Hyde Park, N.Y., 1957). I, 138–151; George Rawick, "The New Deal and Youth" (unpublished Ph.D. dissertation, University of Wisconsin, 1957), Ch. 5; Tucker Smith to Jane Addams, March 8, 1933, Addams MSS., Box 22; "Memorandum for the Secretary of war: Civilian Conservation Corps," April 3, 1933, Louis Howe MSS., Box 59.

[34] Reminiscence by John Carmody, "Ex-President Hoover in New York," Carmody MSS., Box 57.

[35] Schlesinger, *Coming of New Deal,* p. 297; Lorena Hickok to Harry Hopkins, August 6, 1933, Hopkins MSS.

[36] C. Lowell Harriss, *History and Policies of Home Owners' Loan Corporation* (New York, 1951), pp. 35, 38–39, 127–128; John Blum, *From the Morgen-*

As Roosevelt followed one startling recommendation for reform legislation with yet another, the progressive bloc in Congress came to the pleasant realization that all kinds of proposals that had been doomed to defeat for more than a decade now had an excellent chance of adoption. Of all the projects espoused by the Republican progressives, one in particular symbolized their frustration during the recent reign of three Republican Presidents. Led by George Norris of Nebraska, the progressives had fought year in and year out for government operation of the Muscle Shoals properties on the Tennessee River, an electric power and nitrogen development built during World War I. Twice Congress had passed a Muscle Shoals bill; twice it had been killed by Republican Presidents. Now the progressives found an ally in the new Democratic President. Even before he took office, he had toured the Tennessee Valley with Norris. "Is he really with you?" reporters asked Norris afterward. "He is more than with me, because he plans to go even farther than I did," the Senator responded.[37]

Norris was right. He himself was mainly interested in the potential of Muscle Shoals for hydroelectric power. Southern farm interests and their congressional spokesmen were excited chiefly by the possibility of using the project to manufacture fertilizers. "I care nothing for the power," observed Alabama's Senator Hugo Black. Still others wanted to secure flood protection and curb soil erosion. Roosevelt, with his long-standing interest in forest, land, and water, saw all of these elements as part of a whole. He proposed not merely unified development of the resources of the valley but a vast regional experiment in social planning which would affect directly the lives of the people of the valley.[38]

On April 10, the President asked Congress to create the Tennessee Valley Authority. The TVA would build multipurpose dams which would serve as reservoirs to control floods and at the same time generate cheap, abundant hydroelectric power. Its power operations were

thau Diaries (Boston, 1959), p. 284; Ruth McKenney, *Industrial Valley* (New York, 1939), p. 110.

[37] Schlesinger, *Coming of New Deal*, pp. 323–324; *The New York Times*, February 3, 1933.

[38] Schlesinger, *Coming of New Deal*, pp. 321–323; C. Herman Pritchett, *The Tennessee Valley Authority* (Chapel Hill, 1943), pp. 27–29; Howard K. Menhinick, L. L. Durisch, and Tracy B. Augur, "Origin of the Regional Planning and Development Concept in TVA Legislation," February 6, 1943; Augur to Menhinick, March 1, 1943, TVA Archives, Knoxville (file copy).

designed to serve as a "yardstick" to measure what would be reasonable rates for a power company to charge. The Authority, which would be a public corporation with the powers of government but the flexibility of a private corporation, would manufacture fertilizer, dig a 650-mile navigation channel from Knoxville to Paducah, engage in soil conservation and reforestation, and, to the gratification of the planners, co-operate with state and local agencies in social experiments. Utility executives argued that there would be no market for the power TVA dams would produce; Representative Joe Martin of Massachusetts declared the TVA was "patterned closely after one of the soviet dreams"; and *The New York Times* commented: "Enactment of any such bill at this time would mark the 'low' of Congressional folly."[39] But in the remarkable springtime of the Hundred Days, few heeded. The House passed the measure by a whopping margin; Norris steered it through the Senate; and, after he had first made sure that congressional conferees had approved his more ambitious conception of the project, Franklin Roosevelt signed the Tennessee Valley Authority Act on May 18.[40]

While Roosevelt had sent an impressive number of legislative proposals to Congress, he had still done nothing directly to stimulate industrial recovery. He had, to be sure, encouraged Senator Wagner to look into various schemes for business self-government under federal supervision, and he had requested Raymond Moley to commission the New York financier, James Warburg, to consolidate disparate proposals for industrial co-ordination. Yet, when Warburg reported back, Roosevelt concluded that business thought had not yet crystallized to the point where legislation was feasible. By early April, the President had decided against taking any action.

On April 6, Roosevelt received a rude awakening. In December, 1932, Senator Hugo Black of Alabama had introduced a thirty-hour bill to bar from interstate commerce articles produced in plants in which employees worked more than five days a week or six hours a day. Black, influenced by the English writer G. D. H. Cole and the Americans Arthur Dahlberg and Stuart Chase, claimed that his pro-

[39] *Congressional Record*, 73d Cong., 1st Sess., pp. 2178–2179; *The New York Times*, April 26, 1933; "Reactions of the Country to the Passage of the Muscle Shoals Bill," *Public Utilities Fortnightly*, XI (1933), 724–729. The *Times* was to become one of the most steadfast defenders of the TVA.

[40] Schlesinger, *Coming of New Deal*, pp. 324–326.

posal would create six million jobs. William Green, the mild-mannered president of the A.F. of L., declared the Black bill struck "at the root of the problem—technological unemployment," and startled the country, and perhaps himself, by threatening a general strike in support of the thirty-hour week.[41] On April 6, bowing to pressure from organized labor and the American Legion, the Senate upset all Roosevelt's calculations by passing the Black bill, 53–30.

The President, who believed that the Black bill was unconstitutional, that it was inflexible, and that it would retard recovery, was stung to action. He directed Moley to start the wheels turning once again on a plan for industrial mobilization. Throughout the month of April, various people, holding vague commissions, worked on separate drafts of a recovery bill. They reflected a variety of different demands. New Dealers like Tugwell, writers like Charles Beard and Stuart Chase, and senators like Wagner and La Follette stressed the need for national planning. Wagner and La Follette, together with Senator Costigan, stepped up their campaign for federal public works, with the backing of administration leaders ranging from Secretary of Labor Frances Perkins to Secretary of War George Dern. Old Bull Moosers like the labor attorney Donald Richberg, nourished on the doctrines of the New Nationalism, revived interest in theorists of the Progressive era who had speculated about concentration and control.

The most popular of all proposals arose from the plea of such business leaders as Gerard Swope of General Electric and Henry I. Harriman of the U.S. Chamber of Commerce that the government suspend the antitrust laws to permit trade associations to engage in industrywide planning. While trade associations had adopted "codes of fair competition" in the past, they had had to do so gingerly lest they be slapped with an antitrust suit for collusion, and they had no way to coerce fractious minorities. Some businessmen sought government sanction for business agreements simply in order to fix prices to their own advantage; others, like Swope, who had worked with Jane Addams at Hull-House, conceived of the scheme as a way to benefit labor as well as capital. All shared a common revulsion against the workings of a competitive, individualistic, laissez-faire economy.[42]

[41] U.S. Congress, Senate, *Thirty-Hour Work Week*, Hearings before Subcommittee on the Judiciary, U.S. Senate, 72d Cong., 2d Sess., on S.5267, January 5–19, 1933 (Washington, 1933), pp. 1–23.

[42] J. George Frederick (ed.), *The Swope Plan* (New York, 1931); Gerard

In quest of a precedent for government-business co-operation, the draftsmen of the recovery bill turned to the experience with industrial mobilization in World War I. Swope himself had served in a war agency, and his plan was one of many, like William McAdoo's proposal for a "Peace Industries Board," which drew on recollections of government co-ordination of the economy during the war.[43] Since they rejected laissez faire, yet shrank from embracing socialism, the planners drew on the experience of the War Industries Board because it offered an analogue which provided a maximum of government direction with a minimum of challenge to the institutions of a profit economy.

A number of union leaders, especially men like John L. Lewis and Sidney Hillman in the sick coal and garment industries, had come to believe that only by national action could their industries be stabilized. They were willing to agree to business proposals for a suspension of the antitrust laws because they assumed the Supreme Court would not sanction a federal wages and hours law, and they reasoned that industrial codes offered high-wage businessmen badly needed protection from operators who connived to undersell them by exploiting their workers. But Senator Wagner, the most eloquent champion of the rights of labor in Congress, insisted that if business received concessions labor must have a guarantee of collective bargaining.[44]

On May 10, out of patience with the wrangling over the industrial recovery bill, Roosevelt named a drafting committee and told the draftsmen to lock themselves in a room and not come out until they had a bill. A week later, the President was able to present Congress with an omnibus proposal that had a little for everyone. Business got government authorization to draft code agreements exempt from the

---

Swope, COHC, pp. 123 ff.; Joseph Knapp to Lindsay Warren, February 9, 1933, Warren MSS., Box 9; "A Plea from 123 Representatives of Independent Industrial Units and of Labor for the Trial of a Two Years' Truce in Destructive Competition," The Group to Herbert Hoover, February 11, 1932, copy in Harvey Williams to Robert Wagner, March 24, 1932, Wagner MSS.

[43] Raphael Herman to William McAdoo, June 10, 1931, McAdoo MSS., Box 359; Representative Chester Bolton to Walter Gifford, August 24, 1931, Newton Baker MSS., Box 192.

[44] Matthew Josephson, *Sidney Hillman* (Garden City, N.Y., 1952), p. 357; Hillman, "A Proposal for Labor Boards as an Essential in the Emergency," in Henry Moskowitz to Robert Wagner, April 12, 1933, Wagner MSS.; W. Jett Lauck to John L. Lewis, April 27, May 5, 1933, Lauck MSS., Correspondence.

antitrust laws; the planners won their demand for government licensing of business; and labor received Section 7(a), modeled on War Labor Board practices, which guaranteed the right to collective bargaining and stipulated that the codes should set minimum wages and maximum hours. In addition, the bill provided for $3.3 billion in public works.[45]

The House, largely as a demonstration of its support for Roosevelt, passed the bill, 325–76. In the Senate it had much tougher sledding. The Black group viewed the measure as a "sellout," inflationists favored a different remedy, and conservatives objected to the labor provisions and to the immense power given the federal government. The weightiest objections came from antitrusters such as Senator Borah, who warned that the bill sanctioned cartels and that the big interests would inevitably dominate code-making. Wagner countered by challenging the progressives to find any feasible way other than the codes to outlaw sweatshops and protect labor. "I do not think we will ever have industry in order," Wagner argued, "until we have nationally planned economy, and this is the first step toward it." In the end the Senate approved the bill, but by the narrow margin of seven votes, with both conservatives like Glass and progressives like La Follette and Norris in the opposition on the crucial roll call. On June 16, when President Roosevelt signed the National Industrial Recovery Act, he observed: "Many good men voted this new charter with misgivings. I do not share these doubts. I had part in the great cooperation of 1917 and 1918 and it is my faith that we can count on our industry once more to join in our general purpose to lift this new threat. . . ."[46]

While Roosevelt offered industrialists a partnership in the mobilization for recovery, he insisted that the financiers be disciplined. On March 29, he sent Congress his recommendation for federal regulation of securities, a proposal which added "to the ancient rule of *caveat*

[45] The best accounts of the origins of the industrial recovery bill—a subject on which there is sharply conflicting testimony—are Schlesinger, *Coming of New Deal,* Ch. 6; Moley, *After Seven Years,* pp. 184–190; and James Burns, "Congress and the Formation of Economic Policies: Six Case Studies" (unpublished Ph.D. dissertation, Harvard University, 1947). On section 7(a), see Irving Bernstein, *The New Deal Collective Bargaining Policy* (Berkeley and Los Angeles, 1950), pp. 29–39, and "Docket—Coal & Stabilization," April 24, May 6, 23, 1933, National Recovery Act file, W. Jett Lauck MSS.

[46] *Congressional Record,* 73d Cong., 1st Sess., p. 5245; *Public Papers,* II, 252.

emptor, the further doctrine 'let the seller also beware.' "[47] A short while before, such legislation would have been inconceivable, but the debate on the securities bill took place as the Pecora committee was carrying the popular outcry against Wall Street to a heightened pitch.

A tribune of righteousness, Ferdinand Pecora summoned the nation's financial rulers to the bar. In the twenties, the House of Morgan had been an arcane shrine, Morgan himself a prince of the realm who dwelt somewhere in the recesses of the marble building on Broad and Wall far removed from the common throng. Now Pecora haled Morgan before the Senate committee and put to him questions it had not seemed fit to ask him before. "Pecora has the manner and the manners of a prosecuting attorney who is trying to convict a horse thief," Morgan protested to his friends. "Some of these senators remind me of sex suppressed old maids who think everybody is trying to seduce them." On the witness stand, Morgan appeared to have been resurrected from some Dickensian countinghouse—he called one of his clerks a "clark"—but as Pecora jabbed his stubby finger at him, while Morgan fondled the heavy gold chain across his paunch, he seemed less the awesome figure of Broad and Wall, more like a Main Street banker with his eye on the main chance.[48]

The Pecora probe made possible the adoption of the Securities Act late in May, but the law itself bore the imprint of the Brandeisian faction of the New Dealers. Sharply at variance with the NRA philosophy of government-business co-operation in the interest of national planning, the measure reflected the faith of the New Freedom in regulation rather than direction, in full disclosure rather than coercion. The Securities law, modeled on the British Companies Act, gave the Federal Trade Commission power to supervise issues of new securities, required each new stock issue to be accompanied by a statement of relevant financial information, and made company directors civilly and criminally liable for misrepresentation. Drafted in part by lawyers supplied by the Brandeisian Felix Frankfurter, and shepherded through Congress by the Wilsonian liberal Sam Rayburn, the law delighted progressives who believed that Wall Street wielded too much un-

[47] *Public Papers,* II, 93.
[48] Henry Morrow Hyde MS. Diary, June 13, 1933; Thomas Stokes, *Chip Off My Shoulder* (Princeton, 1940), pp. 345–350; Ferdinand Pecora, *Wall Street under Oath* (New York, 1939); "Confidential Files, Governor Harrison," March 28, 1933, Harrison MSS., Conversations, II.

disciplined power over the economy. On the other hand, spokesmen for co-operation with business and advocates of planning—both of whom shared an organic conception of the economy—thought the measure regressive. William O. Douglas of the Yale Law School dismissed it as a "nineteenth-century piece of legislation" which, instead of aiding national planning, sought to restore the old capitalistic order of competitive small units.[49]

Two days after Pecora's inquiry revealed that the twenty Morgan partners had not paid a penny in income taxes in two years, the Senate passed the Glass-Steagall banking bill without a dissenting voice. The Pecora committee had urged the separation of investment from commercial banking, and this feature of the Glass-Steagall bill, highly popular with investors, had even won the support of Winthrop Aldrich, the new head of the Chase National Bank, although Wall Street hooted that Aldrich's remarks represented an assault by Rockefeller interests on the House of Morgan. Much more controversial was the proposal for federal insurance of bank deposits, a panacea advanced by Senator Arthur Vandenberg of Michigan and by Representative Henry Steagall of Alabama, the leader of the small-bank faction in the House. They were motivated in part by a recognition of the anguish of the small depositor, but also by the conviction that the safety-fund device would protect the circulating medium and would help sustain a system of small unit banks. On the basis of past experience, the insurance proposal seemed highly questionable, and not only bankers but both Senator Glass and President Roosevelt frowned on it. Yet, unlike his more conservative fiscal advisers, such as Dean Acheson, Roosevelt recognized that the demand for federal bank insurance could not be resisted. When Congress adopted the Glass-Steagall Act, it approved not only the separation of investment from commercial banking and certain reforms of the Federal Reserve System but the creation of the Federal Deposit Insurance Corporation. A stepchild of the New Deal, the federal guarantee of bank deposits turned out to be a brilliant achievement. Fewer banks suspended during the rest of the decade than in even the best single year of the twenties.[50]

[49] William O. Douglas, "Protecting the Investor," *Yale Review*, XXIII (1934), 529; Schlesinger, *Coming of New Deal*, pp. 439–442; Moley, *After Seven Years*, pp. 176–183; "SEC," *Fortune*, XXI (June, 1940), 92, 120; Daniel Roper to E. M. House, March 27, 1933, House MSS.

[50] Carter Golembe, "The Deposit Insurance Legislation of 1933: An Ex-

When Congress adjourned on June 16, precisely one hundred days after the special session opened, it had written into the laws of the land the most extraordinary series of reforms in the nation's history. It had committed the country to an unprecedented program of government-industry co-operation; promised to distribute stupendous sums to millions of staple farmers; accepted responsibility for the welfare of millions of unemployed; agreed to engage in farreaching experimentation in regional planning; pledged billions of dollars to save homes and farms from foreclosure; undertaken huge public works spending; guaranteed the small bank deposits of the country; and had, for the first time, established federal regulation of Wall Street. The next day, as the President sat at his desk in the White House signing several of the bills Congress had adopted, including the largest peacetime appropriation bill ever passed, he remarked: "More history is being made today than in [any] one day of our national life." Oklahoma's Senator Thomas Gore amended: "During all time."[51]

Roosevelt had directed the entire operation like a seasoned field general. He had sent fifteen messages up to the Hill, seen fifteen historic laws through to final passage. Supremely confident, every inch the leader, he dumfounded his critics of a few months before. "Roosevelt the Candidate and Roosevelt the President are two different men," observed a veteran newspaperman. "I assert as much, for I am well acquainted with both. In the first role he was famous for his beguiling smile, famous for his soft words, famous for his punchless speeches. . . . The oath of office seems suddenly to have transfigured him from a man of mere charm and buoyancy to one of dynamic aggressiveness."[52] Others conceded that they might have been mistaken.

amination of Its Antecedents and Its Purposes," *Political Science Quarterly*, LXXV (1960), 181–200; J. F. T. O'Connor MS. Diary, May 17, June 1, 2, 7, 1933; "Memorandum from Dr. [H. Parker] Willis to Senator Glass," March 15, 1933, Glass MSS., Box 311; "Memorandum of Conversation Between Mr. Stimson and the President, Franklin D. Roosevelt, March 28, 1933," Henry Stimson MS. Diary; Elmer Thomas to W. E. Hocker, April 4, 1933, Elmer Thomas MSS., Box 467; Schlesinger, *Coming of New Deal*, pp. 442–443. To unify the banking system, the act stipulated that only members of the Federal Reserve System or applicants for membership would be included in the insurance system after July 1, 1936. The FDIC insured depositors in member banks against losses up to $5,000.

[51] J. F. T. O'Connor MS. Diary, June 16, 1933.

[52] J. Fred Essary, "The New Deal for Nearly Four Months," *Literary Digest*, CXVI (July 1, 1933), 4. See Schlesinger, *Coming of New Deal*, p. 21; Oswald

"When Norman Thomas and I wrote a book on New York last year," Paul Blanshard confessed, "we sadly underestimated your courage and intelligence."[53] "How do you account for him?" puzzled William Allen White. "Was I just fooled in him before the election, or has he developed? As Governor of New York, I thought he was a good, two-legged Governor of the type that used to flourish in the first decade of the century under the influence of La Follette and Roosevelt. We had a lot of them but they weren't Presidential size except Hiram Johnson, and I thought your President was one of those. Instead of which he developed magnitude and poise, more than all, power! I have been a voracious feeder in the course of a long and happy life and have eaten many things, but I have never had to eat my words before."[54] Republican editor and Socialist writer, G. O. P. senators and progressive critics, all shared in a common rejoicing. The nation, at last, had found a leader.

Garrison Villard, "Mr. Roosevelt's Two Months," *New Statesman and Nation,* V (1933), 593–594.

[53] Blanshard to F.D.R., July 21, 1933, FDRL PPF 714.

[54] White to Harold Ickes, May 23, 1933, FDRL PPF 714. Hiram Johnson himself commented: "The admirable trait in Roosevelt is that he has the guts to try. . . . He does it all with the rarest good nature. I am fond of saying we have exchanged for a frown in the White House a smile. Where there were hesitation and vacillation, weighing always the personal political consequences, feebleness, timidity, and duplicity, there are now courage and boldness and real action." Johnson to Katherine Edson, April 20, 1933, Edson MSS., Box 3. Cf. Arthur Capper to Stanley Resor, June 17, 1933, Capper MSS.

CHAPTER 4

Over the Top

WELL before Congress had recessed, Roosevelt confronted a stupendous task in state-building. Like the leader of a colony which had just achieved independence, the President had to create an entirely new government. The Republicans, after twelve years in office, held a near monopoly of skilled administrators, and most of these viewed the New Deal experiment with dismay. "We stood in the city of Washington on March 4th," recalled Raymond Moley, "like a handful of marauders in a hostile territory." Only the veterans of the war mobilization had had much experience with the kind of massive undertaking Roosevelt had inaugurated. "One cannot go into the Cosmos Club without meeting half a dozen persons whom he knew during the war," wrote one New Dealer.[1]

In the spring of 1933, Washington quickened to the feverish pace of the new mobilization. From state agricultural colleges and university campuses, from law faculties and social work schools, the young men flocked to Washington to take part in the new mobilization. Wholly apart from their beliefs or special competences, they imparted an enormous energy to the business of governing and impressed almost everyone with their contagious high spirits and their dedication. After a visit to the Department of Agriculture, Sherwood Anderson wrote: "I stood there in the office a few minutes and at least ten of my old western friends came in, old radicals, young ones, newspaper men etc.

[1] Raymond Moley, *After Seven Years* (New York, 1939), p. 128; Isador Lubin to Louis Brandeis, August 25, 1933, Brandeis MSS. G5.

. . . There is certainly a curiously exilerating feeling. You cannot be there now without a feeling of the entire sincereity of many of these men."[2]

No sooner would the public grasp the name of one adviser than another would be gaining headlines. By midsummer, they were coming to Washington in droves. Whereas formerly they arrived one at a time, reported the *Literary Digest* in July, "they are now beginning to arrive in pairs. We find ourselves bumping into them wherever we turn. . . ." "A plague of young lawyers settled on Washington," carped one administrator later. "They all claimed to be friends of somebody or other and mostly of Felix Frankfurter and Jerome Frank. They floated airily into offices, took desks, asked for papers and found no end of things to be busy about. I never found out why they came, what they did or why they left."[3] Critics of the administration learned to single out Frankfurter, who sent a great number of his most promising students at Harvard Law School to Washington, as the symbol of the intellectual in government, as, in the words of one writer, "a kind of alderman-at-large for the better element." Hugh Johnson later named Frankfurter "the most influential single individual in the United States"; Hearst's New York *American* called him "the IAGO of this Administration."[4] By September, college deans of admission were reporting that students were turning away from subjects that prepared them for business and were requesting courses "which would point them toward the 'brain trust.'"[5]

The New Deal's most important task in the summer of 1933 was to negotiate NRA code agreements with the major industries. To head the Recovery Administration, Roosevelt named General Hugh Johnson, who had helped organize the draft in World War I, and had

[2] Sherwood Anderson to Burton and Mary Emmett, October 8, 1933, Emmett MSS. I have retained Anderson's spelling.

[3] *Literary Digest,* CXVI (July 22, 1933), 10; Denver *Post,* May 21, 1933; George Peek, with Samuel Crowther, *Why Quit Our Own* (New York, 1936), p. 20.

[4] Eliot Janeway, *The Struggle for Survival* (New Haven, 1951), p. 140; "Felix Frankfurter," *Fortune,* XII (1936), 63. Cf. Henry Stimson MS. Diary, November 2, 1935.

[5] Forrest Davis, "The Rise of the Commissars," *New Outlook,* CLXII (December, 1933), 23–24. Harvard's Department of Government showed an increase in concentrators from 5.5 to 12.3 per cent between 1931 and 1936. *New Republic,* XC (1937), 251–252.

served as a liaison between the Army and the War Industries Board. A gruff, pugnacious martinet with the leathery face and order-barking rasp of a former cavalry officer, Johnson had no illusions about the dimensions of the job. "It will be red fire at first and dead cats afterward," he remarked. "This is just like mounting the guillotine on the infinitesimal gamble that the ax won't work."[6]

Johnson flung himself into code-making with the frontier abandon of a boy raised in the Cherokee Strip. The NRA codes of fair practices, Johnson explained, would "eliminate eye-gouging and knee-groining and ear-chewing in business. Above the belt any man can be just as rugged and just as individual as he pleases."[7] For all of his ardor, Johnson won the immediate consent of only one of the big ten industries. At the dramatic cotton-code hearing, the room burst into cheers when textile magnates announced their intention to abolish child labor in the mills. In addition, the cotton textile code stipulated maximum hours, minimum wages, and collective bargaining. But to recompense the textile industrialists for increased labor costs, the NRA agreed to permit them to limit production, a precedent which other manufacturers quickly demanded for themselves.

By July, factory production had shot up from an index of 56 in March to 101, industrial stocks from 63 to 109. On July 19, the market broke. The boom, which had encouraged the belief that the country was on its way to recovery, had turned out to be largely an attempt to beat the gun before code wage provisions and other New Deal devices could take effect. Alarmed by the threat to the recovery drive, and vexed at the procrastination of the remaining big nine industries, Johnson sought a way to place a floor under wages until codification was completed. He proposed a national campaign, similar to the wartime Liberty Loan drives, to get individual employers to pledge to the President that they would accept a blanket agreement to uphold NRA standards on wages and hours.

Having roundly denounced Hoover for his faith in exhortation, the New Dealers launched an unabashed revivalism that exceeded anything Hoover had done. To dramatize the campaign, Johnson sketched the symbol of a Blue Eagle with the legend "We Do Our Part." Over-

[6] Hugh Johnson, *The Blue Eagle from Egg to Earth* (Garden City, N.Y., 1935), p. 208.
[7] *Ibid.*, p. 282.

night, the NRA eagle appeared everywhere—on newspaper mast-heads, on store windows, even on girls in the chorus. At a Blue Eagle parade in New York—the greatest in the city's history—a quarter of a million people marched down Fifth Avenue. They paraded all day, and when evening came, they kept coming, thousands more under the amber lights, tramping to the step of "Happy Days Are Here Again." In the end, more than two million employers signed the pledge. "There is a unity in this country," observed Franklin Roosevelt, "which I have not seen and you have not seen since April, 1917. . . ."[8]

As the country rocked with the tumult of the Blue Eagle campaign, Johnson mounted a fresh offensive to win acceptance of the codes. By early August, he had half the big ten industries, as shipbuilding, woolens, electricals, and the garment industry signed up. Johnson strong-armed the oil industry into a code; got steel to consent to $7(a)$, only after making a raft of concessions on price policy; and blustered and wheedled lumber into an agreement. He sped from town to town in an Army plane, issuing marching orders and reporting on action at the front. One day, Johnson tore into the White House, handed the President three codes to approve, and, as Roosevelt was signing the last one, raced to catch his Army plane for points unknown. "He hasn't been seen since," the President remarked.[9] On August 27, the automobile manufacturers, all save Henry Ford, came to terms. Ford charged that the NRA was a plot of his competitors, linked with the "bankers' international."[10] When the soft-coal operators fell in line on September 18, Johnson had won the last of the big ten industries to the NRA in a period of only three months.[11]

Johnson endeavored to enforce the codes not by direct coercion but by appealing to the patriotism of businessmen, and by relying on the pressure of social disapproval of men who broke agreements. What would happen to a man who violated a code? Johnson elucidated:

[8] Samuel Rosenman (ed.), *The Public Papers and Addresses of Franklin D. Roosevelt* (13 vols., New York, 1938–50), II, 345.

[9] James Burns, *Roosevelt: The Lion and the Fox* (New York, 1956), p. 192.

[10] Ford is reported to have said: "Hell, that Roosevelt buzzard! I wouldn't put it on the car." Sidney Fine, "The Ford Motor Company and the N.R.A.," *Business History Review*, XXXII (1958), 362.

[11] Arthur Schlesinger, Jr., *The Coming of the New Deal* (Boston, 1959), pp. 116–118; Harold Ickes, *The Secret Diary of Harold Ickes* (3 vols., New York, 1954), I, 72; Thomas Stokes, *Chip Off My Shoulder* (Princeton, 1940), p. 371.

"As happened to Danny Deever, NRA will have to remove from him his badge of public faith and business honor and 'takin of his buttons off an' cut his stripes away' break the bright sword of his commercial honor in the eyes of his neighbors—and throw the fragments—in scorn —in the dust at his feet. . . . It will never happen. The threat of it transcends any puny penal provision in this law." Johnson alternately cajoled and blustered. He recited Tennyson's "Maud" to the N.A.M. and treated opponents to choice barracks invective. Arthur Krock later reflected: "Hugh Johnson was like McClellan to me. He organized the N.R.A. just as McClellan organized the army of the Potomac, but, once they had done that, both of them should have retired."[12]

By the beginning of 1934, the NRA had acquired a formidable aggregation of critics. Housewives complained about high prices, businessmen about government edicts, and workers about the inadequacy of 7(a). Old-school Democrats like Carter Glass deplored "the utterly dangerous effort of the federal government at Washington to transplant Hitlerism to every corner of this nation."[13] Much of the criticism centered on the charge that the NRA promoted monopoly. Progressive Republicans like William Borah and Gerald Nye denounced the NRA as an oppressor of small business, while New Dealers like Leon Henderson feared that the codes retarded recovery by permitting price rises and cutbacks in production. On March 7, 1934, Roosevelt created a National Recovery Review Board to study monopolistic tendencies in the codes. Johnson, in what he later said was "a moment of total aberration," agreed to have the noted attorney and iconoclast Clarence Darrow head the investigation.[14]

After a brief study, the Darrow committee found that giant corporations dominated the NRA code authorities and squeezed small business, labor, and the public. Johnson was furious. "Bloody old Jeffries [sic] at the Assizes never conducted any hearing to equal those for cavalier disposal of cases," he stormed. In fact, the Darrow probe was prejudiced and its findings were inconsistent. A preliminary report urged a return to free competition, but then described competition

[12] Johnson, *Blue Eagle,* p. 265; *Time,* XXIII (January 1, 1934), 9; Arthur Krock, COHC.

[13] Glass to Walter Lippmann, August 10, 1933, Glass MSS., Box 4.

[14] Johnson, *Blue Eagle,* p. 272; Johnson to F.D.R., December 12, 1933, FDRL OF 466; Schlesinger, *Coming of New Deal,* pp. 128–134.

as savage and wolfish. Darrow himself joined with William O. Thompson, a former member of Altgeld, Darrow and Thompson, in a supplementary report which advocated socialized ownership.[15]

Roosevelt, irritated by the bias of the Darrow report, nonetheless had to concede that much was wrong with the operation of NRA. Disturbed by a Brookings study on NRA price policy, and prodded by such Recovery Administration officials as Henderson, he curbed government acquiescence in price fixing.[16] Roosevelt also recognized that the NRA had bitten off too much in attempting to supervise even businesses of little consequence. Code 450 regulated the Dog Food Industry, Code 427 the Curled Hair Manufacturing Industry and Horse Hair Dressing Industry, and Code 262 the Shoulder Pad Manufacturing Industry. In New York, I. "Izzy" Herk, executive secretary of Code 348, brought order to the Burlesque Theatrical Industry by insisting that no production could feature more than four strips. Beginning in May, 1934, Roosevelt exempted employers in peripheral areas.[17]

By the summer of 1934, Hugh Johnson had become an unsupportable burden, and the President began to ease him out. The General drank too much, lost his temper, made rash decisions, and even rasher statements.[18] In September, Roosevelt secured his resignation. As Johnson's influence waned, Donald Richberg, whom Roosevelt had named director of several co-ordinating agencies in June, took his place. Even more resolutely opposed to direct government coercion than Johnson had been, Richberg, whom newsmen called "Assistant President," urged "co-operation" with business in enforcing the NRA. Increasingly, it became clear that by "co-operation" Richberg meant

[15] Johnson, *Blue Eagle*, p. 272. There were three reports in all, together with supplementary reports and responses from the National Recovery Administration. National Recovery Review Board, Reports of May 4, June 14, 28, 1934, mimeographed.

[16] George Terborgh, *Price Control Devices in NRA Codes* (Washington, 1934); Blackwell Smith to General Johnson, June 2, 1934, Leon Henderson MSS., Box 10. Ironically, it was New Deal bureaucrats who sought to restore the free market, while businessmen insisted on retaining regimented price and production controls.

[17] "The Business of Burlesque, A.D. 1935," *Fortune* XI (February, 1935), 140; Schlesinger, *Coming of New Deal*, pp. 135, 159–162.

[18] "Insanity in public performance, added to insanity in private relations, is pretty dangerous," Donald Richberg expostulated. Richberg to Marvin McIntyre, September 15, 1934, Richberg MSS., Box 1.

that the government should refrain from action business did not approve.[19]

By the time Johnson left office, many critics had judged the NRA a failure, a verdict that most economists since that time have reaffirmed. Much of this censure has been too severe. Too great fault has been found with the NRA's price policies; the Recovery Administration has been blamed unfairly for the acceleration of industry toward concentration; and commentators have exaggerated the degree to which the NRA hurt small business. The main grievance of small businessmen against the Recovery Administration lay not in the fact that they were oppressed by monopoly but that the government would not permit them to continue to exploit labor. Finally, the NRA could boast some considerable achievements: it gave jobs to some two million workers; it helped stop a renewal of the deflationary spiral that had almost wrecked the nation; it did something to improve business ethics and civilize competition; it established a national pattern of maximum hours and minimum wages; and it all but wiped out child labor and the sweatshop.[20]

But this was all it did. It prevented things from getting worse, but it did little to speed recovery, and probably actually hindered it by its support of restrictionism and price raising. The NRA could maintain a sense of national interest against private interests only so long as the spirit of national crisis prevailed. As it faded, restriction-minded businessmen moved into a decisive position of authority. By delegating power over price and production to trade associations, the NRA created a series of private economic governments. The codes, as Walter Lippmann observed, were, in effect, charters like those granted the East India Company. The large corporations which dominated the code authorities used their powers to stifle competition, cut back production, and reap profits from price-raising rather than business expansion. Since Roosevelt, Johnson, and Richberg hesitated to use punitive

[19] Donald Richberg, *The Rainbow* (Garden City, N. Y., 1936), pp. 115–121; Christopher Lasch, "Donald Richberg and the Idea of a National Interest" (unpublished M.A. essay, Columbia University, 1955), pp. 64–65; Felix Frankfurter to Louis Brandeis, April 27, 1935, Brandeis MSS., G9.

[20] Schlesinger, *Coming of New Deal,* pp. 167–176; George Galloway *et al., Industrial Planning Under Codes* (New York, 1935); Arthur R. Burns, Review of Leverett Lyon *et al., The National Recovery Administration, Political Science Quarterly,* L (1935), 598–602; "NRA Examined," *American Economic Review,* XXV, Supplement (March, 1935), 4–6.

powers, and had little intention of undertaking extensive national planning, the private interests of business corporations overwhelmed the public interest.[21]

Although the trend of NRA policies was mischievous, the weakness of the early New Deal lay less in the NRA itself than in the absence of any consistent strategy to boost purchasing power or step up investment. Roosevelt, committed to budget-balancing and skeptical of the value of federal construction, never saw public works as an important lever for industrial recovery. Secretary of the Interior Harold Ickes, whom Roosevelt named to head the Public Works Administration, operated the agency with such extreme caution that it did next to nothing to stimulate the economy.[22] Ickes had a long memory—one that embraced the Ballinger-Pinchot affair and Teapot Dome—and he was determined that neither Interior nor the PWA would repeat this earlier history. Moreover, Ickes feared, on good grounds, that local politicians would waste money on useless projects when they did not pilfer it. Planners like Lewis Mumford warned him that public works should be undertaken only after careful planning, or the country would be pockmarked with jerry-built monstrosities. Ickes feared that either a new scandal or inept construction would set back the cause of public development a generation.

Energetic, courageous, incorruptible, "Honest Harold" was the ideal man to undertake a program of beautifying the nation with well-conceived buildings. Vain, quarrelsome, suspicious of mankind, a man who saw little but a void of wickedness beyond the arc of his own

[21] Walter Lippmann, *The Good Society* (Boston, 1937), p. 123; F.D.R. to Carter Glass, February 21, 1935, FDRL PPF 1435; Memorandum, "Codes in Crying Need of Change Because of Misuse or Abuse of Code Provisions," James Hughes to Leon Henderson, March 14, 1935, Henderson MSS., Box 3; Leverett Lyon *et al., The National Recovery Administration* (Washington, 1935); Lionel Robbins, *The Great Depression* (London, 1934), pp. 139, 147; Arthur R. Burns, "The First Phase of the National Industrial Recovery Act, 1933," *Political Science Quarterly*, XLIX (1934), 193.

[22] The editors of the *Economist* concluded in 1936: "Its net result has been to restore the average expenditure on public construction to about 60 per cent. of its pre-depression level. In these circumstances, it would be idle to look for the effects which have been theoretically predicated for a policy of increasing public works expenditures in times of depression. The plain truth is that in the United States public works have not been increased; they have merely been prevented from fading altogether away." "The New Deal: An Analysis and Appraisal" (October 3, 1936), p. 5.

righteousness, Ickes proved an unfortunate choice to head an operation which required fast, even reckless, spending if it were to succeed. Yet the real responsibility for failure was less Ickes' than Roosevelt's. He not only saw little need for urgency, but repeatedly raided Ickes' funds for money for other projects; he even appeared to hope that recovery would come swiftly enough so that some of the PWA grant could be turned back to the Treasury.[23]

Roosevelt made markedly greater progress in adapting the Reconstruction Finance Corporation to his recovery drive. Under Hoover, the RFC had been an instrument of the established businessmen, particularly of the eastern *rentiers*.[24] Roosevelt put the agency under a Texas banker, Jesse Jones, who represented the southwestern boomers' desire for expansion, and who spoke for businessmen less interested in protecting existing holdings than in fresh ventures. Jones converted the RFC into a vastly different organization from what it had been under Hoover. Instead of lending money to banks, and thereby increasing their debt, as had been done in the Hoover regime, Jones sought to enlarge their capital. By buying bank preferred stock, he bolstered the capital structure of banks, created a base for credit expansion, and made it possible for the deposit insurance system to function. Under Jones, the Corporation became not only the nation's largest bank but its biggest single investor. Eventually, Jones ruled an empire of RFC subsidiaries: federal mortgage agencies; the Commodity Credit Corporation; the Electric Home and Farm Authority,

[23] Schlesinger, *Coming of New Deal*, pp. 284–287; Marquis Childs, *I Write from Washington* (New York, 1942), p. 60; John T. Flynn, "Who's Holding Back Public Works?" *New Republic*, LXXVI (1933), 145–148; "Other People's Money," *idem.* (1933), 99–100; Harold Ickes, "My Twelve Years with F.D.R.," *Saturday Evening Post*, CXX (June 12, 1948), 111–113. Cf. Arthur Gayer, *Public Works in Prosperity and Depression* (New York, 1935), pp. 121–122; J. Kerwin Williams, *Grants-in-Aid under the Public Works Administration* (New York, 1939).

[24] Chartered by Congress in 1932, the RFC had been authorized to lend money to banks, railroads, and other institutions threatened by destruction. In part because of Hoover's misgivings about federal intervention, the agency made so little use of its powers that it frustrated the intent of Congress. Gerald Nash, "Herbert Hoover and the Origins of the Reconstruction Finance Corporation," *Mississippi Valley Historical Review*, XLVI (1959), 455–468; Eugene Meyer, COHC, pp. 612–613; Jackson Reynolds, COHC, pp. 152–153; J. Franklin Ebersole, "One Year of the Reconstruction Finance Corporation," *Quarterly Journal of Economics*, XLVII (1933), 464–492; *Business Week* (March 23, 1932), pp. 6–7.

which sparked the purchase of electrical appliances; and the Export-Import Bank, which spurred foreign trade.[25]

Although the activities of the industrial recovery agencies—NRA, PWA, and RFC—were critically important, Roosevelt and many of his advisers believed the New Deal would stand or fall on the success of its farm program. In his new Secretary of Agriculture, Henry Agard Wallace, Roosevelt had a man of parts. A famed geneticist, known particularly for his experiments in breeding hybrid corn, a first-rate farm economist, skilled in mathematics, an editor who had thought deeply about farm problems, Wallace attracted ardent admirers and just as passionate detractors. Many were impressed by his knowledgeability and won by his shy yet candid manner, his Lincolnian appearance, and his rumpled simplicity. "I think Will Rogers is the type," wrote Sherwood Anderson. "There is the same little smile . . . an inner rather than an outward smile . . . perhaps just at bottom sense of the place in life of the civilized man . . . no swank . . . something that gives us confidence." Yet others were dismayed by his reserve, by his flights of mysticism, even of gibberish, and by a faddism that on one occasion led him to try to live on corn meal and milk after learning that this had been the food of Caesar's forces in Gaul.[26]

Wallace's first task was to carry out a policy which he accepted as necessary but did not relish: crop destruction. When Roosevelt sent his farm message to Congress, he warned that he was racing against the sun. By the time the Senate acted, he had lost the race. Kansas was green with spring wheat; sows had littered in the corn belt; cotton had long since been planted in Georgia. Even if the country did not harvest a single boll of cotton in 1933, it had enough on hand from previous years to sate the world market. Faced with the disastrous

[25] Schlesinger, *Coming of New Deal,* pp. 425–433; Jesse Jones, with Edward Angly, *Fifty Billion Dollars* (New York, 1951) ; Cyril Upham and Edwin Lamke, *Closed and Distressed Banks* (Washington, 1934), pp. 145–207; Bascom Timmons, *Jesse H. Jones* (New York, 1956), pp. 187, 198–233; "The House of Jesse," *Fortune,* XXI (May, 1940), 45–51, 122–144; Jesse Jones to Harvey Gibson, October 27, 1933; Jones to N. R. Kobey, October 27, 1933, Jones MSS., Box 38.

[26] Anderson, "No Swank," *Today,* I (November 11, 1933), 23; John Blum, *From the Morgenthau Diaries* (Boston, 1959), p. 39; Schlesinger, *Coming of New Deal,* pp. 28–34; Louis Bean, COHC, pp. 116–130; Russell Lord, *The Wallaces of Iowa* (Boston, 1947), pp. 107–341. Wallace, the son of Harding's Secretary of Agriculture, had been reared in a sanctum of the farm cause.

prospect of a new bumper crop, the Agricultural Adjustment Administration sent county agents through the South to urge planters to uproot cotton already in the field. That summer mules that had been lambasted for years if they trod on a row of cotton were taught to trample down the stalks they had learned to protect. For plowing up ten million acres of cotton, farmers collected over $100 million in benefit payments.[27]

To tide the Corn Belt over the winter, Wallace reluctantly agreed to a proposal of farm leaders to forestall a glut in the hog market by slaughtering over six million little pigs and more than two hundred thousand sows due to farrow. While one million pounds of salt pork was salvaged for relief families, nine-tenths of the yield was inedible, and most of it had to be thrown away. The country was horrified by the mass matricide and infanticide. When the piglets overran the stockyards and scampered squealing through the streets of Chicago and Omaha, the nation rallied to the side of the victims of oppression, seeking to flee their dreadful fate. Wallace responded sardonically: "I suppose it is a marvelous tribute to the humanitarian instincts of the American people that they sympathize more with little pigs which are killed than with full-grown hogs. Some people may object to killing pigs at any age. Perhaps, they think that farmers should run a sort of old-folks home for hogs and keep them around indefinitely as barnyard pets."[28]

The cotton plowup and the slaughter of the little pigs fixed the image of the AAA in the minds of millions of Americans, who forever after believed that this was the agency's annual operation. In fact, almost all the actual destruction ended in 1933, and thereafter the government restricted yields in a more orderly fashion.[29] In October, 1933,

[27] Henry I. Richards, *Cotton Under the Agricultural Adjustment Act* (Washington, 1934), pp. 46–66; Schlesinger, *Coming of New Deal*, pp. 59–61; *New Republic*, LXXVI (August 23, 1933), adv.; W. L. Clayton to George Peek, May 7, 1933, Peek MSS.

[28] Henry Wallace, *Democracy Reborn* (New York, 1944), p. 104; Theodore Saloutos and John Hicks, *Agricultural Discontent in the Middle West 1900–1939* (Madison, Wis., 1951), pp. 475–476; D. A. FitzGerald, *Corn and Hogs under the Agricultural Adjustment Act* (Washington, 1934), pp. 17–49.

[29] Even in 1933, since bad weather was expected to reduce the wheat yield, it was not necessary to plow up wheat. Instead, the government undertook a three-year program in which it offered benefit payments to wheat growers in

Jesse Jones set up the Commodity Credit Corporation to lend money to farmers who agreed to take land out of production the following year. Since the loans were set at a figure higher than the market price, they served as an ingenious kind of price support. Through such devices as CCC loans and planned cuts in hog breeding, the Administration mounted a long-range offensive to raise farm income.[30]

Yet most of this would benefit the farmer in the future; he wanted relief now. On July 18, wheat broke sharply, carrying other crops down with it, and when hot prairie winds scorched Minnesota and the Dakotas, farm organizations demanded action. Led by the spellbinder Milo Reno of the Farmers' Holiday Association and Minnesota's Farmer-Labor Governor Floyd Olson, midwestern farmers insisted on compulsory production control and price-fixing, with a guaranteed cost of production. When, in early November, Roosevelt, on Wallace's insistence, refused to endorse a plan which Wallace warned would mean licensing every plowed field in the country, the Holiday Association responded by going ahead with its long-delayed strike. Strikers dumped kerosene in cream, broke churns, and dynamited dairies and cheese factories. In Lac qui Parle, Chippewa, and Yellow Medicine counties in Minnesota, in the dairy country of Iowa and Wisconsin, milk spilled over the highways.[31]

By late November the farm violence was sputtering to an end, but the demand for compulsory regulation of agriculture would not be silenced. Southern planters wanted to ride herd on mavericks who had cultivated their land so intensively that, even though a quarter of the cotton crop was plowed under, the 1933 yield was greater than it had been the previous year. Reports that the Southern Railway was hauling four times as much fertilizer as a year before indicated that farmers would explode the cotton price structure in 1934 by planting up to the

---

return for pledges to restrict sowing. Joseph S. Davis, *Wheat and the AAA* (Washington, 1935).

[30] FitzGerald, *Corn and Hogs,* pp. 56–85, and *Livestock under the AAA* (Washington, 1935); Henry I. Richards, *Cotton and the AAA* (Washington, 1936), pp. 194–337; Dean Albertson, *Roosevelt's Farmer* (New York, 1961), pp. 85–93; Murray Benedict and Oscar Stine, *The Agricultural Commodity Programs* (New York, 1956).

[31] George Mayer, *Floyd B. Olson,* pp. 149–157; Schlesinger, *Coming of New Deal,* pp. 64–67; Saloutos and Hicks, *Agricultural Discontent,* p. 485; Donald McCoy, *Angry Voices* (Lawrence, 1958), pp. 34–35; Lord, *Wallaces of Iowa,* pp. 361–366.

fence rails. Secretary Wallace strongly opposed a bill filed by Senator John Bankhead of Alabama for compulsory crop reduction with ginning quotas; but when he sent out questionnaires to cotton farmers, he found more than 80 per cent favored the plan, and only 2 per cent were clearly opposed. The Bankhead Cotton Control Act, which one senator said he voted for with all his fingers and toes crossed, marked the death of voluntarism in the cotton states. That same year, Congress responded to the entreaties of tobacco growers by passing the coercive Kerr-Smith Tobacco Control Act. In 1935, it yielded to the demands of "individualistic" senators like Idaho's William Borah and the cry of farmers from "conservative" Maine for a compulsory Potato Control Act.[32]

When Wallace was not fighting off demands for regimented agriculture, he was called on to mediate quarrels within his own ranks. Most of them revolved around the administrator of the AAA, George Peek. Like General Johnson a Baruch man who had served with the War Industries Board and had been an evangelist of McNary-Haugenism, Peek had one interest, the American farmer, and cared not a whit for anything else. He argued that there was no farm surplus, only a disparity in price, and that the country needed more people in agriculture rather than fewer. Peek reasoned: "It is not a healthful thing from any angle to have people crowded together in industrial centers where they may become the victim of every 'ism' of any agitator. The healthsome atmosphere of the country is far better for all of us." He had a pioneer's hatred of the "un-American policy" of restricting production. He would let the farmer grow all he wanted, negotiate marketing agreements to push up farm prices, and dump the surplus overseas.[33]

Peek was opposed by a liberal faction in the Department of Agriculture headed by the assistant secretary, Rexford Tugwell, and by the volatile general counsel, Jerome Frank, author of the brilliant *Law*

[32] Senator John Bankhead to Representative John Kerr, January 22, 1935, Kerr MSS., Box 6; Representative Ralph Brewster to Lindsay Warren, August 14, 1935, Warren MSS., Box 12; Josiah Bailey to James Pou, March 1, 1934, Bailey MSS., Personal File; Schlesinger, *Coming of New Deal*, p. 73; Harold Rowe, *Tobacco under the AAA* (Washington, 1935); *Time*, XXIII (April 9, 1934), 16; XXV (February 18, 1935), 17–18.

[33] Peek to Charles Brand, March 7, 1933, Peek to Randall Brown, March 4, 1933, Peek MSS.; Gilbert Fite, *George N. Peek and the Fight for Farm Parity* (Norman, Okla., 1954), pp. 251–252.

and the Modern Mind. Frank recruited an astonishing array of legal talent: Adlai Stevenson from Chicago, Thurman Arnold and Abe Fortas from Yale Law School, Alger Hiss, Lee Pressman, John Abt, and Nathan Witt from Harvard Law School. The Tugwell-Frank faction could count on the support of the office of the Consumers' Counsel, headed by Frederic Howe, a veteran of Tom Johnson days in Cleveland, and his aide, Gardner "Pat" Jackson, who had battled to save the lives of Sacco and Vanzetti.

These men wanted to exploit the sense of crisis to push through long-needed reforms to relieve the poverty of sharecroppers, tenant farmers, and farm laborers, and to crack down on packers, millers, and big milk distributors to make sure that increased farm prices came out of middlemen and not the consumer. "Henry's father's gang," as they called the old hands in the Department, scorned the Tugwell-Frank crowd as a strange crew of urban intellectuals whose knowledge of agriculture, such as it was, came not from tilling the upper forty but from books. According to one tale, Lee Pressman, at a meeting on the macaroni code, truculently asked what the code would do for the macaroni growers.[34]

Wallace found it increasingly difficult to reign over a house divided. He questioned the advisability of subsidizing exports, and resented Peek's assumption that he was not Wallace's subordinate but a rival potentate. Roosevelt, who oscillated between the two factions, finally succumbed to Tugwell's proposal that he get rid of Peek by giving him some assignment developing foreign markets. "Lordy, Lordy," the President retorted, "how Cordell Hull will love that!" On December 11, 1933, it was announced that Peek had been eased out of AAA to become Roosevelt's special adviser on foreign trade. It was a typically Rooseveltian solution. He tossed the economic nationalist Peek into a new combat with Hull, his internationalist Secretary of State. The liberals had killed off their arch foe, but at the same time many of the processing industries were taken out of the hands of the Tugwell-Frank group and turned over to NRA.[35]

---

[34] Schlesinger, *Coming of New Deal,* pp. 50–51, 74–75: John Hutson, COHC, pp. 114 ff., 175 ff.; Fite, *Peek,* pp. 258–262; Florence Kiper Frank, Review of *The Coming of the New Deal, Yale Law Journal,* LXVIII (1959), 1724.
[35] Schlesinger, *Coming of New Deal,* p. 58; Peek to Wallace, May 12, Novem-

Although stormy seas lay ahead, Wallace and Roosevelt now had a good idea of the course they intended to navigate. As Wallace explained a year later: "Well, you see we don't like to have a ship that lists stronger to the left or the right, but one that goes straight ahead." Going straight ahead meant subsidizing the farmer, but opposing price-fixing; inducing farmers to curtail production, but resisting proposals to compel them; economic nationalism, but no dumping of surpluses abroad; bettering the lot of the sharecropper, but clamping down on reformers who would revolutionize class relations. For Wallace especially, going straight ahead meant turning one's thoughts to meet the criticism that the New Deal was solving the paradox of want in the midst of plenty by doing away with the plenty. The Secretary believed that as an emergency measure sabotaging production could be justified, but he was perpetually astonished that hard work and productivity could yield a harvest of disaster, and he sought ways to modify the AAA's scarcity economics. "Only the merest quarter-turn of the heart separates us from a material abundance beyond the fondest dream of anyone present," he told a Des Moines gathering. Wallace felt relieved when the drought of 1934 made it possible to slacken crop restriction. Yet even that year, he scheduled the removal from cultivation of an area larger than the entire state of Illinois.[36]

However questionable in its conception, the Roosevelt-Wallace farm program worked out reasonably well. During Roosevelt's first term, gross farm income rose 50 per cent, crop prices climbed, and rural debts were reduced sharply. Government payments to farmers benefited merchants and mail order houses. Even the stanchly conservative Sewell Avery, head of Montgomery Ward, conceded that AAA had been the single greatest cause of the improvement in Ward's position. On the other hand, rising prices of farm equipment wiped out some of these gains, and much of the credit for the rise in crop prices should probably be attributed not to the Triple A but to the drought. Since payments to farmers came from processing taxes passed on to the consumer, economists questioned whether they were of much benefit

ber 15, 25, December 2, 1933; Wallace to Peek, May 12, 1933; Peek to F.D.R., December 15, 1933, Peek MSS.; Fite, *Peek,* pp. 264–265.

[36] Lord, *Wallaces of Iowa,* pp. 407, 363, 384–385; Henry Wallace, *New Frontiers* (New York, 1934), pp. 138–139, 200; Wallace, *America Must Choose* (New York, 1934).

to the economy as a whole. In sum, it appears that the AAA was a qualified success; not until 1941 did farm income exceed the returns for 1929, an unsatisfactory year for the farmer.[37]

Roosevelt ultimately derived a good measure of satisfaction from the achievements of his farm program, but the Triple A did not do much to meet the crisis he confronted when he first took office. By the fall of 1933, Roosevelt was in deep trouble. *The New York Times* Weekly Business Index, which had climbed from 60 in March to 99 in June, had dropped to 72 in October.[38] The Farm Belt, disgusted with the slow start of AAA, threatened rebellion. If the President did not do something quickly to bail out the debt-ridden West and South by inflating crop prices, he faced the prospect of a march of irate farmers on Washington.[39]

For some time, the President had been fascinated by the commodity-dollar theories of Irving Fisher and been attracted to the novel currency doctrines of the "Gold Dust Twins," Professors George Warren and Frank Pearson. Perhaps their ideas were unsound, but they at least offered him an opportunity to act, unlike his orthodox advisers who told him all he could do was wait. He did not know anything about economics, he told a White House conference of fiscal experts, but he did know something about politics and he must demonstrate that he was responsive to the cries of the prairie farmers.[40]

[37] "The Stores and the Catalogue," *Fortune*, XI (January, 1935), 74; Murray Benedict, *Can We Solve the Farm Problem?* (New York, 1955), pp. 443–446; Edwin G. Nourse, *Marketing Agreements under the AAA* (Washington, 1935); Edwin Nourse, Joseph Davis, and John Black, *Three Years of the Agricultural Adjustment Administration* (Washington, 1937), Chs. 10–14; Schlesinger, *Coming of New Deal*, pp. 70–71.

[38] "Historians for the period have generally overlooked how close we came to a second economic collapse in the fall of 1933," Marriner Eccles observed subsequently. Eccles, *Beckoning Frontiers* (New York, 1951), p. 126.

[39] John Simpson, president of the National Farmers' Union, wrote the President: "I know the people are overwhelmingly disappointed and disgusted with your program." Unless Roosevelt gave real inflation "within a very short time," Simpson insisted, the President's administration would be "an absolute failure." Simpson to F.D.R., September 14, 1933, Simpson MSS., Special Correspondence, 1932–34.

[40] Irving Norton Fisher, *My Father Irving Fisher* (New York, 1956), pp. 274–283; Joseph Reeve, *Monetary Reform Movements* (Washington, 1943), pp. 162–184; Charles Hardy, *The Warren-Pearson Price Theory* (Washington, 1935); J. E. Crane, "Confidential Files," October 30, 1933, George Leslie Harrison MSS., Conversations, II; Felix Frankfurter to Louis Brandeis, November 7, 1933, Brandeis MSS., G9; Schlesinger, *Coming of New Deal*, pp. 236–238.

On October 19, the President decided to brush aside the objections of Under Secretary of the Treasury Dean Acheson and embrace the Warren-Pearson tactic of gold buying.[41] The economists contended that if the government purchased gold at increasing prices, not only would the country win a greater share of the world's trade, but domestic commodity prices would soar as the gold value of the dollar fell. Devaluing the dollar would restore the balance between the prices of consumers' goods and raw materials. Their larger ambition was to achieve a commodity dollar—one which would have a constant buying power for all commodities—by keying the dollar to the price index through manipulating the gold content.[42]

Starting on October 25, Warren, Jesse Jones, and Henry Morgenthau, Jr., would meet each morning at the White House to set the price of gold for the day. While the President ate his soft-boiled eggs, the others would give him the latest reports on gold and commodity price trends. One morning, Roosevelt, after taking a quick glance at the lugubrious Morgenthau, suggested a rise of 21 cents. "It's a lucky number," the President said, "because it's three times seven." Morgenthau set down in his diary: "If anybody ever knew how we really set the gold price through a combination of lucky numbers, etc., I think they would be frightened." Only afterward did he realize that Roosevelt had been teasing him. It did not matter which price they set so long as it was a little above the world price and the increment varied enough so that speculators could not guess it and make a killing.[43]

Most of the country was mystified by the intricacies of gold buying, but they sensed Roosevelt must be on the right track because of the enemies he was making. Financial journals and eastern sound-money Democrats belabored the President for manipulating the currency; Al Smith announced: "I am for gold dollars as against baloney dollars."[44] But inflationist agrarians and the business promoters of the

[41] Blum, *Morgenthau Diaries,* pp. 65–68.

[42] George Warren and Frank Pearson, *Prices* (New York, 1933); Henry Stimson MS. Diary, July 21, October 24, 1933; "Copy of a letter from Prof. Irving Fisher to his wife, describing a talk with President Franklin D. Roosevelt at Hyde Park, August 9, 1933," Irving Fisher MSS.

[43] Blum, *Morgenthau Diaries,* pp. 69–70. In the midst of the gold-buying experiment, Roosevelt named Morgenthau Acting Secretary of the Treasury to succeed Woodin, who was too ill to return.

[44] Smith, "Sound Money: An Open Letter to the Chamber of Commerce of the State of New York," *New Outlook,* CLXII (December, 1933), 9–11.

South and West, who wanted a supply of cheap money for new investment, rallied to Roosevelt. The President himself believed he confronted a conspiracy of bankers who hoped to coerce him into adopting the financial views of Wall Street. He wrote Colonel House: "The real truth . . . is, as you and I know, that a financial element in the larger centers has owned the Government ever since the days of Andrew Jackson—and I am not wholly excepting the administration of W.W. The country is going through a repetition of Jackson's fight with the Bank of the United States—only on a far bigger and broader basis."[45]

Roosevelt's gold-buying venture produced the first serious breach within the administration. Not only Acheson but two of his ablest financial advisers, James Warburg and the President's old Harvard professor, Oliver M. W. Sprague, opposed the policy. Roosevelt came to feel that such men were either tools of Wall Street or impractical academic economists, or both; he thought them callous to the suffering of debtors and indifferent to the need for compromise with angry farmers. As a result, he dealt with these men, who in fact sympathized with much of the New Deal and had legitimate grounds for doubting the wisdom of gold buying, severely and unfairly. "Sprague is a nuisance," Roosevelt wrote Colonel House on November 21. "He carries no weight except with the Bank of England crowd and some of our New York City bankers. . . ." That same day, Sprague announced his resignation. James Warburg followed by breaking with Roosevelt. The President fired Acheson, who he wrongly thought had been disloyal, although Acheson had, in fact, been faithfully carrying out a program he detested.[46]

Gold buying was one of the most ill-considered moves Roosevelt ever made. Warren's program failed to work as he had predicted. Wholesale commodity prices actually fell slightly during November and December, and farm prices dropped most of all. The President, perplexed

[45] F.D.R. to House, November 21, 1933, House MSS. A week earlier he had written Robert Bingham: "All goes well here, though some of the inevitable sniping has commenced, led by what you and I would refer to as the Mellon-Mills influence. . . ." F.D.R. to Bingham, November 13, 1933, FDRL PPF 716. Cf. Katherine Blackburn, "The Public Reaction to the Gold Buying Policy as Reflected in the Newspapers," Louis Howe MSS., Box 64.

[46] F.D.R. to House, November 21, 1933, House MSS.: Schlesinger, *Coming of New Deal*, pp. 236–243; Henry Morrow Hyde MS. Diary, Nov. 22, 1933; "Confidential Memorandum," May 24, 1933, Emanuel Goldenweiser MSS., Box 1; James Warburg, *The Money Muddle* (New York, 1934), pp. 144–159.

by the way prices were behaving, remarked that "it was funny how sometimes they seemed to go against all the rules."[47] The most that could be credited to gold buying was that it helped arrest a deflationary trend. Moreover, Roosevelt kept control of monetary policy in his own hands instead of yielding either to the inflationists or the goldbugs. The real damage wrought by gold buying was that for several crucial weeks the President placed his hopes in a panacea which did not and, professional economists warned him, could not work. Keynes thought Warren's theories "puerile" and hooted that his *Prices* would not be tolerated in an undergraduate. The gyrations of the dollar, he observed, "looked to me more like a gold standard on the booze than the ideal managed currency of my dreams."[48] Gold buying cost valuable time which might have been better spent in speeding public works, and it lost to the New Deal men who might well have been held.[49]

At the beginning of 1934, Roosevelt decided to halt gold buying. He wanted to reap the profits from devaluation and to establish an equalization fund to diminish the supposed influence of the British on international exchange. Roosevelt's proposal to set an upper limit of 60 per cent for the gold value of the dollar ran into opposition from inflationists, who denounced it as a concession to the conservatives, and from conservatives, who protested that Roosevelt was sanctioning "the clipping of the national coin in exactly the fashion of the medieval kings." Despite these animadversions, both houses whipped through the bill in a few days. On the President's birthday, January 30, he signed the Gold Reserve Act—"the nicest birthday present I ever had."[50] The very next day, the President set the price of gold at $35 an ounce, which reduced the dollar to 59.06 per cent of its pre-1933 gold content, and placed the country on what Morgenthau called "the 1934 gold bullion standard . . . streamlined . . . air flow . . . and knee action." If the Gold Reserve Act, as some economists insisted,

[47] "Conversations," Nov. 16, 1933, George Leslie Harrison MSS.

[48] Felix Frankfurter to Louis Brandeis, December 9, 1933, Brandeis MSS. G9; Keynes, "Open Letter to President Roosevelt," *The New York Times,* December 31, 1933.

[49] "Confidential Memorandum," Oct. 23, 1933, Emanuel Goldenweiser MSS., Box 1; Schlesinger, *Coming of New Deal,* p. 246; Blum, *Morgenthau Diaries,* pp. 74–75. It is arguable, however, that even a misguided venture in managing money was an important step toward managing money more sensibly.

[50] New Haven *Journal-Courier,* quoted in *Literary Digest,* CXVII (January 27, 1934), 5; *The New York Times,* January 31, 1934.

made Roosevelt's monetary policy less flexible, it freed him temporarily from some of the pressure of the inflationist lobby as well as from the burden of constant preoccupation with the price of gold, and it enhanced Treasury control of credit and currency.[51]

While the end of gold buying appeased some conservatives, it only intensified the demands of the inflationists, and particularly of the silverites. Silver was an industry less important than chewing gum, spaghetti, or the output of "Eskimo Pies," yet the cohesive bloc of silver senators from the sparsely settled mountain states who often held the balance of power in the Senate was able to win concessions for the industry far beyond its actual significance. Some of the silver senators, notably their leader, Nevada's Key Pittman, simply wished to benefit silver interests; others, such as Oklahoma's Elmer Thomas, cared little about the metal but hoped to use silver to inflate prices.[52] Both groups were fanatical in their devotion to the silver cause. When Secretary Morgenthau, baffled by the intransigence of the silverites, tried to reason with them, Senator Henry Fountain Ashurst of Arizona replied: "My boy, I was brought up from my mother's knee on silver and I can't discuss that any more with you than you can discuss your religion with me."[53]

In a sober interlude, Pittman had extracted from the London Economic Conference of 1933 an international agreement which committed the United States to purchase what amounted to almost the entire annual production of American silver for the next four years. Pittman persuaded Roosevelt to effect ratification not by Senate action but by a presidential edict, using his powers under the Thomas amendment. On December 21, 1933, the President announced that for the next four years the government would buy the annual output of silver at 64.5 cents an ounce, more than 21 cents above the current market price. When news of Roosevelt's proclamation reached the West,

[51] Blum, *Morgenthau Diaries,* pp. 120–124; Arthur Whipple Crawford, *Monetary Management under the New Deal* (Washington, 1940), pp. 78–89; Gustav Cassel, *The Downfall of the Gold Standard* (Oxford, 1936), pp. 181–190; Irving Fisher to F.D.R., December 1, 1933, Fisher MSS.

[52] "Silver, Just Silver," *Fortune,* VIII (July, 1933), 103; *Time,* XXXV (March 11, 1940), 12; Fred Israel, "Nevada's Key Pittman" (unpublished Ph.D. dissertation, Columbia University, 1959), pp. 164–178; Earl Wixcey to Elbert Thomas, November 15, 1935, Elbert Thomas MSS., Box 23.

[53] Allan Everest, *Morgenthau the New Deal and Silver* (New York, 1950), p. 39.

miners shot off revolvers in the streets of Leadville and danced with girls atop the bars of Tonopah.[54]

Even this did not satisfy the silverites. In January, 1934, Senator Wheeler filed a bill to compel the purchase of another billion ounces of silver, and to issue currency against it. When twenty-eight Democrats broke ranks, the Senate came within two votes of adopting the Wheeler proposal. On March 19, on the seventy-fourth anniversary of the birth of William Jennings Bryan, the House passed an ingenious bill framed by Representative Martin Dies of Texas which provided for the sale of farm surpluses abroad in return for silver at a rate above the world price; it united inflationists, agrarians, and silverites. On April 10, a Senate committee reported out the Dies bill with still more bounties for the silver states.[55]

Roosevelt fought a fierce rear-guard action. He called the silverites to the White House and lectured them. "In the case of silver you don't know one damn thing—Key to the contrary, notwithstanding," the President said testily. After more than an hour of argument, one senator reported wearily: "We discussed silver from the time of the finding of the first nugget." When Roosevelt made no headway, he permitted Secretary Morgenthau to publish a list of silver speculators, but this had no effect either. The silver bloc had the votes, and the President was forced to compromise.[56] The Silver Purchase Act of June, 1934, directed the Secretary of the Treasury to buy silver until it reached one-fourth of the country's monetary reserve or until the world price of silver climbed to $1.29 an ounce. To cover the cost, the Treasury would issue silver certificates. The act gave the Secretary wide discretion as to how to bring this about. In the next fifteen years, the silver industry, employing fewer than five thousand persons, would wring from the government almost a billion and a half dollars—much more than the government would pay to support farm prices in that same period. These purchases, which jeopardized the

[54] Fred Israel, "The Fulfillment of Bryan's Dream: Key Pittman and Silver Politics, 1918–1933," *Pacific Historical Review*, XXX (1961), 380; Everest, *Morgenthau*, pp. 26–32; *Time*, XXIII (January 1, 1934), 5–6.

[55] Everest, *Morgenthau*, pp. 36–42; Blum, *Morgenthau Diaries*, pp. 184–185; Henry Rainey to H. G. Pine, March 21, 1934, Rainey MSS., Box 1.

[56] "The President's conference with the Senators, Saturday afternoon, April 14, 1934," FDRL PSF 37; *Time*, XXIII (April 30, 1934), 9; Key Pittman to Henry Morgenthau, April 13, May 17, 1934, Pittman MSS., Box 142; Everest, *Morgenthau*, pp. 42–46.

currencies of China, Mexico, and Peru, failed either to remonetize silver or promote inflation.[57]

By 1934, the pattern of the early New Deal was beginning to emerge. Its distinguishing characteristic was the attempt to redress the imbalances of the old order by creating a new equilibrium in which a variety of groups and classes would be represented. The New Dealers sought to effect a truce similar to that of wartime, when class and sectional animosities abated and the claims of partisan or private economic interest were sacrificed to the demands of national unity. Roosevelt presented himself not as the paladin of liberalism but as father to all the people, not as the representative of a single class but as the conductor of a concert of interests. A man above the political battle, the President aimed to serve as the unifier of interests and the harmonizer of divergent ideologies. In March, 1934, he declined to take part in Jefferson Day celebrations, and even suggested it would be a "fine thing" if nonpartisan Jefferson dinners were held with as many Republicans as Democrats on the banquet committee. "Much as we love Thomas Jefferson we should not celebrate him in a partisan way," the President wrote.[58]

Roosevelt incorporated a wide spectrum of views in the government and tolerated deep divisions within his administration over which policies he should sponsor. At the head of one faction, Rexford Tugwell, the debonair Assistant Secretary of Agriculture, advocated extensive government planning, heavy spending for relief and public works, and curbs on profiteers. He rued the failure to take advantage of the banking crisis to set up a national bank and take over "large blocks of paralyzed industries."[59] Tugwell spoke for the ardent New Deal reformers and for congressional progressives like Bob La Follette. At the head of the opposing faction, Budget Director Lewis Douglas, Roosevelt's "Minister of Deflation," opposed government spending, abhorred currency tinkering, and believed that prosperity would come by balancing the budget and leaving capital investment to private in-

[57] Schlesinger, *Coming of New Deal*, p. 252; Blum, *Morgenthau Diaries*, pp. 184, 199–228; Herbert Bratter, "The Silver Episode," *Journal of Political Economy*, XLVI (1938), 609–652, 802–837; Ray Westerfield, *Our Silver Debacle* (New York, 1936), pp. 94–207.

[58] F.D.R. to E. M. House, March 10, 1934, House MSS.

[59] Tugwell, "Notes from a New Deal Diary," April 21, 1933. Cf. Aline Chalufour and Suzanne Desternes, "Deux théoriciens de la crise américaine," *Revue des Sciences Politiques*, LVII (1934), 221–235.

dustry. "The condition precedent to the establishment of a communistic or fascistic order of society," Douglas warned the President, "is the destruction of the middle class by paper inflation or unbearable taxation." Around Douglas gathered the "Treasury crowd," business-oriented New Dealers such as Raymond Moley and Donald Richberg, and the conservative southern Democrats, many of whom felt a halt should be called to further reforms.[60]

Roosevelt spurned the opportunity to be broken on the rocks of factional warfare. At heart a man of orthodox fiscal ideas who dreaded unbalanced budgets, he wanted Douglas around to check the spenders. Douglas, the President confided, was "in many ways the greatest 'find' of the administration." At the same time, he recognized that relief appropriations would be necessary during the emergency, and, while he nodded reassuringly to Douglas, he went ahead with plans which upset the budget. Although he shunned massive deficit spending, the President refused to restrict expenditures to estimated receipts. He favored a middle course of providing for the needy, even at the cost of an unbalanced budget, convinced that in the near future the economy, partly in response to different New Deal enterprises, would take an upward turn and the budget could be balanced out of tomorrow's surpluses.[61]

The Roosevelt administration spun an intricate web of procedures to permit the will of different interest groups to be expressed. Millions of farmers voted in AAA crop referendums, including thousands of southern Negroes who had never before cast a ballot, and the Triple A administered the farm law through production-control associations manned by more than 100,000 committeemen.[62] The Taylor Grazing Act of 1934, which provided for the segregation of up to eighty million acres of public grasslands, stipulated that the Secretary of the Interior co-operate with local associations of stockmen in administering the

[60] Douglas to F.D.R., December 30, 1933, FDRL PSF 28; Arthur Schlesinger, Jr., *The Politics of Upheaval* (Boston, 1960), p. 212.

[61] F.D.R. to Edward House, April 5, 1933, House MSS.; Blum, *Morgenthau Diaries*, pp. 248–249.

[62] Robert Martin, "The Referendum Process in the Agricultural Adjustment Programs of the United States," *Agricultural History*, XXV (1951), 34–47; Ralph Bunche, "The Negro in the Political Life of the United States," *Journal of Negro Education*, X (1941), 577. For another phase of agrarian democracy, see H. C. M. Case, "Farm Debt Adjustment During the Early 1930's," *Agricultural History*, XXXIV (1960, 173–181.

grazing districts. At a powwow in the Black Hills, Commissioner John Collier, speaking through his Sioux interpreter, Young Red Tomahawk, explained the New Deal for the Indians to Blackfeet from Montana, Winnebagoes from Iowa, and Arapahoes from Wyoming. In the Tennessee Valley, the TVA raised the principle of local participation to a fine art. By the end of the thirties, the New Deal had extended the idea of popular involvement to a host of different agencies. Millions of workmen cast ballots in Labor Board collective bargaining elections; district boards of operators and miners set quotas for bituminous coal; local advisory committees were consulted in the national youth and arts projects; and farmers balloted to create soil conservation districts. As the Mescalero Apaches prepared to vote under the authority of the Indian Reorganization Act of 1934, their tribal leaders presented plays to illustrate each section of the proposed constitution.[63]

TVA Director David Lilienthal hailed these procedures as "grassroots democracy," Congressman Taylor spoke of "home rule on the range," and Henry Wallace wrote of "a hierarchy of New England town meetings."[64] Lilienthal claimed that they resolved the dilemma of how to achieve effective government without creating a leviathan state that would destroy democracy, for they combined national power with local "grassroots" administration. While such sentiments were sincerely expressed, they obscured the fact that "grassroots democracy" frequently meant a surrender of the national interest to local power groups. The AAA, by vesting authority in county production associations, turned the power of decision over to the Farm Bureau Federation, the Extension Service, and the land-grant colleges—in short, the larger landholders.[65] Conservationists objected that the Taylor

---

[63] *Literary Digest,* CXVII (April 7, 1934), 21; Elizabeth Green, "Indian Minorities under the American New Deal," *Pacific Affairs,* VIII (1935), 420–427; Oliver La Farge, *As Long as the Grass Shall Grow* (New York, 1940), p. 96.

[64] Lilienthal, *TVA: Democracy on the March* (New York, 1944); *Congressional Record,* 76th Cong., 3d Sess., p. A4199; Henry Wallace, *New Frontiers* (New York, 1934), p. 200.

[65] Grant McConnell, *The Decline of Agrarian Democracy* (Berkeley and Los Angeles, 1953), pp. 72–83; James West, *Plainville, U.S.A.* (New York, 1961), p. 217. In Texas, county agents were instructed that the county AAA committees were to be composed of a banker, a businessman, and a farmer. Henry I. Richards, *Cotton under the Agricultural Adjustment Act* (Washington, 1934), pp. 18–19.

Act gave stockmen an opportunity to exploit the domain for their advantage. "If they really believe the Taylor Grazing Act is a blow at 'special privilege,'" one editor observed, "then Heaven help democracy."[66] Lilienthal's mystical prose about "democracy on the march" glossed over the political bargain he had struck with TVA Director Harcourt Morgan. He had neutralized opposition to public power at the high cost of acquiescence to Morgan's administration of the TVA's farm program in the interests of the more prosperous white planters.[67] Fifteen years later Rexford Tugwell penned the biting verdict: "TVA is more an example of democracy in retreat than democracy on the march."[68]

Interest-group democracy had the tactical advantage of weakening opposition by incorporating potential opponents within the administration, but it also served to make the Roosevelt administration the prisoner of its own interest groups.[69] The President often shied away from decisions that might antagonize one or another of the elements in the coalition. Government sanction greatly enhanced the power of such groups as the Farm Bureau Federation and the trade associations. Like the mercantilists, the New Dealers protected vested interests with the authority of the state. "What Colbert did under Louis XIV," wrote Walter Lippmann, "was precisely what General Johnson and Secretary Wallace did under President Roosevelt." By 1942, the farm bloc would be able to impose a demand for 110 per cent of parity; it had developed, observed A. Whitney Griswold, "from a ward of charity

[66] *Nature Magazine,* XXXI (1938), 46; Roy Robbins, *Our Landed Heritage* (Princeton, 1942), pp. 421–422; Wesley Calef, *Private Grazing and Public Lands* (Chicago, 1960), pp. 52–77. Cf. Phillip O. Foss, *Politics and Grass* (Seattle, 1960), p. 69. While some conservationists thought of the Taylor Act as a way to save the cattle lands from overgrazing, many stockmen viewed the measure as a way to stabilize their industry by curbing overproduction, a view of "conservation" similar to that held by the oilmen.

[67] Philip Selznick, *TVA and the Grass Roots* (Berkeley and Los Angeles, 1953). Cf. Norman Wengert, *Valley of Tomorrow* (Knoxville, 1952). The TVA even opposed or ignored the programs of New Deal farm agencies—notably the Farm Security Administration and the Soil Conservation Service—that had no such commitments.

[68] R. G. Tugwell and E. C. Banfield, "Grass Roots Democracy—Myth or Reality?" *Public Administration Review,* X (1950), 47–55.

[69] Murray Edelman, "New Deal Sensitivity to Labor Interests," in Milton Derber and Edwin Young (eds.), *Labor and the New Deal* (Madison, Wis., 1957), p. 180; Jonathan Mitchell, "Mr. Roosevelt on Stilts," *New Republic,* LXXVII (1933), 69.

into a political force capable of pursuing its own interests even to the point of defying the head of the nation in wartime." As early as 1934, the Harvard political scientist E. Pendleton Herring inquired: "Can the presidential system continue as a game of touch and go between the Chief Executive and congressional *blocs* played by procedural dodges and with bread and circuses for forfeit?"[70]

Roosevelt's predilection for balanced government often meant that the privileges granted by the New Deal were in precise proportion to the strength of the pressure groups which demanded them. "Because NRA policy formulation involved the pitting of private economic groups against each other, the outcome could never be too different from the parties' estimate of their chances of winning in a direct conflict in the economic arena," one scholar has observed. "Unions won important gains in NRA codes only where they were economically strong."[71] Conversely, causes which were not sustained by powerful interest groups frequently made little headway. Those which relied on a mythical "consumer interest" fared poorly, as the history of the NRA Consumers' Advisory Board and the sad saga of Tugwell's futile attempt to get an effective pure food and drugs bill amply demonstrated.[72]

If the state was to be no more than "a parallelogram of pressures," any attempt at reform which required a direct challenge to vested interests was doomed. Joseph Eastman had little success in co-ordinating the railroads because the railroads would not let him. When carriers and unions withdrew their support, the program died. Advocates of slum clearance made only superficial progress because the administration would not tackle those elements—the real estate bloc, insurance companies, and building trades unions—which blocked large-scale public housing. NRA policy, noted Arthur R. Burns, "was directed to

[70] Lippmann, *Good Society,* p. 10; Griswold, *Farming and Democracy* (New York, 1948), pp. 156–157; E. Pendleton Herring, "Second Session of the Seventy-third Congress, January 3, 1934, to June 18, 1934," *American Political Science Review,* XXVIII (1934), 866.

[71] Edelman, "New Deal Sensitivity," p. 167. See Fred Greenbaum, "A 'New Deal' for the United Mine Workers of America," *Social Science,* XXXIII (1958), 154–155; Sidney Fine, "The N.I.R.A. and the Automobile Industry," *Michigan Alumnus Quarterly Review,* LXVI (1960), 249–259.

[72] Persia Campbell, *Consumer Representation in the New Deal* (New York, 1940). When the pure food and drugs bill, first introduced in 1933, was finally passed as the Wheeler-Lea Act of 1938, it was a toothless crone.

considerations no more far reaching than an estimate of what representatives of industry would accept without protest. Essential reform must necessarily involve from time to time the imposition of policies distasteful to the class that is most highly organized and articulate."[73] An early commentator on the New Deal advanced a view which later historians have adopted: "Roosevelt is essentially a broker. He has neither the will nor the power to move against the political currents of the day. He does not and cannot invent his policies."[74]

Critics have been quicker to see the limitations than the benefits of "broker state" government. It marked an important advance from the "single interest" administrations of the 1920's. It gave to such interests as the farm groups a voice in government that had been largely denied them in the twenties. If industrial unionists were disadvantaged in the early New Deal years, they also had new forums for expressing their grievances and champions like Senator Wagner who argued that interest-group government implied public encouragement of unionization. "In order that the strong may not take advantage of the weak," Wagner reasoned, "every group must be equally strong."[75] Moreover, the New Deal was never simply a broker state. The need to create overnight a new civil service brought to power men who had not been routinized by years in a bureaucracy and who were not easily disciplined; instead of viewing themselves simply as agents of the President, they often pushed programs of their own which the President was compelled to accept. Finally, Roosevelt himself insisted that there was a public interest paramount to that of any private group. He scolded bankers in 1934: "The old fallacious notion of the bankers on one side and the Government on the other side as being more or less equal and independent units, has passed away. Government by the necessity of things must be the leader, must be the judge of the conflicting interests of all groups in the community, including bankers."[76]

[73] Max Lerner, *Ideas for the Ice Age* (New York, 1941), p. 379; Earl Latham, *The Politics of Railroad Coordination, 1933–1936* (Cambridge, 1939); Arthur R. Burns, *The Decline of Competition* (New York, 1936), p. 519.

[74] "Unofficial Observer [John Franklin Carter]," *The New Dealers* (New York, 1934), p. 25. The "broker state" conception of the New Deal is most lucidly developed in John Chamberlain, *The American Stakes* (Philadelphia, 1940), and Burns, *Roosevelt*, pp. 183–202.

[75] Irving Bernstein, *The New Deal Collective Bargaining Policy* (Berkeley and Los Angeles, 1950), p. 101.

[76] *Public Papers*, III, 436. Tugwell recalls that Roosevelt, incensed by the

Roosevelt's ire at the bankers sounded the fundamental note of discord in the early New Deal. The President's desire to run a pluralistic government in which he would serve chiefly as a mediator of interests clashed with the fact that he was the agent, both willingly and unwillingly, of forces of reform that business found unacceptable. As eager to hold the allegiance of businessmen as of any other group, Roosevelt was also determined to regulate the practices of high finance, and it was inevitable that this would cost him business support. No single act did more to mobilize the business community against Roosevelt than his message of February 9, 1934, asking Congress for legislation to regulate the Stock Exchange. Financiers protested that the Fletcher-Rayburn bill, introduced the same day, would turn Wall Street into a "deserted village."[77] "You gentlemen are making a great mistake," scolded Richard Whitney, the supercilious president of the New York Stock Exchange. "The Exchange is a perfect institution."[78]

Roosevelt, who never endorsed the stringent Fletcher-Rayburn draft, was willing to compromise, especially if this would detach such moderates as James Forrestal of Dillon, Read from the Whitney faction. But after the bill had been overhauled to ease the more rigid features of the original draft, the President would give no more ground. The bill sped through both houses in May, and on June 6, 1934, the President signed the Securities Exchange Act. The act, which created a new agency, the Securities and Exchange Commission, to administer the law and the Securities Act of 1933 as well, aimed to prevent manipulation of the securities market by insiders by placing trading practices under federal regulation. It sought to end misrepresentation by requiring registration and full disclosure of information on all securities traded on the exchanges, although, as Professor Galbraith later ob-

---

draft he had read of the speech of introduction at the bankers' convention, exploded: "Imagine referring to a representative of the American Bankers Association and the President of the United States as equals! And imagine what national policy would be like if, instead of requiring such groups to conform to the public interest, they were free to bargain about what they would or would not consent to accept. Outrageous!" Tugwell, *Roosevelt*, p. 381.

[77] *Literary Digest*, CXVII (April 7, 1934), 7. Cf. Russell Leffingwell to F.D.R., January 4, March 3, May 15, 1934, FDRL PPF 866; New York *Journal of Commerce*, March 9, 1934.

[78] Joseph Alsop and Robert Kintner, "The Battle of the Market Place," *Saturday Evening Post*, CCX (June 11, 1938), 9. Cf. Richard Whitney to Carter Glass, October 19, 1933, Glass MSS.

served, "no way was found of making would-be investors read what was disclosed." By regulating margin requirements, it hoped to reduce irresponsible speculation. After months of revelations by the Pecora committee, the demand for regulation of Wall Street had been met, and the men who had been cleaned out in 1929 felt avenged.[79]

By the summer of 1934, Roosevelt's coalition of all interests was breaking up. Businessmen came more and more to feel that the New Deal would mean extensive regulation of business. The stock exchange law would have been bad enough, but that was not all. Congress in 1934 passed the Communications Act, which created a Federal Communications Commission to regulate the radio, telegraph, and cable businesses; the Railroad Retirement Act, which set up pensions for railway workers; and the Air Mail Act, which stipulated rigid controls in awarding federal contracts. Even more disturbing was the mounting deficit. Not only did Congress restore the cuts made in the Economy Act, but Roosevelt himself sponsored multibillion-dollar expenditures for relief and public works. On August 30, 1934, deeply troubled by Roosevelt's fiscal policies, Lewis Douglas resigned. Some weeks later, he wrote the President: "I hope, and hope most fervently, that you. will evidence a real determination to bring the budget into actual balance, for upon this, I think, hangs not only your place in history but conceivably the immediate fate of western civilization." With Douglas gone, Henry Morgenthau, Jr., became the Roland of the balanced budget; but Morgenthau, unlike Douglas, opposed regressive tax policies and would tolerate deficits if this was the price of adequate relief.[80]

In August, 1934, the same month Douglas resigned, the District of Columbia granted a charter to a new organization of disaffected businessmen, the American Liberty League. Five months earlier, R. R. M. Carpenter, a retired Du Pont official, had written John J. Raskob, a high Du Pont officer and former chairman of the Democratic party: "Five Negroes on my place in South Carolina refused work this Spring

[79] John Kenneth Galbraith, *The Great Crash* (Boston, 1955), p. 171; Schlesinger, *Coming of New Deal,* pp. 465–470; "SEC," *Fortune,* XXI (June, 1940), 120–121; Alfred Bernheim and Margaret Grant Schneider (eds.), *The Security Markets* (New York, 1935).

[80] Douglas to F.D.R., November 28, 1934, FDRL PPF 1914; Lewis Douglas, *The Liberal Tradition* (New York, 1935); Blum, *Morgenthau Diaries,* pp. 230, 246, 388–389. In January, 1935, Morgenthau, who had been Acting Secretary, was appointed Secretary of the Treasury when Woodin resigned.

. . . saying they had easy jobs with the government. . . . A cook on my houseboat at Fort Myers quit because the government was paying him a dollar an hour as a painter. . . ." Touched by Carpenter's plight, Raskob urged him and his friend Irénée du Pont to form an organization to educate the nation "to the value of encouraging people to work; encouraging people to get rich."[81]

While its founders insisted that the Liberty League welcomed all classes of people, northern industrialists, notably executives of Du Pont and General Motors, dominated the organization. Although its president, Jouett Shouse, insisted it was not anti-Roosevelt, it served as a vehicle for anti-New Dealers of both parties. Four of its six officers were Democrats: Al Smith, Irénée du Pont, Shouse, and Raskob. Yet they were conspicuously conservative Democrats who had not been able to reconcile themselves to Roosevelt's nomination in 1932 and the displacement of the Raskob group from control of the party. Raskob himself wrote Jouett Shouse a few days after the 1932 convention: "When one thinks of the Democratic Party being headed by such radicals as Roosevelt, Huey Long, Hearst, McAdoo, and Senators Wheeler and Dill, as against the fine, conservative talent in the Party as represented by such men as you, Governor Byrd, Governor Smith, Carter Glass, John W. Davis, Governor Cox, Pierre S. Du Pont, Governor Ely and others too numerous to mention, it takes all one's courage and faith not to lose hope completely."[82] They still considered themselves Democrats, but their main link to Roosevelt had been severed when the Twenty-first Amendment had been ratified. As in the fight against prohibition, the campaign against the encroachments of the welfare state appeared to involve an issue of personal liberty, and the League embraced many of the old repeal leaders. The President scoffed at the claims of the new organization. The League reminded him, he told newsmen, of an organization formed to uphold two of the Ten Commandments.[83]

[81] *The New York Times,* December 21, 1934; Frederick Rudolph, "The American Liberty League, 1934–1940," *American Historical Review,* LVI (1950), 19–33.

[82] Raskob to Shouse, July 7, 1932, Shouse MSS.

[83] *The New York Times,* August 25, 1934; Memorandum, "Relating to the Substance of Conversation Had with President Roosevelt at the White House, Wednesday Afternoon, August 15, 1934," Shouse MSS.; George Wolfskill, *The Revolt of the Conservatives* (Boston, 1962). Herbert Hoover scornfully rejected Raskob's bid to join. He did not like the "Wall Street model of human liberty"

Roosevelt continued to woo businessmen, but he would not jettison his program to hold their favor. "With the passage of the Stock Exchange bill, the New Deal is practically complete," wrote Raymond Moley in May, 1934. "There is nothing that the President or any responsible member of the Administration has said to indicate that any important further development of governmental authority is contemplated." The President had other ideas. "During June and the summer there will be many new manifestations of the New Deal, even though the orthodox protest and the heathen roar!" Roosevelt wrote Colonel House that same month. "Do you not think I am right? We must keep the sheer momentum from slacking up too much and I have no intention of relinquishing the offensive in favor of defensive tactics."[84] In June, Roosevelt announced he would recommend to Congress in 1935 an ambitious program of social insurance, housing, and the development of natural resources.

Roosevelt's determination to go ahead with reforms was, in part, a response to the state of the economy. By the spring of 1934, the country had pulled out of the near-crisis recession of the autumn of 1933. *The New York Times* Weekly Business Index climbed from 72 in October, 1933, to 86 in May, 1934. Detroit had twice as many men at work as a year before. After farmers cashed their AAA checks, Des Moines department stores were jammed like the week before Christmas. Bridge players in Kenosha raised the rate from a tenth of a cent to an eighth of a cent a point. South of Fulton Street, restaurants in New York's financial district which had been half-empty a year before buzzed with lunchtime chatter. Broadway, which had dimmed its lights in September, 1933, rejoiced in its best season in five years after, wholly unexpectedly, Sidney Kingsley's surgical drama *Men in White* scored a smash hit. Before the season ended, more than one hundred and fifty productions brightened the city, including Eugene O'Neill's *Ah, Wilderness!;* Sinclair Lewis' *Dodsworth,* which brought Walter Huston back from Hollywood; Sidney Howard's *Yellow Jack;* Maxwell Anderson's *Mary of Scotland;* and Jerome Kern's *Roberta.*[85]

---

any more than the "Pennsylvania Avenue model," he wrote. Hoover *Memoirs* (3 vols., New York, 1951–52), III, 454–455. See Hoover to Henry Prather Fletcher, August 25, 1934, Fletcher MSS., Box 4.

[84] Moley, "And Now Give Us Good Men," *Today,* II (May 19, 1934), 12; F.D.R. to House, May 7, 1934, House MSS.

[85] Rud Rennie, "Changing the Tune from Gloom to Cheer," *Literary Digest,*

Even more quickly than it began, the surge of recovery subsided. From the spring of 1934 to the spring of 1935, the country rode at anchor. Month after month, the indicator of business activity pointed to the same monotonous number. National income for 1934 totaled $10 billion less than in 1931, a little better than half that of 1929. In the autumn of 1934, many millions were still jobless. No one doubted that conditions had improved—the ranks of the unemployed had been reduced by over two million and national income stood almost a quarter higher than in 1933—but it was no longer clear that Roosevelt could solve the riddle of bad times. One of Hopkins' aides reported on sentiment in Houston: "Nobody seems to think any more that the thing is going to WORK."[86]

In a novel published that year, one character bursts out: "What the hell has happened to everything? You read in the paper things have never been so good; there's never been so much prosperity; the God-damned stock market is booming; and then you find out you can't get work, everybody's losing his job, or their wages are being cut—Christ, there are thousands of people on the roads, thousands, men looking for work—sometimes whole families—all up and down the country—thousands! You see the poor devils hiking along the highways, even down in the desert, even up in them God-awful mountains in California, miles from nowhere, way out in the woods—everywhere you go."[87] As the country weathered the crisis but failed to achieve prosperity, dissatisfaction mounted. As the New Deal bogged down, demagogues and radicals who feasted on the popular discontent of 1934 challenged Roosevelt for national leadership.

CXVII (June 16, 1934), 25; *Literary Digest,* CXVII (April 14, 1934), 34, 40, 44; (April 28, 1934), 21, 46; (May 12, 1934), 46; Lorena Hickok to Harry Hopkins, April 25, 1934, Hopkins MSS.

[86] Lorena Hickok to Hopkins, April 11, 1934, Hopkins MSS.; Henry Rainey to Joseph Guffey, March 10, 1934, Rainey MSS., Box 1; Eleanor Roosevelt to Mary Dewson, March 9, 1934, Dewson MSS., Box 16.

[87] Robert Cantwell, *The Land of Plenty* (New York, 1934), pp. 286–287.

CHAPTER 5

Waiting for Lefty

THE revolutionary spirit burgeons not when conditions are at their worst but as they begin to improve. In America, the radical tempo quickened not at the bottom of the depression in 1932, when many were dispirited, but in 1934, when things had taken a turn for the better. In Minneapolis in April, a mob of six thousand fired sticks, stones, bottles, and lumps of coal through the windows of City Hall, where the city council was meeting. The rioters, some of whom wore red armbands, demanded government work at a decent wage and an increase in relief payments. That fall, one thousand Denver relief workers, angered at a cut in relief, struck to gain removal of the state administrator, threw the tools of nonstrikers into the Platte, and rioted when police arrested their leader. By 1934, many progressives and most radicals had come to question whether Roosevelt's "broker state" policy held any hope for success. "Failure is a hard word," observed the radical periodical *Common Sense*. "It is too early to convince Roosevelt's supporters that he is a failure. Yet we believe the record indicates that nothing but failure can be expected from the New Deal."[1]

The most forthright of the radical leaders was Minnesota's Governor Floyd Olson. When conservatives tied up relief bills in the legislature in April, 1933, Olson warned: "I shall declare martial law. A lot of people who are now fighting the measures because they happen to

[1] *Time,* XXIII (April 16, 1934), 18; XXIV (November 12, 1934), 17; *Common Sense,* III (September, 1934), 2.

possess considerable wealth will be brought in by the provost guard. They will be obliged to give up more than they are giving up now." If capitalism could not prevent a recurrence of depressions, he cried, "I hope the present system of government goes right down to hell." He wished Roosevelt's recovery program to succeed, he told the Eagles' National Convention in Cleveland that summer, but he did not believe security could be achieved "unless the key industries of the United States are taken over by the government." At the biennial convention of the Minnesota Farmer-Labor party, Olson confessed: "Now I am frank to say that I am not a liberal. I enjoy working on a common basis with liberals for their platforms, etc., but I am not a liberal. I am what I want to be—I am a radical. . . . I am not satisfied with hanging a laurel wreath on burglars and thieves and pirates and calling them code authorities or something else. . . . What is the ultimate we are seeking? . . . The ultimate is a Cooperative Commonwealth."[2]

Of all his political rivals in these years, none gave Franklin Roosevelt such deep concern as Louisiana's flamboyant Senator Huey Pierce Long. At the Democratic convention in Chicago in 1932, Long had been one of the leaders of the Roosevelt forces, and it is even arguable that without him Roosevelt would not have been nominated. Yet the President distrusted Huey instinctively; he told Tugwell that Long— who had given Louisiana badly needed reforms but who also had flouted and caricatured the processes of parliamentary democracy— was one of the two most dangerous men in the country.[3]

For a season, Roosevelt managed to keep Long in line. A few weeks after the election, Huey stormed into Roosevelt's suite at the Mayflower, telling reporters he was going to "talk turkey." He reappeared a half hour later to announce blandly: "He is the same old Frank. He is just like he was before the election, all wool and a yard wide." When asked if Roosevelt planned to "crack down" on him, Long retorted: "Crack down on me? He don't want to crack down on me. He told me, 'Huey, you're a-goin' to do just what I tell you'; and that's just

[2] George Mayer, *The Political Career of Floyd B. Olson* (Minneapolis, 1951), pp. 133, 149; Charles Rumford Walker, *American City* (New York, 1937), p. 67; Donald McCoy, *Angry Voices* (Lawrence, Kan., 1958), p. 55.

[3] Was the other Father Coughlin? Tugwell inquired. "Oh, no," Roosevelt replied, "the other is Douglas MacArthur." Tugwell, *The Democratic Roosevelt* (Garden City, N.Y., 1957), p. 349, and "Notes from a New Deal Diary," January 17, 1933. Cf. T. Harry Williams, "The Gentleman from Louisiana: Demagogue or Democrat," *Journal of Southern History*, XXVI (1960), 3–21.

what I'm a-goin to do." Yet within a month after Roosevelt's inauguration, Long and the President had split. Long disliked the proposals of the Hundred Days—"We took four hundred millions from the soldiers and spent three hundred millions to plant saplings"—but more important than any ideological disagreement was the fact that Roosevelt stood squarely in the path of Long's ambition for national power.[4]

Contemptuous of Roosevelt and the well-educated urban New Deal reformers, Long used the barb of ridicule against them as he had against planter culture and planter pretensions. He had found Louisiana planters and oil corporations vulnerable to a style of politics which exposed their naked exploitation of the state, and he had captivated his rural followers by guying upper class political leaders. He now hoped to win national power in the same fashion—by revealing that Roosevelt's claims to moral leadership of the reform forces shielded a national ruling class which held a monopoly of wealth and access to culture.

Huey clowned his way into national prominence. A tousled redhead with a cherubic face, a dimpled chin, and a pug nose, he had the physiognomy of a Punchinello. He wore pongee suits with orchid-colored shirts and sported striped straw hats, watermelon-pink ties, and brown and white sport shoes. He delighted in the name of "Kingfish," a title borrowed from the popular radio serial "Amos 'n' Andy." He had mastered a brand of humor which pricked the pretenses of respectability, which appealed to men convinced that beneath the cloth of the jurist sat the fee-grabbing lawyer, beneath the university professor the carnival faker, and, most of all, beneath the statesman the self-serving politician.

A shrewd, intelligent lawyer, Huey cultivated the impression he was an ignoramus.[5] Yet at the same time he lampooned the serious social thinkers of the day. He argued in favor not merely of curtailing the working day but of shortening the working year. Asked if he had gotten the idea from Einstein, he replied: "It's all in the Scriptures," and cited Isaiah and Deuteronomy. He wanted plowed lands turned into golf links and everyone with enough time to read *The Count of Monte Cristo*. He paid his disrespects to Roosevelt's intellectual advisers. Roosevelt's Secretary of Agriculture he dismissed as Lord Corn Wallace.

[4] Jackson (Miss.) *Daily Clarion-Ledger,* January 20, 1933; Forrest Davis, *Huey Long* (New York, 1935), pp. 192–193.
[5] Arthur Krock, COHC.

When he came to see Roosevelt, he deliberately defied the amenities. He presented his demands for patronage, all the while keeping a sailor straw hat on his head; he never removed it save to tap it against the President's knee to drive home a point. A short while after he arrived in Washington, the Kingfish had become an internationally known figure who was crowding Roosevelt off the center of the stage. When H. G. Wells arrived in the United States, he announced he was going to Washington to see the new American curiosity, Huey Long. Huey, declared Gertrude Stein, had "a sense of human beings, and is not boring the way Harding, President Roosevelt and Al Smith have been boring."[6]

In January, 1934, Long, who had advocated redistribution of wealth for many years, founded a national political organization; its slogan was "Share Our Wealth," its national organizer the shrill Shreveport minister, Gerald L. K. Smith.[7] Although his program went through several versions, he proposed, in essence, to liquidate all personal fortunes above a certain amount and to give to every family enough to buy a home, an automobile, and a radio; old folks would receive pensions, and worthy boys would be sent to college. In addition, he advocated vast public works spending, a national minimum wage, a shortened workweek, a balanced farm program, and immediate cash payment of bonuses. In *My First Days in the White House,* Long announced he would name a National Share Our Wealth Committee headed by John D. Rockefeller, Jr., and numbering other prominent men of wealth. "I was surprised, too, by the views of old Andy Mellon," Huey wrote. "He told us that his fortune had become a tremendous bother and nuisance. . . . He wanted above all to have the affection of the American people, which he believed he had lost. . . . I named him Vice-Chairman of the Committee, and his face lighted up with real pleasure. . . ."[8]

[6] *The New York Times,* January 27, 1933; James A. Farley, *Behind the Ballots* (New York, 1938), pp. 240–242; Leon Pearson, "Huey Long contre Roosevelt," *L'Europe Nouvelle,* XVIII (1935), 352–353; Davis, *Huey Long,* p. 125.

[7] H. L. Mencken called Gerald L. K. Smith "the greatest rabble-rouser seen on earth since Apostolic times." Mencken to Henry Morrow Hyde, August 29, 1933, Hyde MSS.

[8] Long, *My First Days in the White House* (Harrisburg, Pa., 1935), pp. 89–90.

Share Our Wealth had an immense attraction. By February, 1935, Long claimed more than 27,000 clubs, and his staff said it had a mailing list of more than 7,500,000 persons. No one doubted the Kingfish's political strength in neighboring states such as Mississippi and Arkansas, but the allure of Share Our Wealth was not limited to the bayou or the forks of the creek.[9] Huey had the backing of a Georgian who had served twice as president of the Atlanta Board of Education, of a University of Illinois professor of engineering, of a Montana bank president.[10] Philadelphia cheered him to the rafters when he presented his Share Our Wealth program; thousands who fought for admission to the hall had to be turned away.[11] At a time when millions were destitute, Long made a deep appeal to the northern worker. A Chicago shoe salesman wrote Huey: "I voted for Pres. Roosevelt but it seems Wall St. has got him punch drunk. What we need is men with guts to go farther to the left as you advocate."[12]

In the spring of 1935, Long reached out for support from discontented farmers in the mercurial prairie states. At the State Fair Grounds in Des Moines, Milo Reno, leader of the radical farmers·in Iowa, introduced Long as "the hero whom God in his goodness has vouchsafed to his children," in compensation for "Roosevelt, Wallace, Tugwell and the rest of the traitors." "Maybe somebody says I don't understand it," Long told the farmers in Des Moines. "Well, you don't have to. Just shut your damned eyes and believe it. That's all." Afterwards he commented: "You know that was one of the easiest audiences I ever won over. I could take this State like a whirlwind. What I did in Louisiana is nothing compared to what I could do in Iowa."[13]

Huey gave Democratic party leaders a bad fright. A poll conducted by the Democratic National Committee revealed that Long might win

[9] Allan Sindler, *Huey Long's Louisiana* (Baltimore, 1956), p. 85; Jackson (Miss.) *Daily News*, November 8, 1934; Hermann Deutsch, "Huey Long, the Last Phase," *Saturday Evening Post*, CCVIII (October 12, 1935), 86.

[10] Ira Harrelson to F.D.R., February 19, 1935; Paul Black to Long, May 3, 1935; W. E. Warren to F.D.R., February 14, 1935, FDRL OF 1403. Warren was president of the Big Horn County Bank. "It is symptoms like this I think we should watch very carefully," Louis Howe noted in passing the letter on to Roosevelt.

[11] See *The American Progress,* II (May, 1935), 8.

[12] C. H. Eckstrom to Long, March 8, 1935, FDRL OF 1403.

[13] *Time,* XXV (May 6, 1935), 19; Robert Morss Lovett, "Huey Long Invades the Middle West," *New Republic,* LXXXIII (1935), 11.

between three and four million votes on a third-party ticket in 1936.[14] His strength was not confined to the South; particularly alarming was the fact that he could draw at least one hundred thousand votes in New York State, most of them from Democrats. Perhaps the party leaders were too frightened. Not only was the poll crude, but Long's very success had antagonized powerful Democrats in the Senate, and driven conservatives to take shelter with Roosevelt.[15] People fearful of Long and Father Coughlin, noted the publisher Frank Gannett, "would quickly swallow Roosevelt at his worst in preference to a Kingfish or a Padre."[16] Yet if Long would probably have been no more than a nuisance in 1936, no one could shrug off the threat Huey would pose four years later if the country was still in the grip of the depression. In 1940, Long would be only forty-six years old.

Great as his following was, Long had to share the spotlight with another Pied Piper: the eloquent Reverend Charles Coughlin, priest of the parish of Royal Oak, Michigan. Father Coughlin, who had given his first radio sermons over WJR in Detroit in 1926, had proved such a popular speaker that by the fall of 1930 CBS was carrying his talks over a national network. One address, "Hoover Prosperity Means Another War," drew 1,200,000 letters. By the end of 1932, he had a weekly audience estimated at from thirty to forty-five million listeners. His radio voice—rich, melodic, authoritative—had unbelievable charm and persuasiveness. He appealed to an audience bewildered by the conflicting theories about the nature of the depression and hungry for some comprehensible explanation. Born the same year that Leo XIII issued *Rerum novarum,* Coughlin offered a combination of advanced Catholic doctrine and *argumenta ad hominem.* He simplified and dramatized the depression by pointing out by name the villains who had caused it. At the same time, he suggested that the simplicity with which he spoke reflected his mastery of the intricacies of finance that baffled ordinary minds. Coughlin told the country what it already half-believed: that the bankers were the source of evil, that international complications were the work of a bankers' conspiracy, that

[14] James A. Farley, *Jim Farley's Story* (New York, 1948), p. 51. Cf. Harold Ickes, *The Secret Diary of Harold Ickes* (3 vols., New York, 1954), I, 462.

[15] Josiah Bailey to Mrs. J. W. Bailey, June 13, 1935, Bailey MSS., Personal File. "Five thousand dollars for every man? Huh!!" wrote a Californian. "Five thousand dollars for John-the-Parasites? For Pete-the-Lazies? Or for Tom-the-Kidnappers? and the Killers?" Andres L. Mijares to Long, March 6, 1935, FDRL OF 1403.

[16] Gannett to Oswald Garrison Villard, March 12, 1935, Villard MSS.

communism was equally bad, and that there was magic in money.[17]

At first, Coughlin gave Franklin Roosevelt warm support, although he did not identify himself directly with the administration. "You won't see me any more," the Radio Priest said he told Roosevelt on March 4. "This Roman collar won't do you any good around the White House." Through most of 1933, Coughlin assured his radio audiences that Roosevelt stood on the side of right. In June, 1933, he declared: "Gabriel is over the White House, not Lucifer." In December, he asserted that the only choice was "Roosevelt or ruin, Roosevelt or Morgan."[18]

Yet increasingly he coupled his encomiums with criticisms of specific features of the New Deal. When he wired Roosevelt, "Am with you a million percent," he also protested the "asinine philosophy" of Henry Wallace. Most of all he was irritated because Roosevelt did not move swiftly enough in the direction of inflation. Coughlin made the purchase and monetization of silver his rallying cry. In November, 1933, he announced: "Forward to Christ all ye people! God wills it—this religious crusade against the pagan god of gold." "Silver," he asserted, "is the key to world prosperity.—Silver that was damned by the Morgans!"[19]

By 1934, Father Coughlin had the heaviest mail of any person in the United States, more even than the President. Each week he spoke to the largest steady radio audience in the world. A poll taken by WOR in New York revealed that an overwhelming majority of its listeners believed the Radio Priest to be the country's most "useful" citizen. Coughlin himself seemed unsure of what to do with his potential political power. He oscillated between defense of the New Deal and sorties against it. In the spring of 1934, he promised he would "never change my philosophy that the New Deal is Christ's Deal," but a week later he called the NRA and AAA "abortive."[20] When the

---

[17] Ruth Mugglebee, *Father Coughlin* (Garden City, N.Y., 1933); Louis Ward, *Father Charles Coughlin* (Detroit, 1933); Wallace Stegner, "The Radio Priest and His Flock," in Isabel Leighton (ed.), *The Aspirin Age* (New York, 1949), pp. 232–238. For Coughlin's charm, see Rupert Hart-Davis, *Hugh Walpole* (London, 1952), pp. 386–387.

[18] Marquis Childs, "Father Coughlin," *New Republic*, LXXVIII (1934), 327; *Time*, XXI (June 26, 1933), 11; Rev. Charles E. Coughlin, *Money Control* (Lecture, December 3, 1933 [Detroit, 1933]).

[19] Coughlin to F.D.R., September 24, 1933, FDRL OF 306; Coughlin to Marvin McIntyre, September 12, 1933, FDRL OF 306.

[20] *The New York Times*, February 25, April 9, 16, 1934.

Treasury revealed that Coughlin's secretary had been investing funds of the Radio League of the Little Flower in silver futures, the priest upbraided Secretary Morgenthau for his assault on the "gentile" metal, but still refrained from a direct attack on Roosevelt.[21] Yet there was no political home for him in the New Deal, and by late 1934 he was moving toward a split. He denounced the AAA as a "Pagan Deal," railed against Roosevelt's advisers as the "Sycophants" of the "Drain Trust," and scored the Treasury for playing into the hands of "international bankers." On November 11, 1934, Father Coughlin announced the formation of the National Union for Social Justice. Convinced that capitalism was finished, he proposed a system of "social justice" to take its place. With the encouragement of his bishop, the Right Reverend Michael Gallagher, a friend of Regent Horthy of Hungary and of Austrian premier Engelbert Dollfuss, Coughlin advocated a political order strikingly similar to that of Italian corporatism.[22]

In early March, 1935, the rupture broke into the open when General Hugh Johnson execrated Coughlin and Long as "a couple of Catilines." "We expect politics to make strange bedfellows," Johnson gibed, "but if Father Coughlin wants to engage in political bundling with Huey Long, it is only a fair first move to take off his Roman cassock."[23] Many of the Coughlinites, bewildered by the breach between a Roosevelt aide and the Radio Priest, beseeched Roosevelt "to take the place of a father in a home and call those people together and bring the quarrel to an end at once." Others resolved the dilemma by deciding that Roosevelt, a great leader embattled in a hopeless struggle against a cabal of bankers, had been deceived by conspiratorial subalterns. "Many of your advisers who mislead you are the advance agents of the money changers in the temple," an Indiana doctor wrote Roosevelt. "Many of your advisers are and is a Judas." Attacks on bankers like the Rothschilds and advisers like Felix Frankfurter had anti-Semitic overtones which grew stronger year by year.[24]

[21] Schlesinger, *Coming of New Deal,* p. 251.

[22] Coughlin to Forrest J. Alvin, September 1934, FDRL OF 306; Rev. Charles E. Coughlin, "How Long Can Democracy and Capitalism Last?" *Today,* III (December 29, 1934), 6–7.

[23] *The New York Times,* March 5, 1935.

[24] S. S. Sager to F.D.R., n.d.; Bruce Bartholomew to F.D.R., March 12, 1935, FDRL OF 306; M. W. Yencer to F.D.R., March 12, 1935, FDRL OF 306. While Coughlin's anti-Semitism escaped public attention in the early New Deal

General Johnson's blast succeeded only in driving Coughlin and Long closer together. Coughlin, who could not be a rival for the White House because of his Canadian birth and his clerical collar, could provide Long with support from the money-crank sections of the rural Middle West, and even more from his real source of strength: lower-middle-class, old immigrant Catholics in the urban Northeast.[25] He challenged the New Deal by his appeal to a wide spectrum of reformers: the California progressive Rudolph Spreckels, who joined his organization; the labor attorney Frank Walsh, who reported himself "absolutely enthusiastic" about Coughlin's speeches; and the union leader Homer Martin, who urged auto workers to enroll in the National Union for Social Justice.[26] If Coughlin and Long were to unite with the western pension campaigners led by Dr. Francis Townsend and the rebellious farmers headed by Milo Reno, they might split the Democratic coalition in 1936 and permit the election of a Republican.

Of the three most important challengers Roosevelt confronted, Dr. Francis Townsend was much the most benign. An unsuccessful country practitioner, Doctor Townsend had migrated to Long Beach, California, in 1919. When various promotional schemes failed, he took a job as a public health official. A change of administration cost Townsend his position. He was sixty-seven years old, and had less than a hundred dollars in savings. Disturbed not only by his own plight but by that of others like him—elderly people from Iowa and Kansas who had gone west in the 1920's and now faced the void of unemployment with slim resources—Townsend came up with a plan which he believed would get the country once more on the road to recovery and which would, incidentally, provide security for the aged.

In January, 1934, Townsend and Robert Clements, a real estate promoter, set up Old Age Revolving Pensions, Limited. They proposed

---

years, it was expressed privately in the most extreme form as early as February, 1933. Coughlin to Elmer Thomas, February 15, 1933, Thomas MSS., Box 459.

[25] Joseph Thorning, S.J., "Senator Long on Father Coughlin," *America*, LIII (1935), 8–9; Raymond Gram Swing, "The Build-up of Long and Coughlin," *The Nation*, CXL (1935), 325–326; James Shenton, "The Coughlin Movement and the New Deal," *Political Science Quarterly*, LXXIII (1958), 360.

[26] Rudolph Spreckels to Franklin Hichborn, November 19, 1934, Hichborn MSS.; Frank Walsh to Judson King, November 1, 1932, Walsh to Louis Ward, December 22, 1934, Walsh MSS., Correspondence; Homer Martin to Coughlin, February 16, April 25, 1935, Homer Martin MSS.

to pay a pension of $200 a month to every citizen over sixty (save for habitual criminals), on condition that he or she retire from all gainful work and promise to spend the sum within one month in the United States. The pension would be financed by a 2 per cent tax on business transactions which would be paid into a "revolving fund."[27] The Townsendites argued that their plan would end mass joblessness both because older people would be compelled to surrender their positions to the younger unemployed and because the rapid spending of pension checks would produce a demand for goods and services that would create still more jobs.[28]

Although Townsend's plan was essentially a recovery program rather than a pension scheme, it was the pension promise that caught on. By September, 1934, Townsend and Clements required a staff of ninety-five to handle their mail, and by the end of the year they claimed some twelve hundred clubs, mostly in the western states. In San Diego, customers sought to make purchases which they would pay for when their Townsend checks arrived. One old couple ordered a stove, a refrigerator, kitchen cabinets, a radio, twin beds, and other furnishings; they said they would settle their account when the government sent the check they expected next month. In some stores, when anyone past sixty entered, all the salesmen disappeared. Irate Townsendites boycotted western merchants if they refused to extend credit on the basis of future pension checks. In San Francisco, elderly aliens, intent on becoming naturalized before the Townsend plan became law, crowded immigration offices.[29]

Critics tried in vain to show that the plan was based on fallacious assumptions. They pointed out that it would at most redistribute purchasing power by taking money from the young and giving it to the old, and that, with a national income of $40 billion a year, Townsend expected to spend $24 billion to compensate 9 per cent of the people.[30]

[27] Townsend later came to recognize that the transactions tax was regressive and advocated instead a levy on the gross incomes of individuals, businesses, and corporations. Dr. Francis Townsend, *New Horizons* (Chicago, 1943), p. 140.

[28] Abraham Holtzman, "The Townsend Movement: A Study in Old Age Pressure Politics" (unpublished Ph.D. dissertation, Harvard University, 1952), pp. 1–80; Luther Whiteman and Samuel Lewis, *Glory Roads* (New York, 1936).

[29] "If Money," *Saturday Evening Post*, CCVII (May 11, 1935), 12–13, 121–127; Richard Neuberger and Kelley Loe, "The Old People's Crusade: The Townsend Plan and Its Astonishing Growth," *Harper's*, CLXXII (1936), 431, and *An Army of the Aged* (Caldwell, Ida., 1936); *The New York Times*, February 4, 1935.

[30] Nicholas Roosevelt, *The Townsend Plan* (Garden City, N.Y., 1936), and

Walter Lippmann observed: "If Dr. Townsend's medicine were a good remedy, the more people the country could find to support in idleness the better off it would be." "My plan is too simple to be comprehended by great minds like Mr. Lippmann's," the doctor retorted. When Congressman Robert Doughton, chairman of the House Ways and Means Committee, asked Townsend whether he was sure the tax he proposed would raise the necessary funds, he answered: "I'm not in the least interested in the cost of the plan."[31]

The Townsend Plan represented the conviction of millions of Americans that they had to take action themselves in order to restore the country to its former prosperity, and that, if the right formula could be found, they could lift themselves by their own bootstraps. Distrustful both of the sophistication and the big-city paternalism of the New Deal, they sought more "American" answers. Townsend's idea of "revolving" pensions drew on the prevalent assumption that the country had great wealth if only some device could be found to tap it.[32]

The Townsend crusade reached out to people raised in the traditions of rural Protestantism; assembled, they acted like a gathering of Methodist picnickers. Townsend meetings featured frequent denunciations of cigarettes, lipstick, necking, and other signs of urban depravity. Townsendites claimed as one of the main virtues of the plan that it would put young people to work and stop them from spending their time in profligate pursuit of sex and liquor. It appealed to men and women gullible enough to accept the pitch of patent medicine vendors; the *Townsend National Weekly,* in fact, featured ads from patent medicine firms hawking remedies for rheumatism and "bladder weakness." Strongly antiradical, the Townsendites insisted their plan would preserve the profit system and avert alien collectivism. His movement, Townsend explained, embraced people "who believe in the Bible, believe in God, cheer when the flag passes by, the Bible Belt solid Americans."[33]

Twentieth Century Fund, *The Townsend Crusade* (New York, 1936), are among the many critiques of the Townsend Plan.

[31] *Time,* XXV (January 14, 1935), 15; XXV (February 18, 1935), 15; New York *Herald-Tribune,* January 4, 1935.

[32] The Townsend Plan made its appearance the very same year that the chain-letter mania swept the nation. Like the chain-letter vogue, Townsendism embodied the ideas of individual contribution, an anticipated windfall, and a self-generating, perpetual-motion economics.

[33] Pittsburgh *Press,* quoted in *New Republic,* LXXXVII (1936), 242; Had-

Dr. Townsend unleashed a new force in American politics: the old people. Thanks to gains in medicine, more of them lived longer; thanks to changes in the organization of the family, many of them no longer felt wanted. Buffeted by the depression, often made to feel a burden to their own children, which, indeed, they often were, they found in the Townsend clubs a home; in the movement, a sense of unity with other elderly people; in the plan, a cause; and in the crusade, a newfound sense of their political power. Townsend claimed twenty-five million Americans had signed his petitions, and his critics conceded he had at least ten million signatures. West Coast politicians quaked at the threat of Townsend reprisal unless they endorsed the Doctor's program. On one key vote, sixteen out of twenty California congressmen supported a Townsend bill.[34]

The same winds of discontent that gave Townsend, Long, and Coughlin their opportunity blew through the streets of industrial America. In the grimy alleys of Birmingham, in the drab bars of Akron, on Seattle's wharves, workers were saying what for so long they had not dared to say: "Organize!" In the summer of 1933, John L. Lewis, his once sizable United Mine Workers shrunk to little more than 150,000 members, gambled his entire treasury on a bold campaign in the coal fields. When his organizers invoked the magic name of Roosevelt, tens of thousands of members signed up. Within a year, the U.M.W. had more than 500,000 members. In Camden, New Jersey, shipbuilders raised on Scotland's Clydeside created the Industrial Union of Marine and Shipbuilding Workers. Across the river in Philadelphia, twenty-one-year-old James Carey organized workers at the Philco Radio plant into a Walking, Hunting and Fishing Club, captured control of a company union, and won a closed-shop agreement and substantial wage increases. Some 100,000 auto workers

---

ley Cantril, *The Psychology of Social Movements* (New York, 1941), pp. 190–209; Neuberger and Loe, "Old People's Crusade," pp. 435–436.

[34] In Oregon, Howard Merriam, a state legislator who refused to vote for a memorial to Congress endorsing the Townsend Plan, was recalled from office. Portland *Oregonian*, Mar. 21, 1935. "About 120 people in my district have signed a Townsend petition," explained one perplexed Oregon congressman. "When I first heard of this I laughed at it. Then I got the smile off my face. It's like a punching bag—you can't dodge it. If we don't pass it this session we'll have to meet it when we get back home." *Time,* XXV (January 14, 1935), 14. Cf. James Murray to Mrs. Howard McIntyre, January 21, 1936, Murray MSS.; Twentieth Century Fund, *Townsend Crusade,* pp. 9–12.

signed union cards; almost as many steel workers organized lodges which they named "New Deal," "NRA," and "Blue Eagle."[35]

Franklin Roosevelt was somewhat perturbed at being cast in the role of midwife of industrial unionism. When Roosevelt agreed to Section 7(a) of the National Industrial Recovery Act, he had little idea what he was letting himself in for. Labor insisted the President should enforce its right to collective bargaining, while employers either defied the government or resorted to the subterfuge of setting up company unions. The administration was uncertain about what 7(a) meant or how it could be enforced, but when strikes threatened the recovery program President Roosevelt and General Johnson created a series of labor boards to bring industrial peace and, incidentally, to give some substance to 7(a). Under the vigorous leadership first of Senator Wagner and later of Lloyd Garrison, dean of the Wisconsin Law School, and the Philadelphia attorney, Francis Biddle, these tribunals, almost powerless, and with the vaguest of assignments, worked out a rough body of labor law. They claimed for the government the power to conduct secret elections to permit workers to decide whether they wanted to be represented by a nationally affiliated union, a company union, or no union at all. Increasingly, they also came to hold that the union chosen by a majority of the workers would have exclusive bargaining rights for all workers. If this principle were established, the company union speaking for a minority of workers in a unionized plant would be doomed. Finally, the board insisted that 7(a) required employers to bargain with unions in good faith, and that bargaining must lead to an agreement.[36]

In making these rulings, labor board officials had to fight a running war with Hugh Johnson and Donald Richberg. Both Johnson and

---

[35] Harris, *American Labor,* p. 140; Levinson, *Labor on March,* pp. 52–54; Schlesinger, *Coming of New Deal,* pp. 137–140; "Docket—Coal & Stabilization, May 22, 1933," W. Jett Lauck MSS., National Recovery Act file; David McCabe, "The Effects of the Recovery Act upon Labor Organization," *Quarterly Journal of Economics,* XLIX (1934), 52–78. The drive for industrial unionism was also abetted by the Norris-La Guardia Act of 1932, which created safeguards for the issuance of injunctions in labor cases and made the yellow-dog contract unenforceable in any federal court. Edwin Witte, "The Federal Anti-Injunction Act," *Minnesota Law Review,* XVI (1932), 638–658.

[36] Irving Bernstein, *The New Deal Collective Bargaining Policy* (Berkeley and Los Angeles, 1950), Chs. 5–7; Lewis Lorwin and Arthur Wubnig, *Labor Relations Boards* (Washington, 1935); Schlesinger, *Coming of New Deal,* pp. 146–149; Gerard Swope, COHC, pp. 145 ff.

Richberg viewed strikes as pointless "aggression" which delayed the achievement of early recovery. "The trouble with Hugh," remarked Secretary Perkins, "is that he thinks a strike is something to settle." Johnson even went so far as to conclude that a strike against an industry which had adopted a code was a strike against the government. Neither Johnson nor Richberg accepted the principle of the right of a union which had majority support to speak for all, despite evidence that the alternative principle of proportional representation served as a union-busting tactic; neither favored using the power of the government to compel businessmen to bargain; and both resented what they regarded as a usurpation of authority by the labor boards.[37]

President Roosevelt shared Wagner's indignation at the intransigence of employers, but he shared too Johnson's perturbation that mass labor organizing might impede the recovery drive. Essentially a "patron" of labor, Roosevelt had far more interest in developing social legislation to help the worker than in seeing these gains secured through unions. Moreover, while the details of agricultural policy fascinated him, the minutiae of collective bargaining vexed him. The President either could not or would not grasp the critical importance of the issue of employee representation. He told reporters irritatedly in May, 1934, that workers could choose anyone they wished to represent them, including a union, the Ahkoond of Swat, or the Royal Geographic Society.[38] When Senator Wagner sought to give the labor boards statutory power, Roosevelt forestalled him and put a mild congressional resolution in its place. "The new deal," protested Senator Cutting, "is being strangled in the house of its friends."[39]

[37] Hugh Johnson, *The Blue Eagle from Egg to Earth* (Garden City, N.Y., 1935), p. 311; Schlesinger, *Coming of New Deal*, pp. 149–150; Jonathan Mitchell, "Grand Vizier: Donald R. Richberg," *New Republic*, LXXXII (1935), 301–304; Donald Richberg to George Berry, January 9, 1935, Richberg MSS., Box 2; Frances Perkins, COHC, VII, 139–140.

[38] Raymond Moley, *After Seven Years* (New York, 1939), p. 13; *Public Papers*, III, 301.

[39] *Congressional Record*, 73d Cong., 2d Sess., p. 12052. Cf. Bernstein, *New Deal Collective Bargaining*, pp. 76–83; Schlesinger, *Coming of New Deal*, pp. 150–151, 397–400; Sidney Fine, "President Roosevelt and the Automobile Code," *Mississippi Valley Historical Review*, XLV (1958), 23–50; "The President's conference with the Senators Saturday afternoon, April 14, 1934," FDRL PSF 37; Edward A. Filene, "A Report of a Study Tour of Business Conditions in Fourteen Large Cities in the United States," March 1, 1934, FDRL PPF 2116.

The crux of the difficulty lay in Roosevelt's determination to serve as a balance wheel between management and labor, when management was mighty and labor, in industries like steel and automobiles, still pathetically weak. Roosevelt's defenders argued that it was not the President's job to organize the auto workers, and, so long as the union represented a minority of the workers in the industry, no amount of government good will would help very much. Unless the A.F. of L. could build powerful industrial unions, it could not compel much respect from either industry or government.

Under the cautious leadership of Samuel Gompers' successor, William Green, and his circumspect lieutenant, Matthew Woll, the A.F. of L. had treated the influx of industrial workers as an embarrassment. It placed tens of thousands of these new members in federal labor unions, "a kind of Ellis Island where the immigrants into the AFL were to be kept until the crafts had decided into what pen they were to be driven."[40] Many of the Federation's leaders, primarily skilled workers of northern European stock, displayed open contempt toward the new industrial unionists. Dan Tobin, head of the Teamsters, dismissed the factory workers as "rubbish."[41] William Collins, the A.F. of L. representative in New York State, confided: "My wife can always tell from the smell of my clothes what breed of foreigners I've been hanging out with." The Federation handled its new recruits in steel so ineptly that out of some one hundred thousand members who had signed up in 1933, only six thousand were still paying dues in late 1934.[42]

At the annual convention of the A.F. of L. in San Francisco in 1934, a faction representing about 30 per cent of the membership demanded that the Federation conduct a drive to organize factory workers into industrial unions. Under John L. Lewis, whose commitment to in-

[40] Denis Brogan, *The Era of Franklin D. Roosevelt* (New Haven, 1950), p. 175.

[41] *Report of Proceedings of the Fifty-fourth Annual Convention of the American Federation of Labor* (Washington, [1935]), p. 453.

[42] Levinson, *Labor on March,* p. 60; Edwin Young, "The Split in the Labor Movement," in Milton Derber and Edwin Young, *Labor and the New Deal* (Madison, Wis., 1957), p. 53; James Morris, *Conflict within the AFL* (Ithaca, 1958), pp. 150–170; Ed Hall, Wayne Oral History Interview, pp. 3 ff.; Max Handman, "Conflicting Ideologies in the American Labor Movement," *American Journal of Sociology,* XLIII (1938), 525–538; Robert Christie, *Empire in Wood* (Ithaca, 1956), pp. 285–287.

dustrial unionism meshed with his interest in the welfare of his own particular union and his vaulting personal ambitions, the industrial unionists won a surface victory which netted them little. The Executive Council delayed issues of charters to industrial unions for many months, watered down the charters by clauses protecting craft unions, and did little to organize industrial labor. Workers who had joined the A.F. of L. in high hopes tore up their union cards in disgust.[43]

At the Atlantic City convention in October, 1935, the industrial union forces determined on a showdown. One labor paper headlined: "The A.F. of L. Must Stop Raping the Brewery Workers Union!" In his typically high-eloquent Welsh style, Lewis thundered: "A year ago at San Francisco, I was a year younger and naturally I had more faith in the Executive Council. I was beguiled into believing that an enlarged Executive Council would honestly interpret and administer this policy—the policy . . . of issuing charters for industrial unions in the mass-production industries. . . . I know better now. At San Francisco they seduced me with fair words. Now, of course, having learned that I was seduced, I am enraged and I am ready to rend my seducers limb from limb, including Delegate Woll. In that sense, of course, I speak figuratively. . . . Heed this cry from Macedonia that comes from the hearts of men."[44]

Lewis' eloquence made little impression on craft union leaders such as Dan Tobin, who subscribed to the creed Samuel Gompers and other founders of the Federation had given the present generation: "Upon the rock of trades autonomy, craft trades, you shall build the church of the labor movement, and the gates of hell nor trade industrialism shall not prevail against it." By a vote of 18,464 to 10,897, the convention adopted the craft unionists' resolution. Near the end of the tumultuous convention, Lewis, who symbolized the new unionism, locked in a duel with Big Bill Hutcheson of the Carpenters. Lewis threw a punch which knocked Hutcheson sprawling, and, after both tumbled to the floor fighting like two clumsy grizzlies, Hutcheson retreated, his face

[43] Levinson, *Labor on March,* pp. 82–98; Christie, *Empire in Wood,* pp. 290–291; Sidney Fine, "The Toledo Chevrolet Strike of 1935," *Ohio Historical Quarterly,* LXVII (1958), 326–356.

[44] *Brewery Worker,* L (March 9, 1935); *Report Proceedings of the Fifty-fifth Annual Convention of the American Federation of Labor* (Washington, [1936]), pp. 538, 542; John Brophy, COHC, pp. 552 ff. The crucial question was not whether the Federation would favor the craft or industrial union structure, but whether it would take a live interest in organizing the unorganized.

bloodied. The country interpreted Lewis' blow for what it was—the determination of the industrial union leaders to smash the Federation's monopoly of organized labor.[45]

The next day, Lewis breakfasted with industrial union supporters at the Hotel President in Atlantic City. In addition to his Mine Worker associates, the breakfasters included Sidney Hillman, David Dubinsky, and Max Zaritsky from various garment unions, Thomas McMahon of the Textile Workers, and the able, disinterested Charles Howard of the Typographers. Three weeks later, the Committee for Industrial Organization was established. Ostensibly, the C.I.O. was not a rival organization but a committee within the A.F. of L. working to organize the unorganized and to educate the remainder of the Federation on the wisdom of their policies. Actually, the cleavage was total. In the summer of 1936, after a heresy trial at which C.I.O. leaders refused to appear, the Federation formally suspended the truant unions for the crime of dual unionism. For a time, the C.I.O. continued to protest the legality of the Council's action, but in October, 1938, it formally recognized the split between the two groups by transforming itself into the *Congress* of Industrial Organizations.[46]

In 1934, a series of violent strikes, many of them led by avowed radicals, shook the country. Milwaukee streetcar workers brought their employers to terms after an uprising in which, aided by a Socialist organization of unemployed, they assaulted car barns, pulled off trolley poles, and crippled dozens of streetcars.[47] In Philadelphia, striking hackies burned a hundred taxicabs, and rioting New York cabbies, impressed by their bloody victory, drove most of the city's 15,000 taxis off the streets.[48] Communists led strikes of farm workers from the let-

[45] *Proceedings Fifty-fifth Convention A.F. of L.,* p. 659; Levinson, *Labor on March,* pp. 99–117; John Frey, COHC, pp. 633 ff.; Christie, *Empire in Wood,* pp. 291–293. After the fight, President Green admonished: "You shouldn't have done that, John." "He called me a foul name," Lewis replied. "Oh, I didn't know that," Green answered agreeably. Levinson, *Labor on March,* pp. 116–117.

[46] Walter Galenson, *The CIO Challenge to the AFL* (Cambridge, 1960), pp. 1–56; Schlesinger, *Coming of New Deal,* pp. 413–415; Philip Taft, *The A.F. of L. from the Death of Gompers to the Merger* (New York, 1959), pp. 146–180; "C.I.O.—Statement of Charges Against C.I.O. (John L. Lewis Group) presented by John P. Frey," Frey MSS., Box 5.

[47] Ruben Levin, "Mass Action Works in Milwaukee," *New Republic,* LXXIX (1934), 261–263.

[48] *The New York Times,* February 3–8, 1934, Mar. 10–April 13, 1934. In

tuce sheds of California's Salinas Valley to the tomato fields of southern New Jersey. Union electrical workers pulled a control switch which plunged 160,000 people in Des Moines into darkness. Plants in the paralyzed city could not operate, streetcars stayed in their barns, and nurses patrolled hospital corridors in darkness. Workers tied up Terre Haute with a general strike, Butte copper miners closed the Anaconda pits for months, and striking cooks, waiters, and busboys from the Waldorf-Astoria paraded the sidewalks of Park Avenue singing the "Internationale."[49] At the Electric Auto-Lite plant in Toledo, where A. J. Muste and Louis Budenz of the American Workers' party sought to direct the walkout, mobs battled police and helmeted National Guardsmen. They defied bayonets, clouds of tear gas, and even volleys of rifle fire, and, after threatening a general strike, won most of the union's demands. The American correspondent of the London *Daily Herald* cabled: "Toledo (Ohio) is in the grip of civil war."[50]

On Labor Day, 1934, textile workers began the largest single strike ever undertaken in this country. For the next sixteen days, Italian silk workers from Paterson joined with French weavers in Rhode Island, Portuguese millhands from New Bedford, and lintheads from the Carolinas to shut down the industry in twenty states. "Flying squadrons" of hundreds of pickets raced from town to town in car caravans. At Honea Path, South Carolina, six members of such a squadron were killed in a fight with deputies. In the Moshasauck Cemetery in Saylesville, Rhode Island, fifty were wounded in a running battle between strikers and militia who fought from behind tombstones by the eerie light of Roman candles carried by strikers and flares used by soldiers. The walkout, aimed as much at the NRA's Cotton Code Authority as

---

Clifford Odets' *Waiting for Lefty,* a meeting of cab drivers waits for a leader who never appears. As they wait, the hackies rise to explain why they want to strike. On opening night in January, 1935, when the question was finally put, the audience, caught up in the drama, roared "Strike! Strike!" Harold Clurman, *The Fervent Years* (New York, 1957), pp. 132–133, 138–139.

[49] William Dusinberre, "Strikes in 1934, with a Case Study of New Jersey Farm Workers" (unpublished M.A. essay, Columbia University, 1953); Chicago *Daily Times,* September 19, 1934; Edward Levinson, *Labor on March,* p. 56; Vernon Jensen, *Nonferrous Metals Industry Unionism 1932–1954* (Ithaca, 1954), pp. 13–16; *Time,* XXIII (February 5, 1934), 18.

[50] Schlesinger, *Coming of New Deal,* p. 393; Bernard Karsh and Phillips Garman, "The Impact of the Political Left," in Derber and Young. *Labor and New Deal,* p. 98; *New Republic,* LXXIX (1934), 86–87; London *Daily Herald,* May 29, 1934.

at the operators, ended in failure when the union found itself out-manned by the industrialists and the state governors, who did not hesi-tate to use force to put down the strike. "A few hundred funerals," observed one textile journal, "will have a quieting influence."[51]

The representative prairie metropolis of Minneapolis saw naked class war when the city's truckdrivers sought to crack that open-shop stronghold by crippling its transportation system. They were led by Vincent Raymond Dunne, a teamster at eleven, a Wobbly at fourteen, a radical who had been expelled from the Communist party as a Trotskyite. Business leaders countered by organizing a "citizens' army" of the wealthy, who were sworn in as special deputies to break the strike. On the morning of May 22, 1934, two of the deputies, one a member of one of the town's prominent families, died in a clash in Minneapolis's central market place, where twenty thousand people had massed. In July, when the walkout resumed after a temporary armis-tice, an armed police convoy shot down sixty-seven persons, two of whom died. The teamsters refused to surrender, and in the end won a settlement which made it possible for them to establish themselves as the most powerful union in the Northwest. The open-shop bastion of Minneapolis had been smashed—and by a union under radical leadership.[52]

In San Francisco, as in Minneapolis, bitterly antiunion employers drove workers into the arms of radicals. In the summer of 1934, after a strike of stevedores had tied up Pacific ports from San Diego to Vancouver for two months, the Industrial Association decided to open the harbor of San Francisco by force. When two strikers were killed and many injured on the Embarcadero on Bloody Thursday, July 5, Harry Bridges, a hard, sourfaced Australian who consistently followed the Communist line, persuaded conservative unionists to launch a gen-eral strike on July 16. Strikers shut down not only plants but barber

[51] *Literary Digest*, CXVIII (September 15, 1934), 6; Chicago *Daily Times*, September 12, 1934; Schlesinger, *Coming of New Deal*, p. 394; William Pollock, COHC, pp. 11 ff.; Martha Gellhorn to Harry Hopkins, November 11, 1934, Hopkins MSS.; *Textile World*, LXXXIV (1934), 1814–1816; E. David Cronon, "A Southern Progressive Looks at the New Deal," *Journal of Southern History*, XXIV (1958), 159–160.

[52] Walker, *American City*, pp. 87–221; Mayer, *Floyd Olson*, pp. 198–201; Schlesinger, *Coming of New Deal*, pp. 385–389; William Dunne and Morris Childs, *Permanent Counter-Revolution* (New York, 1934); Eric Sevareid, *Not So Wild a Dream* (New York, 1947), pp. 57–58.

shops, laundries, theaters, and restaurants, blockaded highways, and barred incoming shipments of food and fuel oil. With the President cruising in the Pacific, civic leaders and officials of the Roosevelt administration panicked. Roosevelt later told newsmen: "Everybody demanded that I sail into San Francisco Bay, all flags flying and guns double shotted, and end the strike. They went completely off the handle."[53] The President refused to be pushed into precipitate action, and within four days the strike, which had never actually disrupted essential services, was over. Despite Bridges' opposition, the dockers agreed to arbitration, and in the end won recognition for their union and most of their other demands. An ill-conceived move born of momentary anger and desperation, San Francisco's general strike had no revolutionary intent and lacked any real political objective. But it did indicate the depth of class animosity in 1934.[54]

The current of radicalism ran strongly in the 1934 elections, nowhere more strongly than in California. In the fall of 1933, the well-known novelist and reformer Upton Sinclair had launched EPIC (End Poverty in California). Sinclair proposed a direct attack on the crucial problem the New Deal was not solving: want in the midst of plenty. Instead of placing idle workers on relief, he urged that they be given a chance to produce for their own needs. The state would buy or lease lands on which the jobless could grow their own food; it would rent idle factories in which unemployed workers could turn out staples like clothing and furniture. They would be paid in scrip which could be spent only for goods produced within the "production-for-use" system. At the same time, Sinclair announced his candidacy for governor of California in the November, 1934, elections.

In the Democratic primary on August 28, 1934, Sinclair struck terror among conservatives when he polled 436,000 votes to 288,000 for the Wilsonite George Creel, with 89,000 for two lesser-known candidates. Since Sinclair had piled up more votes than Governor Frank Merriam had received in the Republican primary, he appeared to

[53] Samuel Rosenman (ed.), *The Public Papers and Addresses of Franklin D. Roosevelt* (13 vols., 1938–50), III, 399; F.D.R. to Frances Perkins, July 16, 1934, FDRL PSF 18; F.D.R. to John Garner, August 11, 1934, FDRL PPF 1416.

[54] Schlesinger, *Coming of New Deal*, pp. 389–393; Paul William Ryan, *The Big Strike* (Olema, Calif., 1949); Fremont Older to Rose Wilder Lane, July 18, 1934, Older MSS.

have an excellent chance of winning the November election. For the next two months, hysteria rode the state of California. After the primaries, Sinclair allegedly told reporters that he had said to Harry Hopkins: "If I am elected, half of the unemployed will come to California, and you will have to take care of them." Opponents plastered billboards with a distorted version of Sinclair's statement; movie companies turned out "newsreels" showing armies of hoboes crossing the state borders; three Los Angeles newspapers refused to print news about Sinclair. His enemies culled his books for statements denouncing marriage as a bourgeois institution and characterizing the Roman Catholic Church as "The Church of the Servant Girls."[55]

Many New Deal Democrats, forced to choose between the archconservative Merriam, who opposed the reforms they had fought for, and Sinclair, who favored rural communes, chose Merriam. Sinclair banked his hopes for election on Roosevelt's support; but when it was clear Sinclair was losing, the President cut him adrift.[56] Administration Democrats in California like George Creel opposed the novelist, and Senator McAdoo's law partner came out for Merriam because Sinclair's ticket "concealed the communistic wolf in the dried skin of the Democratic donkey." In the closing days of the campaign, New Deal officials forged an alliance between Democratic conservatives and the anti-New Deal Merriam.[57] On Election Day, Merriam trounced Sinclair with 1,139,000 votes to the novelist's 880,000, with 303,000 for a third candidate.

Despite the evidences of radical discontent, Roosevelt remained immensely popular. His popularity cut across all classes. In the homes of Carolina mill workers pictures of the President, sometimes ones torn from a newspaper, hung over the mantel like an Italian peasant's Madonna. In Mississippi, the Jackson Junior Chamber of Commerce

[55] *Time,* XXIV (November 12, 1934), 16; Charles Larsen, "The Epic Campaign of 1934," *Pacific Historical Review,* XXVII (1958), 134.

[56] Upton Sinclair to F.D.R., October 5, 18, 1934, FDRL OF 1165; F.D.R. to Key Pittman, October 9, 1934, FDRL PPF 745; James A. Farley to William McAdoo, October 3, 1934, McAdoo MSS., Box 398; *Upton Sinclair's Epic News,* October 22, 1934; Arthur Schlesinger, Jr., *The Politics of Upheaval* (Boston, 1960), pp. 115–121.

[57] William Neblett to William McAdoo, October 31, 1934; George Creel to Upton Sinclair, October 18, 1934, McAdoo MSS., Box 398; J. F. T. O'Connor to Louis Howe, O'Connor to Marvin McIntyre, October 27, 1934; FDRL OF 300 (California), Box 16.

sponsored a mammoth "New Deal parade" with a portrait of Roosevelt draped in red, white, and blue bunting followed by floats of each of the alphabet agencies. The best-known man in the building industry, Clarence Mott Woolley, declared: "I have always been a rank Republican, but I take my hat off to President Roosevelt. The Administration has performed a miracle."[58] The *Nation* editor, Oswald Garrison Villard, often a stern critic of the President, responded skeptically to an invitation to join a third-party movement; he had just returned from a tour of the West, where he found 90 per cent of the people for Roosevelt.[59]

The outcome of the 1934 elections demonstrated the force both of the radical spirit and of Roosevelt's popularity. Traditionally, the party in power loses a substantial number of seats in an off-year election. As the November, 1934, congressional elections approached, the tide was running so strongly Democratic that John Garner guessed that the Republicans would win only thirty-seven seats in the House, a gain so small it could be regarded as "a complete victory" for the Roosevelt administration.[60] Most observers thought the G.O.P. gains would be more substantial.[61] No one was prepared for the actual results. Expected to pick up a few score seats, the Republicans lost thirteen instead. The voters elected a new House of 322 Democrats, 103 Republicans, and ten Progressives or Farmer-Laborites. Never in the history of the Republican party had its percentage of House seats fallen so low. In the Senate the rout of the G.O.P. was even more devastating. There the Democrats won better than a two-thirds majority, the greatest margin either party had ever held in the history of the upper house. The nine new Democratic senators—one was Missouri's Judge Harry Truman—swelled the party's total to a dazzling sixty-nine seats.

If there was an issue in the campaign, it was Roosevelt: the elec-

[58] Martha Gellhorn to Harry Hopkins, November 11, 1934, Hopkins MSS.; Jackson (Miss.) *Daily News,* October 9, 1934; "Heating Man," *Fortune,* XI (April, 1935), 81. "Through thick and thin I am for you," Russell Leffingwell of the House of Morgan wrote the President. "It is impossible to imagine greater achievements than yours have been in the year just ending." Leffingwell to F.D.R., March 3, 1934, FDRL PPF 866.

[59] Villard to Alfred Bingham, April 10, 1934, Villard MSS. Cf. Rudolph Spreckels to Franklin Hichborn, May 8, 1934, Hichborn MSS.

[60] Garner to F.D.R., October 1, 1934, FDRL PPF 1416.

[61] See, e.g., *Kiplinger Washington Letter,* September 22, 1934.

tion was a thumping personal victory for the President. Even Republicans had invoked Roosevelt's name to get elected. William Allen White commented: "He has been all but crowned by the people."[62] The elections almost erased the Republican party as a national force. They left the G.O.P. with only seven governorships, less than a third of Congress, no program of any substance, no leader with a popular appeal and none on the horizon.[63]

Yet Roosevelt was riding a tiger, for the new Congress threatened to push him in a direction far more radical than any he had originally contemplated. Economy, the staple of past campaigns, was hardly mentioned in 1934, and a number of the new Democrats, elected on public ownership or "production-for-use" platforms, stood to the "left" of Roosevelt.[64] Faced with the threats of Long, Coughlin, and Townsend, aware that millions of Americans were still jobless, confronting drought-stricken farmers and rebellious industrial workers, Roosevelt now had to work with a Congress committed to free spending. At the same time, the President and his advisers knew that the New Deal still had much unfinished business. "Boys—this is our hour," Harry Hopkins exulted. "We've got to get everything we want—a works program, social security, wages and hours, everything—now or never."[65] When Roosevelt faced the newly elected Congress in 1935, the result promised to be a fresh outburst of reform and recovery legislation which would surpass even that of the Hundred Days.

[62] *Time*, XXIV (November 19, 1934), 11. See Raymond Clapper, *Watching the World* (New York, 1944), p. 140; Rene Pinon, "Les elections américaines," *Revue des Deux Mondes*, Series 8, XXIV (December 1, 1934), 719; Edwin Bronner, "The New Deal Comes to Pennsylvania: The Gubernatorial Election of 1934," *Pennsylvania History*, XXVII (1960), 44–68; P. L. Gassaway to James Curley, November 10, 1934, Gassaway MSS., Correspondence 1926–1934.

[63] On Election night, Jim Farley gloated: "Famous Republican figures have been toppled into oblivion. In fact, we must wonder who they have left that the country ever heard of." *Time*, XXIV (November 12, 1934), 14.

[64] "Memorandum on Democratic Representatives and Senators Elected to 74th Congress for First Time," Katherine Blackburn to Louis Howe, November 14, 1934, FDRL PSF 37; Donald McCoy, "The Formation of the Wisconsin Progressive Party in 1934," *Historian*, XIV (1951), 88–89; Key Pittman to Alben Barkley, November 12, 1934, Pittman MSS., Box 21; Rudolph Spreckels to Franklin Hichborn, November 8, 1934, Hichborn MSS.

[65] Robert Sherwood, *Roosevelt and Hopkins* (New York, 1948), p. 65.

CHAPTER 6

One Third of a Nation

IN THE early months of the depression, jobless men made the rounds of factory gates in search of work only to be faced by hastily scrawled signs with the legend "NO HELP WANTED." By the spring of 1934, the legends were painted in gilt letters, as if to make the words more enduring. The embarrassed "Sorry, buddy" had given way to the brusque "Can't you read signs?" Men who had once felt certain they could find a job the next day, the next week, the next month, had lost their self-assurance. One worker remarked: "There is something about the anniversary of your layoff which makes you feel more hopeless." Where once they had been confident of their own powers, they now felt impotent. "What can a man do?" asked a jobless Italian-American in New Haven. "Dey pulla de string; you move. Dey letta de string loose, you drop."[1]

If, in America, every man rises on his own merits, then he falls through his own failings. "Anyone," it was said, "could find a job if he really tried." That people who were economically secure should perpetuate this myth is understandable; what is, at first glance, more surprising is that the jobless themselves should do so. The unemployed worker almost always experienced feelings of guilt and self-depreciation. Although he knew millions had been thrown out of work through no fault of their own, he knew too that millions more were still

[1] E. Wight Bakke, *The Unemployed Worker* (New Haven, 1940), pp. 187, 349.

employed. He could not smother the conviction that his joblessness was the result of his own inadequacy.[2]

To be unemployed in an industrial society is the equivalent of banishment and excommunication. A job established a man's identity —not only what other men thought of him but how he viewed himself; the loss of his job shattered his self-esteem and severed one of his most important ties to other men. Engulfed by feelings of inferiority, the jobless man sought out anonymity. He withdrew from the associations he had had before he lost his job, even tried to escape the company of friends and neighbors whose opinion he respected. Without work, without hope of work, he spent his days in purposeless inactivity. "These are dead men," observed one writer. "They are ghosts that walk the streets by day. They are ghosts sleeping with yesterday's newspapers thrown around them for covers at night."[3]

Like a drowning swimmer struggling to keep his head above water, the middle-class man fought frantically to maintain his social status. One who found a job after being unemployed remarked: "It all turned out O.K., you see, and I kept my collar white." Others did not do as well, like the Dubuque man whose wife burst into tears when, for the first time in his life, he donned overalls to go to work. Blue shirt or white, he exhausted all his resources before he would accept charity or go on relief: he stopped his child's music lessons; sold the wedding furniture; sent his children to early Mass where few would notice their shabby dress.[4]

[2] Benjamin Glassberg, *Across the Desk of a Relief Administrator* (Chicago, 1938), p. 18; Abram Kardiner, "The Rôle of Economic Security in the Adaptation of the Individual," *The Family*, XVII (1936), 187–197; Bakke, *Unemployed Worker*, pp. 25–26; Mirra Komarovsky, *The Unemployed Man and His Family* (New York, 1940), pp. 23–27. Cf. Edward Rundquist and Raymond Sletto, *Personality in the Depression* (Minneapolis, 1936); Philip Eisenberg and Paul Lazarsfeld, "The Psychological Effects of Unemployment," *Psychological Bulletin*, XXXV (1938), 358–390.

[3] E. Wight Bakke, *Citizens Without Work* (New Haven, 1940), p. 14; Sol Wiener Ginsburg, "What Unemployment Does to People," *American Journal of Psychiatry*, XCIX (1942), 439–446; Tom Kromer, *Waiting for Nothing* (Garden City, N.Y., 1935), p. 166. One character in a novel of the period remarked: "It's funny, ain't it, how we don't work and yet we live?" Edward Anderson, *Hungry Men* (Garden City, N.Y., 1935), p. 66.

[4] Bakke, *Unemployed Worker*, p. 212; Works Progress Administration, *The Personal Side*, by Jessie Bloodworth and Elizabeth J. Greenwood, under the supervision of John Webb, mimeographed (Washington, 1939), p. 286; Helen

Yet the day came when even the proudest family had to seek relief. The failure of the First National Bank in one Midwestern town forced twelve families on the county. When the bank collapsed, one woman, shouting and sobbing, beat on the closed plate-glass doors; all of her savings from a quarter of a century of making rag rugs had vanished. A woman who had taught the fourth grade for fifty-two years lost every penny she had set aside for her old age. Thousands of hard-working, thrifty people now had no alternative but to make their way to the relief office. "It took me a month," said an Alabama lumber-man. "I used to go down there every day or so and walk past the place again and again." A Birmingham engineer put it more directly: "I simply had to murder my pride."[5]

On May 12, 1933, Congress authorized a half-billion dollars in relief money to be channeled by the federal government through state and local agencies. Eight days later, the President announced his choice to head the new Federal Emergency Relief Administration: the social worker Harry Hopkins, who had directed relief operations under Roosevelt in Albany. For a social worker, he was an odd sort. He belonged to no church, had been divorced and analyzed, liked race horses and women, was given to profanity and wisecracking, and had little patience with moralists. Joseph E. Davies said of him: "He had the purity of St. Francis of Assisi combined with the sharp shrewdness of a race track tout." A small-town Iowan, he had the sallow complexion of a boy who had been reared in a big-city pool hall. The artist Peggy Bacon set down in her sketchbook: "Pale urban-American type, emanating an aura of chilly cynicism and defeatist irony like a moony, melancholy newsboy selling papers on a cold night." As he talked to reporters—often out of the side of his mouth—through thick curls of cigarette smoke, his tall, lean body sprawled over his chair, his face wry and twisted, his eyes darting and suspicious, his manner brusque, iconoclastic, almost deliberately rude and outspoken, the cocksure Hopkins seemed, as Robert Sherwood later observed, "a profoundly shrewd and faintly ominous man." Virtually unknown when he took office, his name would be a household word by the end of the year. By Roosevelt's second term, he would be the most powerful man in the

Wright, "The Families of the Unemployed in Chicago," *Social Service Review*, VIII (1934), 17–30; Eli Ginzberg *et al., The Unemployed* (New York, 1943).
   [5] Marquis Childs, "Main Street Ten Years After," *New Republic*, LXXIII (1933), 263; Lorena Hickok to Harry Hopkins, April 2, 1934, Hopkins MSS.

administration; in the war years, he would sit next to the throne, the confidant of Churchill, the emissary to Stalin.[6]

A half-hour after Hopkins left the White House, he placed a desk in the hallway of the RFC building. Amidst discarded packing cases, gulping down endless rounds of black coffee and chain-smoking cigarettes, he spent over five million dollars in his first two hours in office. In the next few months, he spent money like a Medici prince. Yet he still was not doing enough. As winter approached, Hopkins recognized that unless the government acted quickly, millions faced extreme privation.[7]

When Hopkins explained to Roosevelt the critical need for emergency measures, the President authorized him to set up the Civil Works Administration. Unlike the FERA, CWA was a federal operation from top to bottom; CWA workers were on the federal payroll. The agency took half its workers from relief rolls; the other half were people who needed jobs, but who did not have to demonstrate their poverty by submitting to a "means" test. CWA did not give a relief stipend but paid minimum wages. Hopkins, called on to mobilize in one winter almost as many men as had served in the armed forces in World War I, had to invent jobs for four million men and women in thirty days and put them to work. By mid-January, at its height, the CWA employed 4,230,000 persons.[8]

In its brief span the CWA built or improved some 500,000 miles of roads, 40,000 schools, over 3,500 playgrounds and athletic fields, and 1,000 airports. Workmen renovated Montana's State Capitol Building and helped erect Pittsburgh's Cathedral of Learning. The CWA employed fifty thousand teachers to keep rural schools open and to teach adult education classes in the cities. It provided enough money to put

[6] Robert Sherwood, *Roosevelt and Hopkins* (New York, 1948), pp. 3, 49; Peggy Bacon, "Facts About Faces: IV. Harry Hopkins," *New Republic,* LXXXII (1935), 7; Arthur Schlesinger, Jr., *The Coming of the New Deal* (Boston, 1959), pp. 265–266; Raymond Clapper, *Watching the World* (New York, 1944), pp. 120–121; "Reminiscences of Will W. Alexander," COHC, II, 359.

[7] Harry Hopkins, *Spending to Save* (New York, 1936), pp. 97–107; Gertrude Springer, "The New Deal and the Old Dole," *Survey Graphic,* XXII (1933), 347–352, 385, 388; Edward Ainsworth Williams, *Federal Aid for Relief* (New York, 1939), pp. 58–66.

[8] "Reminiscences—Civil Works Administration," John Carmody MSS., Box 73; Sherwood, *Hopkins,* pp. 50–52; Louis Brownlow, *A Passion for Anonymity* (Chicago, 1958), pp. 286–288.

back to work every teacher on Boston's unemployed rolls. It hired three thousand artists and writers, and used a variety of other special skills. Opera singers toured the Ozarks; prehistoric mounds were excavated for the Smithsonian Institution; ninety-four Indians restocked the Kodiak Islands with snowshoe rabbits. The CWA pumped a billion dollars of purchasing power into the sagging economy.[9]

The CWA got the country through the winter. In Des Moines, men raced to the store with their first pay, able for the first time in months to shove cash instead of a grocery order across the counter. One Iowa woman reported: "The first thing I did was to go out and buy a dozen oranges. I hadn't tasted any for so long that I had forgotten what they were like." On Christmas Eve, streets were crowded with wives of CWA workers doing last-minute shopping for children who would otherwise have awoke to a bleak Christmas morning. By midnight, CWA employees in Columbus, Ohio, had swept bare the shelves of shoestores. In February, a record-breaking cold wave chilled the country—56 below in northern New England, 14 below in New York City, 8 below in Richmond, snow and hail in Florida—and only the CWA provided the margin of relief from cold and want.[10]

Alarmed at how much CWA was costing, Roosevelt ended it as quickly as he could. He feared he was creating a permanent class of reliefers whom he might never get off the government payroll. If CWA were continued into the summer, the President told his advisers, it would "become a habit with the country. . . . We must not take the position that we are going to have permanent depression in this country, and it is very important that we have somebody to say that quite forcefully to these people. . . . Nobody," he added blandly, "is going to starve during the warm weather."[11] Revelations of corruption reinforced Roosevelt's determination. The CWA proved immensely popular—with merchants, with local officials, and with workers—and

[9] Henry Alsberg, *America Fights the Depression* (New York, 1934); Hopkins, *Spending to Save,* pp. 108–125; Florence Peterson, "CWA: A Candid Appraisal," *Atlantic Monthly,* CLIII (1934), 587–590; Schlesinger, *Coming of New Deal,* p. 270; *Time,* XXIII (February 5, 1934), 14; (February 19, 1934), 12; Corrington Gill, *Wasted Manpower* (New York, 1939), pp. 163–169.

[10] Lorena Hickok to Harry Hopkins, November 25, December 4, 1933, Hopkins MSS.; Louis Brownlow, *Passion for Anonymity,* p. 288; Sherwood, *Hopkins,* pp. 53–55.

[11] "Proceedings of the National Emergency Council, Dec. 19, 1933–April 18, 1936," microfilm, FDRL, session of January 23, 1934.

Senate progressives such as Cutting and La Follette strongly protested the death of the agency. But Roosevelt was adamant. On February 15, Hopkins began demobilization in the South, where weather was warmest, and paced his pink slips with the northward move of the spring. By early April, he had fired four million workers.

In the spring of 1934, the FERA took up the relief burden once more. It continued the CWA's unfinished work projects; dispensed direct relief; bailed out stranded coal towns like Dewmaine and Colp, Illinois, where 95 per cent of the people were on relief; and, in the course of aiding rural areas, experimented with long-term projects such as loans to farmers and even the creation of new communities. Self-help projects employed jobless men and women in idle canning plants or mattress factories. Most of the reliefers on works projects engaged in routine construction; FERA erected five thousand public buildings and seven thousand bridges, cleared streams, dredged rivers, and terraced land. But Hopkins sought too to use the particular skills of white-collar workers. The FERA taught over one and a half million adults to read and write, ran nursery schools for children from low-income families, and helped one hundred thousand students to attend college.[12]

It was this highly imaginative white-collar program that drew the most heated criticism, especially after an aldermanic investigation in New York revealed that an FERA official was teaching 150 men to make "boondoggles," such items as woven belts or linoleum-block printing. Ever after, "boondoggling"—the official claimed the term went back to pioneer days, but others insisted it had been invented by a fifteen-year-old Eagle Scout seven years before—served as the main epithet for critics of federal relief. Such disparagement threatened to make FERA a political liability, at the very same time that it was failing to give adequate relief. While the worker under CWA had averaged $15.04 a week, he received only $6.50 a week under FERA; moreover, he could get on FERA rolls only if he had first identified himself as a reliefer and submitted to the humiliation of a "means" test.

By the end of 1934, the government had spent over two billion dol-

---

[12] Hopkins, *Spending to Save,* pp. 126–166; Corrington Gill, *Wasted Manpower,* pp. 169–175; Edward Ainsworth Williams, *Federal Aid for Relief,* pp. 67–265.

lars on relief, and Roosevelt felt he had little to show for it. The wear of five years of depression was adding new people to the welfare rolls every day; some twenty million were receiving public assistance, and, in drought-stricken South Dakota, one-third of the state depended on the dole. Both Roosevelt and Hopkins believed the states should bear a greater share of the burden, and both thought direct relief sapped the morale of workers. "What I am seeking is the abolition of relief altogether," Roosevelt wrote Colonel House in November, 1934. "I cannot say so out loud yet but I hope to be able to substitute work for relief."[13]

In January, 1935, Roosevelt proposed a gigantic program of emergency public employment which would give work to three and a half million jobless. They would receive more than the relief dole but less than the prevailing wage so as not to "encourage the rejection of opportunities for private employment." He would turn the remaining million and a half on relief back to local charity as unemployables. The dole, Roosevelt told Congress, was "a narcotic, a subtle destroyer of the human spirit. . . . I am not willing that the vitality of our people be further sapped by the giving of cash, of market baskets, of a few hours of weekly work cutting grass, raking leaves or picking up papers in the public parks," he declared. "The Federal Government must and shall quit this business of relief."[14]

While Roosevelt's proposal to spend almost five billion dollars on work relief alarmed conservatives who favored the less costly dole, the stiffest opposition came from men who felt the President was not asking enough. Roosevelt advocated setting men to work at a "security wage" of roughly $50 a month, about twice what they would get as a dole. The A.F. of L. feared that such a sum would undercut union scales, and advocated paying relievers according to the "prevailing wage scale" in the area. The fight for the "prevailing wage" united New Deal liberals such as Wagner and Black with opportunists like Huey Long, Father Coughlin, and Nevada's silver-haired Pat McCar-

[13] *Time*, XXV (April 15, 1935), 14–16; Schlesinger, *Coming of New Deal*, pp. 278–280; F.D.R. to E. M. House, November 27, 1934, FDRL PPF 222.

[14] Samuel Rosenman (ed.), *The Public Papers and Addresses of Franklin D. Roosevelt* (13 vols., New York, 1938–50), V, 19–21; Schlesinger, *Coming of New Deal*, p. 294; Arthur Macmahon, John Millett, and Gladys Ogden, *The Administration of Federal Work Relief* (Chicago, 1941), pp. 44–47; Daniel Roper to E. M. House, June 12, 1934, House MSS.

ran. Not until late March, after the Senate had first voted for the McCarran "prevailing wage" amendment, did it reverse itself and agree to give the President "discretionary control" of relief wages, save on building projects. The Emergency Relief Appropriation Act of 1935 authorized the greatest single appropriation in the history of the United States or any other nation. The law, which permitted Roosevelt to spend this huge sum largely as he saw fit, marked a significant shift of power from Congress to the President.[15]

The passage of the work relief bill touched off a no-holds-barred fight between Hopkins and Ickes for control of the nearly $5 billion. Ickes viewed Hopkins as an irresponsible spender who was not priming the pump but "just turning on the fire-plug." Hopkins, in turn, thought Ickes a crotchety sorehead: "He is also the 'great resigner'— anything doesn't go his way, threatens to quit. He bores me." Ickes hoped to revive the economy by large public works projects which would prime the pump through heavy capital expenditures.[16] Hopkins, on the other hand, aimed at putting to work as many men as he could who were presently on relief. Since Roosevelt had to divide his billions into three and a half million men, Hopkins' approach—which would get more men to work right away—had the greater appeal. When the President arrived at a rule of thumb which split authority between the two men, it was Hopkins who won out.

Perhaps Roosevelt's decision was inevitable, but it was nonetheless regrettable. Since projects were chosen in which the cost of materials was negligible, housing was doomed. Since the Works Progress Administration, as Hopkins' new agency was called, was not permitted to compete with private industry or to usurp regular governmental work, many WPA projects were make-work assignments of scant value.[17] Yet given the problem Hopkins confronted, he displayed remarkable ingenuity in much that he did. The WPA built or improved

[15] James Burns, "Congress and the Formation of Economic Policies: Six Case Studies" (unpublished Ph.D. dissertation, Harvard University, 1947), pp. 55–88.

[16] Schlesinger, *Coming of New Deal,* p. 347; Harold Ickes, "My Twelve Years with F.D.R.," *Saturday Evening Post,* CXX (June 19, 1948), 30–31, 95–99; Searle Charles, "Harry L. Hopkins: New Deal Administrator, 1933–38" (unpublished Ph.D. dissertation, University of Illinois, 1953), p. 152; Harold Ickes to F.D.R., September 7, 1935, FDRL PSF 17.

[17] In 1939, the agency's name was changed to the Work Projects Administration.

more than 2,500 hospitals, 5,900 school buildings, 1,000 airport landing
fields, and nearly 13,000 playgrounds. It restored the Dock Street
Theater in Charleston; erected a magnificent ski lodge atop Oregon's
Mount Hood; conducted art classes for the insane in a Cincinnati
hospital; drew a Braille map for the blind at Watertown, Massachusetts; and ran a pack-horse library in the Kentucky hills.[18]

The WPA's Federal Theatre Project employed actors, directors, and
other craftsmen to produce plays, circuses, vaudeville shows, and mari-
onette performances. To direct the project, Hopkins named the head
of Vassar's Experimental Theatre, Hallie Flanagan. A classmate of
Hopkins' at Grinnell, she had become an assistant to George Pierce
Baker and had won the first Guggenheim Fellowship granted a
woman. Outstanding theater people served as regional directors, including the playwright Elmer Rice, the actor Charles Coburn, and the
critic Hiram Motherwell.

Far from being the rigidly bureaucratic operation critics had predicted, the Federal Theatre proved ready to experiment. It presented
Marlowe's *Dr. Faustus* without scenery, using lighting to achieve effects, and offered *Macbeth* by an all-Negro company in a Haitian lo-
cale. The Poetic Theatre staged T. S. Eliot's *Murder in the Cathedral*
and Alfred Kreymborg's production of W. H. Auden's satirical verse
drama, *Dance of Death*. On the same night, the curtain rose on
twenty-one stages in eighteen cities on the first presentation of Sinclair
Lewis' *It Can't Happen Here;* there were even versions in Yiddish
and Spanish. In New York, Lewis himself performed the role of his
hero, Doremus Jessup. The Federal Theatre's most striking innovation,
the "Living Newspaper," offered a kaleidoscopic dramatization of
contemporary social and political issues. Critics protested, not unjustly,
that "editions" like *Power* presented New Deal propaganda.

The Federal Theatre presented plays to many people who had never
seen a theatrical production. In April, 1936, a touring company gave
Twelfth Night in the town hall of Littleton, Massachusetts, Detroit
witnessed *Liliom,* and Asheville, North Carolina, saw *Camille,* while
Los Angeles had a choice of attractions at six different WPA theaters
ranging from *Six Characters in Search of an Author* to plays in

[18] Federal Works Agency, *Final Report on the WPA Program, 1935–43*
(Washington, 1946); Works Progress Administration, *Inventory* (Washington,
1938); Grace Adams, *Workers on Relief* (New Haven, 1939), p. 101; *Life,* IV
(February 28, 1938), 41–47.

French, Yiddish, and Mexican. On hot summer nights, youngsters perched on benches in city parks to watch touring puppet companies perform *Punch and Judy* or *Jack and the Beanstalk*. New York's Henry Street Playhouse had to call for police reserves to handle the crowds who came to Dance Theatre productions. In four years, the Federal Theatre played to an audience of some thirty million people.[19]

The Federal Writers' Project, under the direction of Henry Alsberg, a former newspaperman and Provincetown Theatre director, turned out about a thousand publications, including fifty-one state and territorial guides; some thirty city guides; twenty regional guides such as *U.S. One, Maine to Florida,* and *The Oregon Trail;* and such special studies as the imaginative picture book *Who's Who in the Zoo.* The 150 volumes in the "Life in America" series ranged from the moving *These Are Our Lives* to *Baseball in Old Chicago* and embraced a notable series of ethnic studies, including *The Italians of New York, The Hopi, The Armenians of Massachusetts,* and *The Negro in Virginia.* The projects, under the guidance of state directors like Vardis Fisher in Idaho, Ross Santee in Arizona, and Lyle Saxon in Louisiana, reflected the fascination of the thirties with the rediscovery of regional lore, the delighted recapture of place names—Corncake Inlet, Money Island, Frying Pan Shoals—and the retelling of long-forgotten tales of Indian raids. The Project made use of the talents of established writers such as Conrad Aiken, who wrote the description of Deerfield for the Massachusetts state guide, and new men like John Cheever and Richard Wright, whose "Uncle Tom's Children" received the *Story* magazine prize for the best story by an FWP writer. Commercial publishers were happy to print most of the guides, and many of them sold exceptionally well. Congress in 1939 abolished the Federal Theatre and allowed the other projects to continue only if they found local sponsors who would bear 25 per cent of the cost. The Federal Writers' Project proved its popularity when, to the amazement of its critics, every one of the forty-eight states put up the required amount.[20]

[19] Hallie Flanagan, *Arena* (New York, 1940); Willson Whitman, *Bread and Circuses* (New York, 1937); "Unemployed Arts," *Fortune,* XV (May, 1937), 110; Mordecai Gorelik, *New Theatres for Old* (New York, 1940), pp. 397–399, 443–445; Grant Code, "Dance Theatre of the WPA: A Record of National Accomplishment," *Dance Observer* (November, 1939), 280–81, 290; "Footlights, Federal Style, *Harper's,* CLXXIII (1936), 626.

[20] Mabel Ulrich, "Salvaging Culture for the WPA," *Harper's,* CLXXVIII

The Federal Art Project, directed by Holger Cahill, gave jobs to unemployed artists, often men of modest talents, but there were gifted painters too—Stuart Davis and Yasuo Kuniyoshi, Willem De Kooning and Jackson Pollock, Charles Alston and Aaron Bohrod. At night, in schools, in settlement houses, and in rural churches, hundreds of teachers taught painting, clay modeling, weaving, and carving. To prepare an *Index of American Design,* water-colorists and draftsmen copied precisely and in vivid hues objects which had been used since the first days of settlement—andirons and skillets, cigar-store Indians and ship's figureheads. Although the distinguished British critic Ford Madox Ford found the level of WPA art work "astonishingly high," much of the original painting was inevitably second rate. Yet the FAP, and the special projects of the Treasury Department, revived the art of mural decoration, which had been dead in the United States for seventy-five years. Henry Varnum Poor reported: "What has happened and in such a short time is almost incredible. From a government completely apathetic to art, we suddenly have a government very art conscious. Public offices are decorated with fresh paintings instead of old prints." American painters were deeply influenced by the Mexican muralists Rivera, Orozco, and Siqueiros—especially in their forthright treatment of political and social subjects. "Mural art can never be important," wrote George Biddle, "unless it is interpreting a great social and collective idea." Applying the Mexican techniques but drawing on native American themes, painters frescoed the walls of schools and post offices across the nation; "some of it good," Roosevelt commented, "some of it not so good, but all of it native, human, eager, and alive—all of it painted by their own kind in their own country, and painted about things that they know and look at often and have touched and loved."[21]

(1939), 656; "Work of the Federal Writers' Project of WPA," *Publishers' Weekly,* CXXXV (1939), 1130–1135; *Time,* XXXI (January 3, 1938), 55–56; XXXVI (August 12, 1940), 64. Yet another WPA agency, the Historical Records Survey, made inventories of manuscripts, public records, and other historical materials and turned out guides like the *Index to Early American Periodical Literature, 1728–1870.* Robert Binkley, "The Cultural Program of the W.P.A.," *Harvard Educational Review,* IX (March, 1939), 156–174; Memorandum to Harry Hopkins, May 12, 1935, Hopkins MSS., "The President" file. Cf. Willard Hogan, "The WPA Research and Records Program," *Harvard Educational Review,* XIII (1943), 52–62.

[21] "Unemployed Arts," p. 172; "Mr. Whiting. Interview given to American Mag. of Art by Henry Varnum Poor," n.d., Forbes Watson MSS.; George Biddle

The National Youth Administration aimed to help the millions of young Americans who found it even harder than other groups to find work. "The young are rotting without jobs and there are no jobs," commented a Hopkins aide. Young men and women who had "grown up against a shut door," they not infrequently felt they had been born into a society where no one either valued them or cared about them.[22] They had a friend in court in the NYA's director, Aubrey Williams of Alabama. The grandson of a planter who had freed a thousand slaves, he had gone to work in a factory at the age of six, had studied with Bergson at the Sorbonne, had had a try as a Lutheran lay preacher, and had found himself in social work. In the next seven years, the NYA gave part-time employment to more than six hundred thousand college students and to over one and a half million high-school pupils. Most of the student jobs were routine clerical assignments, but NYA directors tried to find work directly related to a student's field: many history students found positions in the document divisions of state historical societies. Connecticut College women compiled a code of social welfare laws, and University of Nebraska students built an observatory and telescope. During this same period, the NYA aided over 2.6 million jobless youth who were not in school. They built tuberculosis isolation huts in Arizona; repaired automobiles in NYA workshops; raised a milking barn at Texas A. & M.; landscaped a park in Cheboygan, Michigan; renovated a schoolhouse in Antelope, North Dakota; laid powdered blue-shale tennis courts in Topeka; and received vocational training at a model village in Passamaquoddy, Maine. Although educators protested that the NYA would destroy the independence of the schools, the government never attempted to exercise political control.[23]

By predepression standards, Roosevelt's works program marked a

to Eleanor Roosevelt, June 28, 1933, Biddle MSS.; *Public Papers,* X, 73–74; Holger Cahill, COHC; Cahill, tape recording, Archives of American Art, Detroit Institute of Arts; Grace Overmyer, *Government and the Arts* (New York, 1939), pp. 95–116.

[22] Martha Gellhorn to Harry Hopkins, December 2, 19, 1934, Hopkins MSS.

[23] Jacob Baker to Harry Hopkins, April 29, 1935, Aubrey Williams MSS., Box 13; National Advisory Committee, NYA, Minutes, February 8–9, 1938, Charles Taussig MSS., Box 6; George Rawick, "The New Deal and Youth" (unpublished Ph.D. dissertation, University of Wisconsin, 1957), pp. 64, 218 ff.; Federal Security Agency, *Final Report of the National Youth Administration* (Washington, 1944); Betty and Ernest Lindley, *A New Deal for Youth* (New York, 1938).

bold departure. By any standard, it was an impressive achievement. Yet it never came close to meeting Roosevelt's goal of giving jobs to all who could work. Of the some ten million jobless, the WPA cared for not much more than three million. Workers received not "jobs" but a disguised dole; their security wage amounted to as little as $19 a month in the rural South. The President split up the billions among so many different agencies—the Department of Agriculture got $800 million of it—that Hopkins had only $1.4 billion to spend for WPA. By turning the unemployables back to the states, he denied to the least fortunate—the aged, the crippled, the sick—a part in the federal program, and placed them at the mercy of state governments, badly equipped to handle them and often indifferent to their plight.[24]

Roosevelt's social security program was intended to meet some of the objections to his relief operations. In June, 1934, the President had named a cabinet Committee on Economic Security, headed by Frances Perkins, to develop a workable social insurance system. The Committee, with little argument, approved a national system of contributory old-age and survivors' insurance. It had a much harder time reaching agreement on the vexatious issue of unemployment compensation. New Dealers like Wallace and Tugwell desired a national system which would insure uniform standards in a country where workers moved from state to state and where state governments were often notoriously mismanaged. A group of reformers centered in Wisconsin, which had the only existing unemployment compensation plan, urged instead a joint national-state system, which would permit experimentation on a problem where many questions still had not been answered. This Brandeisian approach, sponsored by one of the authors of the Wisconsin plan, Paul Rauschenbush, and his wife, Elizabeth Brandeis, the Justice's daughter, won the committee's support, both because the President favored it and because it had the further merit of being more likely to meet a constitutional test before the Supreme Court.[25]

24 Grace Adams, *Workers on Relief,* pp. 15–16, 314–320; Marriner Eccles, *Beckoning Frontiers* (New York, 1951), pp. 185–186; Elias Huzar, "Federal Unemployment Relief Policies," *Journal of Politics,* II (1940), 332–335. Cf. Donald Howard, *The WPA and Federal Relief Policy* (New York, 1943).

25 Schlesinger, *Coming of New Deal,* pp. 301–303; Minutes of the Advisory Council of the Committee on Economy Security, Mary Dewson MSS., Box 1; Gerard Swope to F.D.R., March 22, 1934, Swope MSS.; W. Jett Lauck to Thomas Kennedy, November 22, 1934, Lauck MSS., Correspondence; Eveline Burns, "Social Insurance in Evolution," *American Economic Review,* XXXIV,

On January 17, 1935, President Roosevelt asked Congress for social security legislation. That same day, the administration's bill was introduced in both houses of Congress by men who had felt keenly the meaning of social insecurity. Robert Wagner, who steered the social security measure through the Senate, was the son of a janitor; as an immigrant boy, he had sold papers on the streets of New York. Maryland's David Lewis, who guided the bill through the House, had gone to work at nine in a coal mine. Illiterate at sixteen, he had taught himself to read not only English but French and German; he had learned French to verify a translation of Tocqueville. The aged, Davy Lewis declared, were "America's untouchables. . . . Even under slavery, the owner did not deny his obligation to feed and clothe and doctor the slaves, no matter what might happen to crops or to markets."[26]

Conservatives charged that the social security conception violated the traditional American assumptions of self-help, self-denial, and individual responsibility. New Jersey's Senator A. Harry Moore protested: "It would take all the romance out of life. We might as well take a child from the nursery, give him a nurse, and protect him from every experience that life affords." Southerners worried about its implications for race relations. "The average Mississippian," wrote the Jackson *Daily News*, "can't imagine himself chipping in to pay pensions for able-bodied Negroes to sit around in idleness on front galleries, supporting all their kinfolks on pensions, while cotton and corn crops are crying for workers to get them out of the grass."[27]

Yet, as in the case of the relief bill, opposition to the measure centered not in conservatives but in congressmen who argued that the measure was too parsimonious. The Townsendites rallied behind a proposal, filed by Representative John McGroarty, the poet laureate of California, for monthly pensions of $200. They failed to win approval for their proposition, but they impressed Congress with the need to adopt some kind of legislation which would quiet the demand for old-age pensions. After voting down the McGroarty bill and the

Supplement (March, 1944), 199–211; Paul Douglas, *Social Security in the United States* (New York, 1936), 3–68.

[26] *Fortune*, XI (March, 1935), 116; *Congressional Record*, 74th Cong., 1st Sess., p. 5687.

[27] *Literary Digest*, CXIX (June 29, 1935), 5; Jackson (Miss.) *Daily News*, June 20, 1935.

yet more radical demand of Ernest Lundeen of Minnesota for costly unemployment benefits to be distributed by elected committees of workers, the House passed the social security measure in April, 371–33, although only after every Republican save one had voted to recommit it.[28] In June the Senate approved the bill, and on August 15 the President signed the measure.

The Social Security Act created a national system of old-age insurance in which most employees were compelled to participate. At the age of sixty-five, workers would receive retirement annuities financed by taxes on their wages and their employer's payroll; the benefits would vary in proportion to how much they had earned. In addition, the federal government offered to share equally with the states the care of destitute persons over sixty-five who would not be able to take part in the old-age insurance system. The act also set up a federal-state system of unemployment insurance, and provided national aid to the states, on a matching basis, for care of dependent mothers and children, the crippled, and the blind, and for public health services.

In many respects, the law was an astonishingly inept and conservative piece of legislation. In no other welfare system in the world did the state shirk all responsibility for old-age indigency and insist that funds be taken out of the current earnings of workers. By relying on regressive taxation and withdrawing vast sums to build up reserves, the act did untold economic mischief. The law denied coverage to numerous classes of workers, including those who needed security most: notably farm laborers and domestics. Sickness, in normal times the main cause of joblessness, was disregarded. The act not only failed to set up a national system of unemployment compensation but did not even provide adequate national standards.[29]

Yet for all its faults the Social Security Act of 1935 was a new landmark in American history. It reversed historic assumptions about the

[28] The *Daily Worker,* which backed the Lundeen bill, denounced Roosevelt's social security proposal as "one of the biggest frauds ever perpetrated on the people of this country." *Daily Worker,* June 20, 1935.

[29] Schlesinger, *Coming of New Deal,* pp. 308–315; Twentieth Century Fund, *More Security for Old Age* (New York, 1937); Margaret Grant, *Old-Age Security* (Washington, 1939); Abraham Epstein, "Our Social Insecurity Act," *Harper's,* CLXXII (1935), 55–66; Eveline Burns to Edward Costigan, February 20, 1935, Grace Abbott to Costigan, February 25, 1936, Lincoln Filene to Felix Frankfurter, April 29, 1935, Costigan MSS., V.F. 4. Cf. Lewis Meriam, *Relief and Social Security* (Washington, 1946).

nature of social responsibility, and it established the proposition that the individual has clear-cut social rights. It was framed in a way that withstood tests in the courts and changes of political mood. "I guess you're right on the economics," Roosevelt conceded when told that the employee contributions were a mistake, "but those taxes were never a problem of economics. They are politics all the way through. We put those payroll contributions there so as to give the contributors a legal, moral, and political right to collect their pensions and their unemployment benefits. With those taxes in there, no damn politician can ever scrap my social security program."[30]

Not only did Harold Ickes come out second best in his battle with Hopkins for control of relief act funds; he soon found that he had to protect the PWA from the President's urge to wipe it out altogether, since Roosevelt was unpersuaded by Ickes' claims for the indirect economic benefits of large-scale public works. Yet if Ickes got meager economic returns, he proved to be a builder to rival Cheops. From 1933 to 1939, PWA helped construct some 70 per cent of the country's new school buildings; 65 per cent of its courthouses, city halls, and sewage plants; 35 per cent of its hospitals and public health facilities. PWA made possible the electrification of the Pennsylvania Railroad from New York to Washington and the completion of Philadelphia's 30th Street Station. In New York, it helped build the Triborough Bridge, the Lincoln Tunnel, and a new psychiatric ward at Bellevue Hospital. It gave Texas the port of Brownsville, linked Key West to the Florida mainland, erected the superbly designed library of the University of New Mexico, and spanned rivers for Oregon's Coastal Highway. Under PWA allocation, the Navy built the aircraft carriers *Yorktown* and *Enterprise,* the heavy cruiser *Vincennes,* and numerous other light cruisers, destroyers, submarines, gunboats, and combat planes; the Army Air Corps received grants for more than a hundred planes and over fifty military airports.[31]

The PWA's work in slum clearance and low-cost housing proved

---

[30] Schlesinger, *Coming of New Deal,* pp. 308–309. See Frances Perkins, *The Roosevelt I Knew* (New York, 1946), pp. 188–189; 282–285; Alvin David, "Old-Age, Survivors, and Disability Insurance: Twenty-Five Years of Progress," *Industrial and Labor Relations Review,* XIV (1960), 10–23.

[31] Public Works Administration, *America Builds* (Washington, 1939); Harold Ickes, *Back to Work* (New York, 1935); Schlesinger, *Coming of New Deal,* pp. 287–288; Grace Adams, *Workers on Relief,* pp. 100–101.

disappointing. Ickes, hoping to outwit land speculators and greedy corporations, approved only seven projects out of hundreds of applications from limited-dividend corporations. When he decided the only way to get decent housing was for the government to build it itself, he was hamstrung by a Circuit Court ruling that the government could not claim the right of eminent domain to acquire land for federal housing.[32] In its first four and a half years, the PWA built or started only forty-nine projects with a total of less than 25,000 residential units. Ickes himself confided to his diary in June, 1936, after a tour of the PWA housing projects in Chicago: "From what I saw and heard I was very much disappointed with the progress that has been made. There isn't any doubt that something is wrong in the Housing Division, in fact, has been wrong for a long time. We are not getting results."[33]

Unhappy with the PWA's performance, a small corps of reformers centered in New York called for the creation of a separate federal agency which would undertake low-cost housing as a permanent government activity. They got little help from Roosevelt. A man with small love for the city, and less for the housing reformers, with their cold calculations about plumbing requirements and unit costs, he never shared the enthusiasm of the proponents of model tenements. The main drive for a federal housing act came not from the White House but from Senator Wagner, who introduced legislation in the Senate, and from the urban reformers: social workers like Mary Simkhovitch, housing experts like Catherine Bauer, labor leaders like John Edelman, and local officials like New York City's Housing Commissioner, Langdon Post.[34]

Franklin Roosevelt thought of his housing program less as a way to aid slum dwellers than as a way to revive a sick industry—nearly a third of the jobless were in the building trades—and one which would have incalculable effects on other industries from cement to electrical

[32] United States v. Certain Lands in the City of Louisville, 78.F. (2d) 684 (C.C.A. 6th Cir. 1935).

[33] Harold Ickes, The Secret Diary of Harold Ickes (3 vols., New York, 1954), I, 610; John Blum, From the Morgenthau Diaries (Boston, 1959), p. 287; Robert Moore Fisher, 20 Years of Public Housing (New York, 1959), pp. 82–91; Timothy McDonnell, S.J., The Wagner Housing Act (Chicago, 1957), pp. 29–30; John Carson to Harry Slattery, July 31, 1935, Slattery MSS., Letters.

[34] McDonnell, Wagner Housing Act, pp. 47, 51–87, 137; Ernest Lindley, Half Way with Roosevelt (New York, 1936), p. 72.

appliances. He had far less faith in public housing than in government encouragement of private ventures. Under the National Housing Act of 1934, a new agency, the Federal Housing Administration, insured loans made by private lending institutions to middle-income families to repair and modernize their homes or to build new homes. Home and business modernization proved an outstanding success, but the core of the program, which aimed at new home construction, moved sluggishly. Roosevelt made the strange appointment of the conservative oil executive James Moffett to head the agency, and Moffett placed a retired New York banker, who was unsympathetic to the FHA, in charge of new home construction. Not until 1938, after the act had been liberalized and new men appointed, did the potentialities of FHA as a home-building agency begin to be realized.[35] Meanwhile, public housing foundered. When Moffett charged that the program Ickes proposed for massive low-cost housing would "wreck a 21-billion-dollar mortgage market and undermine the nation's real estate values," Roosevelt sided with Moffett.[36]

For two years, Senator Wagner got nowhere in his attempt to win approval for his federal housing bill. In 1936, the Senate passed the measure, but it died in House committee. In 1937, Wagner made a new try, and on August 6, despite objections from rural senators that the bill would benefit chiefly the great northern cities, the Senate once more passed the measure. Once again, it seemed that the House would kill the bill, but now Roosevelt swung his weight behind it, even though he still felt little enthusiasm for the measure and was annoyed by the prickly, squabbling, impolitic housing reformers. The President's support proved decisive. Henry Steagall, chairman of the House committee, thought the measure both socialistic and financially reckless, but he reported the bill out from a sense of party loyalty. The House finally approved the bill, and on September 1 the President signed it. The Wagner-Steagall Housing Act created the United States Housing Authority as a public corporation under the Department of the In-

[35] Pearl Janet Davies, *Real Estate in American History* (Washington, 1958), pp. 178–180; Eccles, *Beckoning Frontiers*, pp. 158–161.

[36] *Time*, XXIV (December 3, 1934), 15; T.R.B., "Washington Notes," *New Republic*, LXXXI (1934), 128. The National Housing Act also sought to make the mortgage market more fluid. Under authority of the act, the RFC set up a Federal National Mortgage Association, and FNMA or "Fanny May" offered a market for FHA mortgage loans.

terior, and made available $500 million in loans for low-cost housing. By the end of 1940, almost 350 USHA projects had been completed or were under construction. But, like many other New Deal operations, the federal housing venture was notable more because it created new precedents for government action than for the dimensions of its achievements. Measured by the needs, or by the potentialities, Roosevelt's public housing program could make only modest claims.[37]

Franklin Roosevelt responded with much greater warmth to chimerical plans to remove slum dwellers to the countryside than he did to schemes for urban renewal. The President, Tugwell has written, "always did, and always would, think people better off in the country and would regard the cities as rather hopeless. . . ." Roosevelt, who had experimented when Governor with marrying agriculture and industry in a "rural industrial group," instructed Senator Bankhead to write an appropriation for subsistence homesteads into the NIRA. The chief of the Subsistence Homestead Division, M. L. Wilson, envisioned the projects as "a middle class movement for selected people, not the top, not the dregs," for Americans who were not "white-lighters," and who felt themselves "outcasts of the jazz-industrial age."[38]

The New Deal built about a hundred communities, mostly all-rural farm colonies like Penderlea Homesteads in North Carolina or industrial subsistence settlements like Austin Homesteads in Minnesota. The one experiment which caught public attention was the Arthurdale project at Reedsville, West Virginia, in the depressed mountain coal country. The personal pet of Louis Howe and of Mrs. Roosevelt, who spent thousands of dollars of her own money, Arthurdale proved an expensive failure. "We have been spending money down there like drunken sailors," Ickes lamented.[39] Although the leaders of the movement believed they were affording people the opportunity to escape the evils of a vulgar industrial society, the subsistence homesteads which proved most successful were precisely those, such as El Monte near

[37] McDonnell, *Wagner Housing Act*, Chs. 4–15; Robert Wagner to Edith Elmer Wood, May 7, 1937, Wagner MSS.; Catherine Bauer, "Now, at Last: Housing," *New Republic*, XCII (1937), 119–121.

[38] Tugwell, "The Sources of New Deal Reformism," *Ethics*, LXIV (1954), 266; Paul Conkin, *Tomorrow a New World* (Ithaca, 1959), p. 83; Russell Lord, *The Wallaces of Iowa* (Boston, 1947), pp. 420–421.

[39] Ickes, *Diary*, I, 207; Millard Milburn Rice, "Footnote on Arthurdale," *Harper's*, CLXXX (1940), 411–419; Alfred Steinberg, *Mrs. R.* (New York, 1958), pp. 209–212.

Los Angeles and Longview in the Columbia River valley, which quickly took on the character of any suburban subdivision. As soon as the pall of the early depression years lifted, people hurried to get back into the "real" world of the bustling city streets. Ostensibly experimental and utopian, the subsistence homesteads movement soon seemed rather a quest for an ark of refuge, an indication of the despair of the early thirties.[40]

Ironically, at the very time the New Deal was celebrating the rural idyll, the country was beginning to learn some of the desperate facts of rural life. The novels of Erskine Caldwell—especially *Tobacco Road,* which became a long-running Broadway play—books like Caldwell and Margaret Bourke White's *You Have Seen Their Faces,* which showed pictures of the sharecropper with his rickety children, his mangy dogs, his prematurely aged wife; James Agee's sensitive *Let Us Now Praise Famous Men,* with photographs by Walker Evans; and such newspaper reports as Frazier Hunt's in the New York *World-Telegram* in 1935 all opened the nation's eyes to the misery of the sharecropper. The croppers he saw in the South, Hunt wrote, reminded him of Chinese coolies working alongside of the South Manchurian railroad, save that in China he never had seen children in the fields. These people "seemed to belong to another land than the America I knew and loved."[41]

The New Deal was not to blame for the social system it inherited, but New Deal policies made matters worse. The AAA's reduction of cotton acreage drove the tenant and the cropper from the land, and landlords, with the connivance of local AAA committees which they dominated, cheated tenants of their fair share of benefits.[42] When

[40] Conkin, *Tomorrow a New World,* pp. 111–112; Schlesinger, *Coming of New Deal,* pp. 371–372. Illuminating studies of the experiments are Edward Banfield, *Government Project* (Glencoe, Ill., 1951); Russell Lord and Paul Johnstone (eds.), *A Place on Earth* (Washington, 1942); Paul Wager, *One Foot on the Soil* (University, Ala., 1945); G. S. Wehrwein, "An Appraisal of Resettlement," *Journal of Farm Economics,* XIX (1937), 190–202; Raymond Duggan, *A Federal Resettlement Project: Granger Homesteads* (Washington, 1937).

[41] New York *World-Telegram,* July 30, 1935. Cf. Herman Clarence Nixon, *Forty Acres and Steel Mules* (Chapel Hill, 1938).

[42] Arthur Raper, *Preface to Peasantry* (Chapel Hill, 1936), pp. 6–7, 243–253; Gunnar Myrdal, *An American Dilemma* (New York, 1944), pp. 254–258; Henry I. Richards, *Cotton and the AAA* (Washington, 1936), pp. 135–162; Alison Davis, Burleigh Gardner, and Mary Gardner, *Deep South* (Chicago,

Norman Thomas and others called these abuses to Roosevelt's attention, he counseled patience. He had enough on his hands trying to administer a vast mobilization to restore farm prosperity, and he did not propose to attempt to revolutionize social relations in the South. Even the bolder New Deal spirits feared to jeopardize the rest of their program by antagonizing powerful conservative southern senators like Joe Robinson of Arkansas.[43]

It was in Arkansas that croppers and farm laborers, driven to rebellion by the hardhanded tactics of the landlords and the AAA committees, established the Southern Tenant Farmers' Union in July, 1934. Under Socialist leadership, the farmers, Negro and white—some of the whites had been Klansmen—organized in the region around Tyronza. The landlords struck back with a campaign of terrorism. "Riding bosses" hunted down union organizers like runaway slaves; union members were flogged, jailed, shot—some were murdered.[44] The wife of a sharecropper from Marked Tree wrote: "We Garded our House and been on the scout untill we are Ware out, and Havent any Law to looks to. thay and the Land Lords hast all turned to nite Riding. . . . thay shat up some Houses and have Threten our Union and Wont let us Meet at the Hall at all."[45] Queried by a reporter about the violence, one Arkansan responded: "We have had a pretty serious situation here what with the mistering of the niggers and stirring them up to think the Government was going to give them forty acres."[46]

Even more tragic was the experience of the migrant farmers. Tractored off their land, some of the "Arkies," and even more of the "Okies," whose farms had blown away in the dust storms, trekked westward to the orange groves and lettuce fields of the Pacific Coast. By the end of the decade, a million migrants, penniless nomads in end-

---

1941), pp. 283–284, 346–347; Harold Hoffsommer, "The AAA and the Cropper," *Social Forces*, XIII (1935), 494–502.

[43] Norman Thomas to Henry Wallace, February 22, 1934, Thomas MSS.; M. S. Venkataramani, "Norman Thomas, Arkansas Sharecroppers, and the Roosevelt Agricultural Policies, 1933–1937," *Mississippi Valley Historical Review*, XLVII (1960), 225–246.

[44] Dorothy Detzer, *Appointment on the Hill* (New York, 1948), pp. 208–210; H. L. Mitchell to Jack Herling, April 1, 1935, Norman Thomas MSS.; H. L. Mitchell, COHC.

[45] Mrs. T. P. Martin to Norman Thomas, April 1, 1935, Norman Thomas MSS.

[46] *The New York Times*, April 16, 1935.

less caravans of overburdened jalopies bursting with half-clad tow-headed children, had overrun small towns in Oregon and Washington and pressed into the valleys of California. Their dust-covered Hudson sputtering westward along Highway 66, the Joads of John Steinbeck's *The Grapes of Wrath* symbolized thousands of Americans uprooted from their farms, turning to the lotus lands of the West only, when they came to the end of their weary odyssey, to be drowned in a sea of cheap labor, exploited by the great orchards, hounded by sheriffs, their poverty a badge of shame.[47]

In the halls of the Department of Agriculture, Jerome Frank and his allies pleaded the cause of the politically voiceless sharecroppers and tenants. They found AAA Director Chester Davis sympathetic to their aims even though, like his predecessor Peek, he thought the main purpose of the Triple A was farm recovery rather than overhauling the rural power structure. Davis even insisted, over the objections of the planters, that the 1934–35 cotton contract stipulate that the landlord retain the same number of tenant units he had in 1933. But this did not go far enough for the Frank group. Alger Hiss of the Legal Division drafted an opinion, approved by Frank, which required planters to retain the same individuals as tenants during the life of the contract, and Frank had the new edict circulated while Davis was out of town. When Davis returned, he canceled the directive and, furious at the reformers, demanded a showdown. With the backing of Roosevelt and Wallace, he wiped out the office of General Counsel and ousted Frank, Gardner Jackson, Lee Pressman, Frank Shea, and Frederic Howe. Hiss, who, oddly, was not on Davis' elimination list, resigned. Liberals viewed the "purge" of the Frank faction as the end of an era, the triumph of the planters and processors over the advocates of a "social outlook in agricultural policy."[48]

[47] Carey McWilliams, *Factories in the Field* (Boston, 1939); Victor Jones, "Transients and Migrants," in University of California, Berkeley, Bureau of Public Administration, *1939 Legislative Problems, No. 4*, February 27, 1939, mimeographed; William and Dorothy Cross, *Newcomers and Nomads in California* (Stanford, 1937); Richard Neuberger, "Refugees from the Dust Bowl," *Current History*, L (April, 1939), 32–35; Paul Taylor and Edward Rowell, "Refugee Labor Migration to California, 1937," *Monthly Labor Review*, XLVII (1938), 240–250; Clarke Chambers, *California Farm Organizations* (Berkeley and Los Angeles, 1952), pp. 32–33.

[48] Schlesinger, *Coming of New Deal*, pp. 74–80; Russell Lord, *The Wallaces of Iowa* (Boston, 1947), pp. 404–409; Louis Bean, COHC, pp. 158 ff.; John Hutson, COHC, pp. 114 ff., 175 ff.; Mordecai Ezekiel, COHC, pp. 74 ff.;

In April, 1935, shortly after the purge of the Frank group, President Roosevelt set up the Resettlement Administration to meet the problem of rural poverty, and named the leader of the agrarian reformers, Rexford Tugwell, to head the operation. The RA took over the rural rehabilitation and land programs that Harry Hopkins had initiated with the FERA, as well as the subsistence homestead projects. Tugwell had scant sympathy with the homestead colonies, which he thought anachronistic; the apple of his eye was the "greenbelt" town. He built three of them—near Washington, D.C., Cincinnati, and Milwaukee—entirely new towns, located close to employment and girdled by green countryside. These greenbelt communities, which reflected the ideas of the English garden-city advocate Ebenezer Howard, and the experiment of American urban planners at Radburn, New Jersey, attracted more attention from foreign visitors than any New Deal project save TVA.[49]

Tugwell's chief assignment was not communitarianism but resettlement; he sought to move impoverished farmers from submarginal land and give them a fresh start on good soil with adequate equipment and expert guidance. But he never had enough money to do very much. The RA, which planned to move 500,000 families, actually resettled 4,441. A program on this scale hardly began to meet the problems of the cropper and tenant farmer. Some of the southern leaders, disturbed in part by the threat posed by Huey Long, had begun to appreciate the need to take more effective action. When a group of humanitarian reformers centered in the Rosenwald Fund came up with a proposal (drafted by Professor Frank Tannenbaum) to benefit the southern tenant, they found a sponsor in Alabama's Senator Bankhead.[50] The Bankhead bill, first introduced in 1935, aimed to help tenant farmers and farm laborers become landowners; it reflected the reformers' faith in the Jeffersonian dream of the yeoman farmer and the Latin Ameri-

Gardner Jackson to Louis Brandeis, December 21, 1934. Brandeis MSS., G8; Edward Costigan to F.D.R., February 7, 1935. FDRL PPF 1971; *New Republic,* LXXXII (1935), 29; *The Nation,* CXL (1935), 216–217.

49 Conkin, *Tomorrow a New World:* Rexford Tugwell, "The Resettlement Idea," *Agricultural History,* XXXIII (1959), 159–164; George Warner, *Greenbelt* (New York, 1954).

50 Theodore Bilbo to F.D.R., August 18, 1935, FDRL OF 1650; "Notes on Conference of Messers. Embree, Alexander, Johnson, Tannenbaum, and Simon on Agricultural Rehabilitation, Atlanta, Georgia—Jan. 6, 1935"; Will Alexander to Tannenbaum, July 15, 1937, Tannenbaum MSS.

canist Tannenbaum's familiarity with Mexico's experiment in converting peons into peasant proprietors. Southern conservatives objected that the proposal was revolutionary, and radicals protested that it would set up a caste of peasantry like Hitler's *nobel Bauern*.[51] But when it received the blessing of the President's Special Committee on Farm Tenancy in 1937, Congress quickly gave its approval.

Under the Bankhead-Jones Farm Tenancy Act of 1937, a new agency, the Farm Security Administration, which replaced the RA, extended rehabilitation loans to farmers, granted low-interest, long-term loans to enable selected tenants to buy family-size farms, and aided migrants, especially by establishing a chain of sanitary, well-run migratory labor camps. The FSA was the first agency to do anything substantial for the tenant farmer, the sharecropper, and the migrant; by the end of 1941, it had spent more than a billion dollars, much of it, to be sure, in the form of loans which it expected to be repaid. It plowed new ground when it helped launch medical-care co-operatives, and reaped the anger of Twin Cities grain dealers when it lent money to co-operatives to buy grain elevators. At the risk of its political life, the FSA was scrupulously fair in its treatment of Negroes. Yet, for all this, it did not measure up to the dimensions of the problem it faced. The main boast of the FSA—that the rate of repayment of its loans was impressively high—suggested that the FSA did not dig very deeply into the problem of rural poverty.[52] The FSA had no political constituency—croppers and migrants were often voteless or inarticulate—while its enemies, especially large farm corporations that wanted cheap labor and southern landlords who objected to FSA aid to tenants, had powerful representation in Congress. The FSA's opponents kept its appropriations so low that it was never able to accomplish anything on a massive scale.

In the winter of 1939, although the great peregrination had passed its peak, one still saw families huddled over fires along western highways. The highway filling station, bringing together settled and mi-

[51] Frank Tannenbaum to Will Alexander, April 26, 1935, Tannenbaum MSS.; Benjamin Marsh to editors of *The Nation*, CXL (1935), 482; T.R.B., "Washington Notes," *New Republic*, LXXXVII (1936), 352.

[52] Jerry Voorhis, *Confessions of a Congressman* (Garden City, N.Y., 1948), pp. 127–132; Myrdal, *American Dilemma*, pp. 274–275; Richard Sterner, *The Negro's Share* (New York, 1943), pp. 295–309; Grant McConnell, *The Decline of Agrarian Democracy* (Berkeley and Los Angeles, 1953), p. 205, n. 16.

gratory characters, was the scene of the depression epic.[53] Here stopped the Joads on their way west on Highway 66, here the transcontinental buses refueled in a cycle of films of lonely pilgrimages. If in the representative novel of the twenties the hero goes to the small-town depot to board the train to the city full of promise, in the representative novel of the thirties he is left alone on a highroad thumbing his way to an unknown destination. At the end of Dos Passos' *U.S.A.*, young Vag stands, famished and footsore, at the edge of a highway, with no hope or aim save to make his way "a hundred miles down the road."[54] As the depression years wore on, the jobless adapted to a kind of purposeless motion, like the music in the favorite song of the day that went 'round and around, like the marathon dancers who shuffled lifelessly around the dance floor barely able to keep their bodies moving, like the six-day bike riders who spun endlessly around an oval track, anesthetizing spectators by the relentless monotony of the circular grind.[55]

By the end of Roosevelt's first term, many had come to believe that there would be a permanent army of unemployed. Even before he took office, skeptics had queried whether full employment would ever return. By 1934, Raymond Moley was writing: "We are going to keep on providing relief—probably permanently," and one of Hopkins' lieutenants reported: "It looks as though we're in this relief business for a long, long time. . . . The majority of those over 45 probably will NEVER get their jobs back." By 1936, Hopkins had accommodated himself to the idea that even if "prosperity" returned, at least four or five million would be jobless.[56] Like Great Britain after World War I, the United States, it seemed, had achieved for millions of its citizens a permanent plateau of unemployment.

[53] John Dos Passos, *Fortune Heights*, in *Three Plays* (New York, 1934); Irving Bernstein, *The Lean Years* (Boston, 1960), p. 324. All the action in Robert Sherwood's *The Petrified Forest* takes place in the Black Mesa Filling Station and Bar-B-Q in eastern Arizona.

[54] John Dos Passos, *The Big Money* (New York, 1936), p. 561.

[55] "For Six Days," *Fortune*, XI (March, 1935), 88; Horace McCoy, *They Shoot Horses, Don't They?* (New York, 1935).

[56] Bernstein, *Lean Years*, p. 479; William Dodd to Richard Crane, August 2, 1932, Dodd MSS., Box 40; Raymond Moley, "Saving Men and Building the Nation," *Today*, I (February 24, 1934), 13; Lorena Hickok to Hopkins, April 25, 1934, Hopkins MSS.; Arthur Schlesinger, Jr., *The Politics of Upheaval* (Boston, 1960), pp. 357–358.

CHAPTER 7

The Second Hundred Days

FROM the very first, the New Dealers were haunted by the fear that the Supreme Court would hold reform legislation unconstitutional. Four members of the bench—James McReynolds, George Sutherland, Willis Van Devanter, and Pierce Butler—were tenacious conservatives. If any one of the remaining five justices joined these four, the whole New Deal program might be impaired. Louis Brandeis, the scholastic Benjamin Cardozo, and the warm, wide-ranging Harlan Fiske Stone seemed safely in the liberal camp, but there were limits to how great an extension of government power even they would approve. More likely to join the conservatives were the two justices who occupied middle ground: the Jovelike Chief Justice, Charles Evans Hughes, and Owen Roberts, at fifty-eight the youngest member of the Court.

If the Court pursued the line of reasoning it had followed for much of the past decade, all was lost. The administration hoped it would be open to a different conception: that the depression constituted, in Justice Brandeis' words, "an emergency more serious than war."[1] Consequently, New Deal lawyers festooned early legislation with "emergency clauses." When the Court in January, 1934, in a 5–4 decision written by Chief Justice Hughes, upheld a Minnesota moratorium on mortgages, it indicated it might be receptive to such a view of the depression. "While emergency does not create power," Hughes stated, "emergency may furnish the occasion for the exercise of power."[2] Two

[1] New State Ice Co. v. Liebmann, 285 U.S. 306 (1932).
[2] Home Building & Loan Association v. Blaisdell et al., 290 U.S. 426. Cf.

months later, in another 5–4 decision, the Court, speaking through Justice Roberts, sustained a New York law which created a board empowered to set minimum and maximum retail prices for milk. To the horror of conservatives, Roberts declared: "Neither property rights nor contract rights are absolute."[3] The two decisions, Justice McReynolds wrote James Beck, meant "the end of the Constitution as you and I regarded it. An Alien influence has prevailed."[4]

Both these cases involved state laws; not until 1935 did a federal statute reach the Court. In January, 1935, the administration paid the price for its sloppy procedures in delegating powers and issuing edicts when the Court, despite a vigorous dissent by Cardozo, invalidated the "hot oil" provisions of the NIRA.[5] In February, the administration had a close call. The Court found the government's repudiation of the gold clause in government bonds both unconstitutional and immoral, yet ruled, 5–4, that bondholders were not entitled to sue for reimbursement. "As for the Constitution, it does not seem too much to say that it is gone," fumed Justice McReynolds, one of the dissenters. "Shame and humiliation are upon us now."[6] The majority, fearful that a decision in favor of the plaintiffs would have meant financial chaos, was vexed at having been coerced into upholding legislation it thought invalid and unjust. Commentators predicted the Court would seize the first opportunity to right the scales.[7] Eleven weeks later, Justice Roberts sent a chill through New Dealers when he joined the conservatives to find the Railroad Pension Act unconstitutional in yet another 5–4 decision.[8] The future of the New Deal appeared to rest on the mind

---

Jane Perry Clark, "Emergencies and the Law," *Political Science Quarterly,* XLIX (1936), 268–283; George Sutherland to W. D. Anderson, September 21, 1935, Sutherland MSS., Box 5.

[3] Nebbia *v.* New York, 291 U.S. 523 (1934).

[4] Morton Keller, *In Defense of Yesterday* (New York, 1958), p. 254.

[5] Panama Refining Co. *v.* Ryan, 293 U.S. 388. As a result of well-founded criticism of administrator-made law, the government launched the *Federal Register,* which printed federal orders with the force of law.

[6] Perry *v.* United States, 294 U.S. 330, and other cases. Cf. Carl Brent Swisher (ed.), *Selected Papers of Homer Cummings* (New York, 1939), pp. 106–121; items in James McReynolds MSS., Folder H-2, Charles Evans Hughes MSS., Box 157; John Dawson, "The Gold-Clause Decisions," *Michigan Law Review,* XXXIII (1935), 647–684.

[7] T.R.B., "Washington Notes," *New Republic,* LXXXII (1935), 100; *Time,* XXV (February 25, 1935), 11.

[8] R.R. Retirement Board *v.* Alton R.R. Co., 295 U.S. 330.

of Mr. Justice Roberts, and many feared he was permanently lost.

On Black Monday, May 27, 1935, the guillotine fell. In a unanimous 9–0 decision in the "sick chicken" case, the Court struck down the National Industrial Recovery Act. In 1934, the Schechter brothers, Brooklyn poultry jobbers, had been convicted of violating the NRA's Live Poultry Code on several counts, including selling diseased chickens and violating the code's wage and hour stipulations. On two separate grounds, the Court questioned the constitutional basis for the prosecution of the Schechters. The NIRA's sweeping delegation of legislative power, declared Justice Cardozo, was "delegation running riot." Furthermore, the Court found the Schechters' business was in intrastate commerce; hence the federal government had no authority to regulate working conditions in the firm.[9] In London, the *Daily Express* headlined: "AMERICA STUNNED; ROOSEVELT'S TWO YEARS' WORK KILLED IN TWENTY MINUTES." Not only had the Court destroyed Roosevelt's industrial recovery program but, by its narrow interpretation of the commerce clause, it threatened the remainder of the New Deal.[10]

The Schechter decision dumfounded Roosevelt. "Well, where was Ben Cardozo?" the President asked. "And what about old Isaiah?"[11] Roosevelt was grieved to learn that both Cardozo and Brandeis had voted with the conservative members of the bench. At an unusual press conference, he unburdened himself to reporters. He called the implications of the Schechter ruling more important than any since the Dred Scott case, and protested that the nation had been "relegated to the horse-and-buggy definition of interstate commerce." Many commentators claimed Roosevelt was relieved by the decision, for it freed

[9] A. L. A. Schechter Poultry Corp. et al. *v.* United States, 295 U.S. 553. Cf. Robert Stern, "The Commerce Clause and the National Economy, 1933–1946," *Harvard Law Review,* LIX (1946), 659–664; E. S. Corwin, "The Schechter Case—Landmark or What?" *New York University Law Quarterly Review,* XIII (1936), 151–190.

[10] *Literary Digest,* CXIX (June 8, 1935), 12; Nicolas Molodovsky, "Grandeur et décadence du New-Deal: II.—La mort de l'Aigle bleu," *L'Europe Nouvelle,* XVIII (1935), 714–718. That same day, the Court found the Frazier-Lemke mortgage act invalid, and ruled, to Roosevelt's intense annoyance, that he could not remove members of independent regulatory commissions save as Congress provided. Louisville Joint Stock Land Bank *v.* Radford, 295 U.S. 555; Humphrey's Executor *v.* U.S., 295 U.S. 602. Cf. Eugene Gerhart, *America's Advocate: Robert H. Jackson* (Indianapolis, 1958), p. 99.

[11] Gerhart, *Jackson,* p. 99. Brandeis' friends commonly referred to him as "Isaiah."

him of an albatross. In fact, the President believed deeply in the NRA approach and never gave up trying to restore it.[12]

The Schechter verdict climaxed a disastrous five months for Roosevelt. Although the President had apparently won a smashing victory in the November, 1934, elections, the new Congress had passed only a single piece of legislation—the work relief bill—and had taken the reins from him in foreign affairs. By March, government was at a standstill. "Once more," wrote Walter Lippmann, "we have come to a period of discouragement after a few months of buoyant hope. Pollyanna is silenced and Cassandra is doing all the talking." Roosevelt seemed uncertain where he was headed, barely able to manage an unruly Congress, and downright contrary in his refusal to indicate his main line of direction. The historian Arthur Schlesinger wrote him plaintively: "I find it more and more difficult to stand by you and your program because I know less and less about what is going on." In 1934, H. G. Wells had been thrilled by Roosevelt's leadership and by the possibilities of a "more rational social order." In 1935, Wells felt that the Coughlins and Longs had taken the play away from him, and the President seemed to be marking time.[13]

Despite the radical character of the 1934 elections, Roosevelt was still striving to hold together a coalition of all interests, and, despite rebuffs from businessmen and the conservative press, he was still seeking earnestly to hold business support. Some observers believed he hoped to ward off the threat of the Longs, Coughlins, and Townsends in 1936 not by reform measures but by encouraging new investment so that he could face the country in the next campaign with a record of having restored prosperity. In a period of a few weeks, he sided with

[12] Samuel Rosenman (ed.), *The Public Papers and Addresses of Franklin D. Roosevelt* (13 vols., 1938–50), IV, 205. Cf. F.D.R. to Hugo Black, June 3, 1935, FDRL OF 372; Frances Perkins, COHC, VII, 54 ff.; Donald Richberg memorandum to F.D.R., May 1, 1935, FDRL PPF 466.

[13] New York *Herald-Tribune*, March 9, 1935; Schlesinger to F.D.R., May 8, 1935, FDRL PPF 2501; H. G. Wells, *The New America: The New World* (London, 1935), pp. 35–51. Another British writer observed: "The first half of 1935 has brought, in relation to Mr. Roosevelt and his personal standing, a revulsion in the public feeling such as could hardly have been anticipated at the time of the election last November." S. K. Ratcliffe, "The New American Demagogues," *Fortnightly*, n.s. CXXXVII (1935), 674–675. Cf. Key Pittman to F.D.R., February 19, 1935, FDRL PPF 745; F.D.R. to E.M. House, February 16, 1935, FDRL PPF 222.

employers in key labor disputes; supported the ouster of agrarian reformers from the AAA; backed the leader of private housing interests in a dispute over public housing; and opposed the "prevailing wage" amendment to the work relief bill.

He seemed to fear the conservatives less than the spenders and inflationists. When he broke precedent to go personally up to Congress to deliver a vigorous veto of the inflationary Patman bonus bill, he sounded for all the world like Grover Cleveland venting his indignation at the Silver Purchase Act. Roosevelt's actions left his progressive followers in a quandary. They could not countenance his conservative posture, yet they feared to break with him because they anticipated that at any moment he might take charge of the reform movement as he had in the past. The liberal publisher J. David Stern wrote the White House: "I don't want to make the same mistake that was made by Horace Greeley, who turned on President Lincoln during the second year of his Administration, because Greeley did not think he was liberal enough,—and look what history did to Lincoln and to Greeley."[14]

It was less the dismay of Roosevelt's progressive supporters than business' own actions that led him to question the viability of the all-class alliance. At its convention in early May, the U.S. Chamber of Commerce broke with Roosevelt and denounced the New Deal. The President, wounded by the attacks, commented: "Of course, the interesting thing to me is that in all of these speeches made, I don't believe there was a single speech which took the human side, the old-age side, the unemployment side."[15] Senate progressives, led by La Follette, seized the opportunity to instruct Roosevelt on the need for a new departure. At a White House conference on the night of May 14, they told the President bluntly that the time had come for him to assert leadership. Early in May, Frankfurter had written Brandeis: "If only Business could become still more articulate in its true feelings toward F.D.R. so that even his genial habits would see the futility of

---

[14] J. David Stern to Marvin McIntyre, March 7, 1935, FDRL PPF 1039; T.R.B., "Washington Notes," *New Republic,* LXXXII (1935), 158; *Time,* XXV (February 18, 1935), 14.

[15] *Public Papers,* IV, 162. Cf. William Wilson, "How the Chamber of Commerce Viewed the NRA: A Re-examination," *Mid-America,* XLIV (1962), 95–108.

hoping anything from Business in '36, so that he would act more 'irrepressible conflict.' " Now Justice Brandeis himself sent the President word that it was "the eleventh hour."[16]

As Roosevelt weighed the advice of the progressives, the Court handed down the Schechter decision. Some of the New Dealers hoped something could be salvaged. One NRA official queried: "If the court said we couldn't do this with a dead chicken, what else could we do?"[17] Such counsel was anathema to the Brandeisians. For three decades Louis Brandeis, the seventy-eight-year-old Supreme Court justice, had argued that big business was economically inefficient and socially dangerous. He repeated in 1935 what he had told Woodrow Wilson many years before: "We must come back to the little unit." He urged a return to a more competitive society, not to the anarchy of unrestricted profiteering, but to "regulated competition." He challenged the contention of the Tugwells that the concentration of economic power was both inevitable and beneficial, and rejected their faith in centralized planning. All men were limited, Brandeis observed, and none could stand up under the burdens of excessive power. Progress would come not from the direct assault of the planners but slowly through institutions which recognized human frailty. "My dear," he once told his daughter Susan, "if you will just start with the idea that this is a hard world, it will all be much simpler."[18]

Brandeis' ideas made their way in Washington under the aegis of his chief disciple, Felix Frankfurter, and through the young men Frankfurter had sent down to be law clerks to the Justice or to staff New Deal agencies. Of all of Brandeis' young men, the most successful were two of Frankfurter's prodigies, Ben Cohen and Tom Corcoran.

[16] Felix Frankfurter to Louis Brandeis, May 6, 1935; Frankfurter to Brandeis, May 21, 24, 1935, Brandeis MSS., G9; Harold Ickes, *The Secret Diary of Harold Ickes* (3 vols., New York, 1954), I, 363. Brandeis' views were relayed to the President by mutual friends like Frankfurter and Norman Hapgood.

[17] Frances Perkins, COHC, VII, 54 ff.

[18] Alpheus Mason, *Brandeis: A Free Man's Life* (New York, 1946), pp. 621, 642; Arthur M. Schlesinger, Jr., *The Politics of Upheaval* (Boston, 1960), p. 220; Charles Sumner Hamlin MS. Diary, June 5, 1935. Although wary of governmental power, Brandeis would tolerate governmental intervention so long as it was carried on by the states. In addition, the Brandeisians favored enhanced power for labor, not only as a matter of justice, but to balance the power of corporations. Hence, although opposed to the leviathan state, they approved the Wagner bill, which would strengthen labor's power. Frankfurter to Brandeis, March 15, 1935, Brandeis MSS., G9.

The shy, modest Cohen, a Jew from Muncie, Indiana, was a former Brandeis law clerk who had developed into a brilliant legislative draftsman. Corcoran, an Irishman from a Rhode Island mill town, had gone from Harvard Law School to Washington in 1926 to serve as secretary to Mr. Justice Holmes; Holmes found him "quite noisy, quite satisfactory, and quite noisy." Roosevelt had discovered Corcoran and Cohen, who had teamed up to draft Wall Street regulatory legislation, to be remarkably resourceful in resolving knotty governmental problems, and by the spring of 1935 "the boys," as Frankfurter called them, were playing key roles in the New Deal.[19]

Corcoran was a new political type: the expert who not only drafted legislation but maneuvered it through the treacherous corridors of Capitol Hill. Two Washington reporters wrote of him: "He could play the accordion, sing any song you cared to mention, read Aeschylus in the original, quote Dante and Montaigne by the yard, tell an excellent story, write a great bill like the Securities Exchange Act, prepare a presidential speech, tread the labyrinthine maze of palace politics or chart the future course of a democracy with equal ease." He lived with Cohen and five other New Dealers in a house on R Street; as early as the spring of 1934, G.O.P. congressmen were learning to ignore the sponsors of New Deal legislation and level their attacks at "the scarlet-fever boys from the little red house in Georgetown."[20]

In the first two years of the New Deal, the Brandeisians chafed as the NRA advocates sat at the President's right hand. They rejoiced only in the TVA, since it both checked monopoly and accentuated decentralization, and in the regulation of securities issues and the stock market, which marked the success of Brandeis' earlier crusade against the money power. Not until 1935 did the Brandeisians make their way. That year saw the triumph of the decentralizers in the fight over the social security bill, and the enhanced power in TVA of Frankfurter's follower, David Lilienthal. Brandeis himself had a direct hand in the most important victory of all: the invalidation of the NRA.

As Roosevelt surveyed the wreckage left by the Schechter decision, and pondered which fork in the road to take next, Felix Frankfurter

[19] Schlesinger, *Politics of Upheaval,* p. 225; Felix Frankfurter to Louis Brandeis, July 9, 10, 1935, Brandeis MSS., G9.

[20] Joseph Alsop and Turner Catledge, *The 168 Days* (Garden City, N.Y., 1938), p. 82; *Congressional Record,* 73d Cong., 2d Sess., p. 7943. Cf. Max Stern, "The Little Red House," *Today,* II (May 19, 1934), 5.

was always by his side, sometimes arguing alone with the President before turning in for the night as a guest at the White House. The Harvard professor insisted that the attempt at business-government co-operation had failed, and urged Roosevelt to declare war on business. Once the President understood that business was the enemy, he would be free to undertake the Brandeisian program to cut the giants down to size: by dwarfing the power of holding companies, by launching antitrust suits, and by taxing large corporations more stiffly than small business.[21]

By Black Monday, the President had already begun to move decisively in a new direction; the Schechter verdict served only as the final goad. In early June, with Congress prepared to adjourn after putting the finishing touches on an unhappy half-year, Roosevelt suddenly galvanized the administration into action. At a time when congressmen contemplated heading home to escape the hot, steamy capital, Roosevelt insisted on the passage of four major pieces of legislation: the social security bill, the Wagner labor proposal, a banking bill, and a public-utility holding company measure. A few days later, he added a fifth item of "must" legislation: a "soak the rich" tax scheme. In addition, he demanded a series of minor measures, some of them highly controversial, which in any other session would have been regarded as major legislation. After months of temporizing, the President displayed a burst of fresh energy. He summoned House leaders to a White House conference and, thumping his desk for emphasis, told them Congress must pass his entire program before it could go home. Thus began the "Second Hundred Days." Over a long torrid Washington summer, Congress debated the most far-reaching reform measures it had ever considered. In the end, Roosevelt got every item of significant legislation he desired.[22]

The first beneficiaries of Roosevelt's change of direction were the supporters of Senator Wagner's labor relations measure. In February, 1935, Wagner had once again introduced his bill, once more without the support of the President or of Johnson, Richberg, or Secretary Perkins.[23] Wagner's bill proposed to set up a National Labor Relations

[21] Raymond Moley, *After Seven Years* (New York, 1939), pp. 306–313; Frankfurter to Louis Brandeis, August 31, 1935, Brandeis MSS., G9.

[22] Memorandum, June 4, 1935, FDRL PSF 26; *The New York Times,* June 14, 1935; "Mr. Roosevelt Drives Ahead," *New Statesman and Nation,* IX (1935), 916–917.

[23] Roosevelt, Miss Perkins recalls, "never lifted a finger" for the Wagner bill.

Board as a permanent independent agency empowered not only to conduct elections to determine the appropriate bargaining units and agents but to restrain business from committing "unfair labor practices" such as discharging workers for union membership or fostering employer-dominated company unions. On May 2, the Senate Labor Committee reported out the bill virtually unchanged, save that, at the urging of Francis Biddle, it made it an unfair practice for employers to refuse to bargain. When conservative senators tried once more to block consideration, Wagner got the President to agree not to intervene this time until the Senate got a chance to vote. Despite conservative protests that the Wagner bill would spawn conflict, foster tyranny by unions, and "re-establish medievalism in industry," the Senate voted down the crippling Tydings amendment framed by the National Association of Manufacturers and went on to pass the bill by the thumping vote of 63–12.[24]

Roosevelt, who still disapproved of the Wagner bill, now faced a situation where he had to act. The Senate had passed the measure, and on May 20 it was reported out by a House committee. After calling in Wagner to work out certain differences with members of the administration, the President abruptly announced, for reasons that are not wholly clear, that he not only favored the Wagner bill but regarded it as "must" legislation. Once Roosevelt gave his blessing, the measure had clear sailing. On June 27, the bill won final congressional approval, and on July 5 the President signed the National Labor Relations Act.

The Wagner Act was one of the most drastic legislative innovations of the decade. It threw the weight of government behind the right of labor to bargain collectively, and compelled employers to accede peacefully to the unionization of their plants. It imposed no reciprocal obligations of any kind on unions. No one, then or later, fully understood why Congress passed so radical a law with so little opposition and by such overwhelming margins. A bill which lacked the support of the administration until the very end, and which could expect sturdy con-

"Certainly I never lifted a finger. . . . I, myself, had very little sympathy with the bill." Frances Perkins, COHC, VII, 138, 147. Cf. Leon Keyserling, "The Wagner Act: Its Origin and Current Significance," *George Washington Law Review*, XXIX (1960), 200–208.

[24] Irving Bernstein, *The New Deal Collective Bargaining Policy* (Berkeley and Los Angeles, 1950), pp. 89–128; William Mangold, "On the Labor Front," *New Republic*, LXXXI (1935), 333–334.

servative opposition, it moved through Congress with the greatest of ease. One observer later wrote: "We who believed in the Act were dizzy with watching a 200-to-1 shot come up from the outside."[25]

On June 19, Roosevelt jolted conservatives by firing at Congress a radical tax message which aimed to redistribute wealth and power. The President's message reflected a variety of influences and motivations: the hectoring of Treasury advisers; the importuning of Felix Frankfurter, who incorporated Brandeisian disapproval of business concentration in a tax memorandum; perhaps, too, vindictive resentment at business criticism and a desire to "steal Huey's thunder." Roosevelt urged increased taxes on inheritances, the imposition of gift taxes, graduated levies on "very great individual net incomes," and a corporation income tax scaled according to the size of corporation income. Leaving a stunned, weary, and irritated Congress behind him, the President blithely departed for the Yale-Harvard boat races.[26]

Roosevelt's tax proposal, the first which reached directly into the pockets of the wealthy, raised an outcry from business and the press such as had greeted none of the President's previous recommendations. William Randolph Hearst directed his editors to use the phrase "Soak the Successful" henceforth in all references to the tax measure and the words "Raw Deal" instead of "New Deal." The Philadelphia *Inquirer* charged the President with "a bald political stroke . . . to lure hosannas from the something-for-nothing followers of Huey Long, 'Doc' Townsend, Upton Sinclair and the whole tribe of false prophets." As the tax message was being read, Long swaggered through the Senate chamber, chortling and pointing to his chest. As the reading ended, Long piped up: "I just wish to say 'Amen.' "[27]

[25] Malcolm Ross, *Death of a Yale Man* (New York, 1939), p. 170; Bernstein, *Collective Bargaining,* pp. 118–128; Eugene Grace to Robert Wagner, May 2, 1935, Wagner MSS. On the motivations of Congress, see Bernstein, *Collective Bargaining,* p. 116; *Literary Digest,* CXX (July 6, 1935), 31; Robert Doughton to Bernhardt Furniture Company, June 8, 1934, Doughton MSS., Drawer 6.

[26] John Blum, *From the Morgenthau Diaries* (Boston, 1959), pp. 298–301; Moley, *After Seven Years,* pp. 308–313; Robert Doughton to Thurmond Chatham, June 27, 1935, Doughton MSS., Drawer 7; Felix Frankfurter to Louis Brandeis, June 14, 1935, Brandeis MSS., G9; *Public Papers,* IV, 270–274; James Burns, *Roosevelt: The Lion and the Fox* (New York, 1956), p. 224.

[27] E. D. Coblentz to editors, August 7, 1935, FDRL PPF 62; *Literary Digest,* CXX (July 6, 1935), 4; *Congressional Record,* 74th Cong., 1st Sess., p. 9659; *The New York Times,* June 20, 1935. Within a short time, Long was finding fault with the President's tax program.

Roosevelt's decision to dispatch his tax message gave him a sense of buoyancy he had not had in weeks. He felt top man once more. He told Moley gleefully: "Pat Harrison's going to be so surprised he'll have kittens on the spot."[28] Yet, despite his vigorous message, it was not clear whether Roosevelt wished Congress to act at that session. Reports circulated that he would be satisfied merely with a nominal inheritance tax, and he seemed wary of pitting himself against the conservative Harrison, chairman of the Senate Finance Committee. It took a bold move by Senator La Follette to turn the tide. When he and twenty other senators signed a round robin announcing they would keep Congress in session until it approved the President's program, Roosevelt and Democratic congressional leaders responded by declaring they sought immediate action.[29] Still the President seemed remarkably reluctant to take the lead. When Morgenthau pressed him at lunch on whether he really wanted an inheritance levy passed at that session, Roosevelt replied: "Strictly between the two of us, I do not know. I am on an hourly basis and the situation changes almost momentarily."[30]

It took most of the summer to get the measure through Congress. As late as mid-July, Roosevelt indicated he would not fight a move to adjourn without enacting his tax proposals.[31] Without White House guidance, Senator La Follette and the levelers fought a losing battle to save the bill from the assaults of irate businessmen and from congressmen who denounced it as class legislation.[32] Congress eliminated

[28] Moley, *After Seven Years*, p. 310.

[29] La Follette wanted action on Roosevelt's tax bill, because he thought taxation should be a social weapon, but even more because he wanted to reduce the national debt. Famed as the leading advocate of spending in the Senate, he sternly opposed deficit finance. Schlesinger, *Politics of Upheaval*, p. 330; Roy and Gladys Blakey, *The Federal Income Tax* (New York, 1940), pp. 369–371.

[30] Blum, *Morgenthau Diaries*, p. 303. An Oklahoma congressman wrote: "I don't know just what the hell is actually happening up here, but it seems to me that the 'GREAT WHITE FATHER' has either gone to bed with Huey Long or stayed out with the girls too long. I'll be damned if I know just what's going to happen next. I have had to change my public speeches six times in the last three days to try to keep up with the administration." P. L. Gassaway to W. M. Gulager, June 26, 1935, Gassaway MSS., Correspondence, 1935.

[31] *The New York Times*, July 16, 1935; Blum, *Morgenthau Diaries*, pp. 303–304.

[32] The Wilsonian Democrat Senator William McAdoo commented on the bill: "Personally, I dislike class discrimination and any other kind of discrimination in our social and economic order, because fundamentally justice is the thing

the inheritance tax and reduced the graduated corporation income levy to no more than symbolic importance. In its final form, the Wealth Tax Act of 1935 destroyed most of the Brandeisian distinction between big and small business, did little to redistribute wealth, and did still less to raise revenue. Yet since the act stepped up estate, gift, and capital stock taxes, levied an excess profits tax, which Roosevelt had not asked, and increased the surtax to the highest rates in history, it created deeper business resentment than any other New Deal measure.[33] The outcry from high-income brackets obscured the fact that much of Roosevelt's tax program was sharply regressive. His insistence on payroll levies to help finance social security cut into low-income groups, and his emphasis on local responsibility for unemployables helped stimulate the spread of the regressive sales tax. The share of upper-income groups remained fairly constant through the thirties, and the share of the top 1 per cent even increased a bit after the passage of the Wealth Tax Act. Not until the war years was there a substantial change in the distribution of income.[34]

Even before the Second Hundred Days, the President had committed himself to a program the Brandeisians cherished still more than wealth taxes: the breakup of holding companies. In his message to Congress in March, Roosevelt had employed the Brandeisian arguments against bigness, and had even used the phrase "other people's money."[35] Roosevelt required no proselytization from the Brandeisians on the evils of holding companies. He shared the popular outrage at the electric power octopuses which had fleeced the consumer, corrupted legislatures, and, by their elusive operations, evaded state regulation.[36] Indeed, he favored even more punitive measures than did the Brandeisians. He wished to destroy the combines, while they desired severe reform rather than extinction. Roosevelt was persuaded to adopt the more moderate course, but he insisted that Cohen and

---

which should direct all policies of government." McAdoo to Colonel Robert Elbert, August 14, 1935, McAdoo MSS., Box 409.

[33] Schlesinger, *Politics of Upheaval*, pp. 331–334. Cf. Sidney Ratner, *American Taxation* (New York, 1942), pp. 468–472.

[34] Simon Kuznets, *Shares of Upper Groups in Income and Savings*, National Bureau of Economic Research, Occasional Paper 35 (New York, 1950).

[35] *Public Papers*, IV, 101.

[36] Frankfurter informed Brandeis early in 1935: "F.D. is really hot on holding cos. & for drastic action." Frankfurter to Brandeis, January 22, 1935, Brandeis MSS., G9.

Corcoran write a bill that breathed fire. They responded with a draft which included the so-called "death sentence" provision, which would empower the SEC after January 1, 1940 to dissolve any utility holding company which could not justify its existence.[37]

The utility companies mounted their heaviest artillery against the holding-company bill. Lobbyists subjected congressmen to greater pressure than on any measure that had been before the House in years; one correspondent estimated there were more utility lobbyists than congressmen. Power companies organized letter-writing campaigns to inundate Congress with protests against the bill. "Millions of letters are being carted into Washington," Senator Murray reported.[38] Under all this hammering, Roosevelt's following in Congress buckled. In June, the Senate approved the death sentence, but by only a single vote. The next month, the House handed Roosevelt a sharp setback when it turned down the death sentence, 216–146. While Democrats in the public-power-conscious Pacific Northwest voted 8–1 for the death sentence, east of the Mississippi, Democrats had deserted the President in droves. Sixteen of twenty-seven New York Democrats broke party lines. The House went on to approve the rest of the holding-company bill, 323–81.[39]

So much attention focused on the rebellion of the House Democrats that the significant point was lost. There was really little difference between the two bills; the House version was still a stringent piece of legislation. The House bill shifted the burden of proof; instead of requiring holding companies to justify their existence, the commission would have to defend an order of dissolution. The Senate bill would make the death sentence mandatory, while the House measure would make it discretionary. The House bill, explained the Boston *Globe,* "substituted only a chance for life imprisonment in the place of capital punishment."[40]

[37] Schlesinger, *Politics of Upheaval,* pp. 302–306; Moley, *After Seven Years,* p. 303; "SEC," *Fortune,* XXI (June, 1940), 123; Philadelphia *Record,* March 8, 1935.

[38] James Murray to Charles H. Moore, April 6, 1935, Murray MSS.; J. Harry LaBrum, "Power Has No Need for Lobbies," *Public Utilities Fortnightly,* XVI (1935), 20–25.

[39] Washington *Daily News,* July 2, 1935; George Cortelyou to Herbert Claiborne Pell, March 19, 1935, Pell MSS., Box 12; John W. Davis to Newton Baker, June 11, 1935, Baker MSS., Box 84.

[40] *Literary Digest,* CXX (July 13, 1935), 3–4, 11–13, 28, 36; "Bills in Congress, 1933–37," FDRL PSF 28.

Yet the defeat was a slap in the face to administration leaders in Congress, and neither they nor the President would give up the fight for the mandatory death sentence. Indignant at the utilities' campaign of "White Rock and propaganda," and suspicious of the character of the hundreds of thousands of telegrams and messages which were bombarding congressmen, both houses launched investigations of lobbying. The probes revealed that power companies had spent well over a million dollars fighting the bill. A Senate investigating committee headed by Hugo Black discovered that the Hopson interests in Pennsylvania had forged names from the city directory to wires to congressmen. Shortly after a Hopson man had told the Western Union office manager in Warren, Pennsylvania, that it would be a good idea if "somebody threw a barrel of kerosene in the cellar" of the telegraph office, the manager found the charred remains of the messages in a basement stove. Black's inquiry also insinuated that a utility executive had bribed a Texas congressman.[41]

Despite Black's findings and new White House overtures, on August 1 the House once more voted down the death sentence by a decisive margin. Defeated in his attempt to gain a more stringent law, Roosevelt now had to compromise. Yet he still won the substance of what he desired. The final version of the Wheeler-Rayburn bill wiped out all utility holding companies more than twice removed from the operating companies and empowered the SEC, with whom all the combines were compelled to register, to eliminate companies beyond the first degree that were not in the public interest. Most of the great utility empires were to be broken up within three years. In addition, the law authorized the SEC to supervise the financial transactions of the companies.

The Public Utilities Holding Company Act was a bold stroke against bigness, and the Brandeisians were delighted. "If F.D. carries through the Holding Company bill we shall have achieved considerable toward curbing Bigness," Brandeis wrote a friend in June. Two months later, he noted: "F.D. is making a gallant fight, and seems to appreciate fully the evils of bigness. He should have more support than his party is giving him; and the social worker-progressive crowd seems as blind

[41] John Frank, *Mr. Justice Black* (New York, 1949), pp. 78–79; Thomas Stokes, *Chip Off My Shoulder* (Princeton, 1940), p. 342; Charlotte Williams, *Hugo L. Black* (Baltimore, 1950), pp. 58–65; Arthur Capper to Henry I. Allen, July 19, 1935, Capper MSS.

as in 1912."[42] With the passage of the Holding Company Act, the Justice came into his own. The Wheeler-Rayburn law marked the most important triumph for the Brandeisian viewpoint in two decades.

Public-power enthusiasts viewed the Holding Company Act as only one more indication that Roosevelt was with them. "The President," exulted Morris Cooke, "is as you know true blue about power."[43] On the Columbia River, the Corps of Engineers was erecting mighty Bonneville. In eastern Washington, the fantastic Grand Coulee project, the greatest man-made structure in the world, would back up water into a lake 150 miles long, generate power to speed the industrialization of the Pacific Northwest, and make possible the reclamation of more than a million acres. Two thousand miles up the Missouri at Fort Peck, workmen living in shanty towns named Wheeler, New Deal, and Delano Heights were throwing up the world's largest earthen dam. Under the vigorous leadership of Director David Lilienthal, the Tennessee Valley Authority was bringing electricity to a valley where, in 1932, only one Mississippi farm out of a hundred had power. "We are working toward no less a goal than the electrification of America," Lilienthal declared.[44]

Perhaps no single act of the Roosevelt years changed more directly the way people lived than the President's creation of the Rural Electrification Administration in May, 1935.[45] Nine out of ten American farms had no electricity. The lack of electric power divided the United States into two nations: the city dwellers and the country folk. "Every city 'white way' ends abruptly at the city limits," wrote one public-power advocate. "Beyond lies darkness."[46] Farmers, without the benefits of electrically powered machinery, toiled in a nineteenth-century world; farm wives, who enviously eyed pictures in the *Saturday Evening Post* of city women with washing machines, refrigerators, and vacuum cleaners, performed their backbreaking chores like peasant

[42] Norman Hapgood to F.D.R., June 16, 1935, FDRL PPF 2278; Mason, *Brandeis,* p. 622.

[43] Morris Cooke to J. D. Ross, January 16, 1935, Ross MSS.

[44] Ernest Abrams, *Power in Transition* (New York, 1940); Charles McKinley, *Uncle Sam in the Pacific Northwest* (Berkeley and Los Angeles, 1952); David Lilienthal, "T.V.A. Seen Only as Spur to Electrification of America," *Electrical World,* CII (1933), 687–690.

[45] Morris Cooke to F.D.R., April 10, 1935, Cooke MSS., Box 51; Cooke to Harry Slattery, November 29, 1939, Cooke MSS., Box 147.

[46] "Notes on REA," penciled memorandum, Judson King MSS., Box 15.

women in a preindustrial age. Under Morris Llewellyn Cooke, a veteran of the public power fight in Pennsylvania, the REA revolutionized rural life. When private power companies refused to build power lines, even when offered low-cost government loans, Cooke sponsored the creation of nonprofit co-operatives. In the next few years, farmers voted, by the light of kerosene lamps, to borrow hundreds of thousands, even millions, from the government to string power lines into the countryside. Finally, the great moment would come: farmers, their wives and children, would gather at night on a hillside in the Great Smokies, in a field in the Upper Michigan peninsula, on a slope of the Continental Divide, and, when the switch was pulled on a giant generator, see their homes, their barns, their schools, their churches, burst forth in dazzling light. Many of them would be seeing electric light for the first time in their lives. By 1941, four out of ten American farms had electricity; by 1950, nine out of ten.[47]

Roosevelt played a rather indifferent role in the adoption of the last of the major achievements of the Second Hundred Days: the Banking Act of 1935. Yet, as in the passage of the Wagner Act, it was the President who had given both a position of power and free rein to the sponsors of the reform. In November, 1934, Roosevelt had named as Governor of the Federal Reserve Board an unorthodox Utah banker, Marriner Eccles, who had been a Mormon missionary in Scotland, had presided over two Ogden banks at thirty, and had advanced "Keynesian" views at a time when neither he nor most other people had heard of Keynes. Eccles had hardly taken office when he helped draft a new banking bill which called for the first radical revision of the Federal Reserve System since its adoption in 1913. Eccles wished to lodge control of the system in the White House; lessen the influence of private bankers, who he believed had taken over the system; and use the Reserve Board as an agency for conscious control of the monetary mechanism. The 20,000-word banking bill introduced in February, 1935, reflected the thinking of Eccles and of certain members of his staff, especially the Keynesian Lauchlin Currie. Since conserv-

---

[47] Morris Cooke, "Early Days of Rural Electrification," *American Political Science Review,* XLII (1948), 431–447; *Time,* XXXIII (July 4, 1938), 12; Marquis Childs, *The Farmer Takes a Hand* (Garden City, N.Y., 1952) Originally a relief measure, the REA received independent status as a loan agency under the Norris-Rayburn Act of 1936. The REA made especially rapid gains under John Carmody, who succeeded Cooke as Administrator early in 1937.

atives within the administration, notably J. F. T. O'Connor, the Comptroller of the Currency, and Leo Crowley, head of the Federal Deposit Insurance Corporation, opposed Eccles' proposals, they were sandwiched between two other titles which would gain conservative backing by protecting the Comptroller from loss of authority and by easing requirements for bank membership in the FDIC.[48]

The House whipped through the Eccles bill in early May virtually unchanged. In the Senate, it encountered the frail but determined figure of Carter Glass. A states' rights Virginian who was alarmed by Eccles' desire to centralize banking control in the federal government, he had cast the lone nay against Eccles' appointment. He viewed men like Eccles as interlopers, since he felt he, as the founder of the Federal Reserve System—as, with no little exaggeration, he regarded himself —alone had the authority to amend the system. He opposed changes which diminished the power of private bankers, and he worked closely with spokesmen for New York bankers in shaping opposition strategy.[49] The Eccles bill also aroused the strong disapproval of economists, ranging from experimenters like Irving Fisher to conservatives like H. Parker Willis. The sometime Roosevelt adviser James P. Warburg derided the proposal as "Curried Keynes." "It is in fact a large, half-cooked lump of J. Maynard Keynes . . . liberally seasoned with a sauce prepared by Professor Lau[c]hlin Currie," Warburg jeered.[50]

For months, Eccles fought almost alone against Glass and the bankers. Roosevelt would not stamp the measure an administration bill, and such administration officials as Jesse Jones worked actively with the opposition.[51] In June, on the eve of the Second Hundred

[48] "Marriner Stoddard Eccles," *Fortune,* XI (February, 1935), 63; Marriner Eccles, *Beckoning Frontiers* (New York, 1951), pp. 166–176, and "The Federal Reserve—1935 Model," *Magazine of Wall Street,* LV (1935), 666–667, 696–697; Eccles to Edward Costigan, May 27, 1935, Costigan MSS., V.F. 3; Blum, *Morgenthau Diaries,* p. 346; Rudolph Weissman (ed.), *Economic Balance and a Balanced Budget* (New York, 1940).

[49] Eccles, *Beckoning Frontiers,* pp. 176–178; Blum, *Morgenthau Diaries,* p. 347; Robert Owen to Carter Glass, March 25, 1935, Owen MSS., Box 4; Carter Glass to M. E. Bristow, February 9, 1935, Glass MSS., Box 317.

[50] H. Parker Willis, "The Eccles Bill and After," *Bankers Magazine,* CXXXI (1935), 176–178; U.S. Congress, Senate, *Banking Act of 1935,* Hearings before Subcommittee of the Committee on Banking and Currency, U.S. Senate, 74th Cong., 1st Sess., on S. 1715 and H. R. 7617, Apr. 19—June 3, 1935 (Washington, 1935), p. 74.

[51] Charles Sumner Hamlin MS. Diary, February 5, 1935; Blum, *Morgenthau Diaries,* pp. 346–348.

Days, the President made the bill a "must" item. In July, he warned Glass that H. Parker Willis belonged "to that little group of Americans who are appendages or appendices—whichever part of the anatomy you prefer—on the large body of international bankers of London, Paris, Shanghai, etc. . . . You have had a long period of intimate contact with banking legislation, but I have seen more rotten practices among the banks in New York City than you have."[52] But Eccles staved off defeat less because of Roosevelt's aid than because he had the support of House leaders like Maryland's T. Alan Goldsborough and of dissident financiers like Amadeo Giannini, the California banking titan, who preferred government control to domination "by the New York banking fraternity."[53]

On August 23, after Glass had completely rewritten much of the bill, the Banking Act of 1935 became law. Glass boasted: "We did not leave enough of the Eccles bill with which to light a cigarette." When Roosevelt gave one of the signature pens to Glass, someone remarked in a stage whisper: "He should have given him an eraser instead."[54] Yet, despite Glass's amendments, the law marked a significant shift toward centralization of the banking system and federal control of banking. The act, observed Walter Lippmann, constituted a victory for Eccles "dressed up as a defeat."[55] The law empowered the President to appoint the seven members of the newly named Board of Governors of the Federal Reserve System for fourteen year terms. The new Board could exercise more direct control over the regional banks since their chief officers could be appointed only with the Board's approval. The act gave the Board greater authority over the rediscount rates and reserve requirements of Reserve banks and expanded the definition of eligible paper on which Reserve banks could make loans to member banks, although Glass frustrated Eccles' wish to make these changes more sweeping. Open-market operations, which had previously been diffused among the Reserve banks, were concentrated in a Federal

[52] F.D.R. to Glass, July 6, 1935, Glass MSS., Box 6.

[53] Giannini to Marvin McIntyre, July 1, 1935, FDRL PPF 1135. "The average country banker would be safer if he knows that too much power is not given to the bankers," wrote the president of another California bank. "I for one would prefer to go along with the Government." R. D. McCook to William McAdoo, June 1, 1935, McAdoo MSS., Box 405.

[54] Glass to Frederick Lee, December 1, 1935, Glass MSS., Box 304; Eccles, *Beckoning Frontiers*, p. 229.

[55] Walter Lippmann, *Interpretations, 1933–35* (New York, 1936), p. 194.

Open Market Committee, composed of seven Board members and five representatives of the Reserve banks. This provision shifted authority over open-market operations to the government, although by a precarious margin. Finally, the law aimed to bring all large state banks under the jurisdiction of the Board by requiring them to join the Federal Reserve System before July 1, 1942, if they wished to enjoy the benefits of the federal deposit insurance system. With the passage of the Banking Act of 1935, Roosevelt had completed his program of establishing government control over currency and credit.[56]

Much of the "minor" legislation of the Second Hundred Days came from attempts to salvage something from the NIRA experience. The Schechter decision had made a shambles of the industrial recovery machinery. Newspapers removed the Blue Eagle from their mastheads; secondhand-furniture men called at the offices of code authorities, and *The New York Times* ran a want ad: "FORMER CODE AUTHORITY DIRECTOR, aggressive, capable, seeks executive position. . . ."[57] The NRA had reached into too many corners of the economy to be dismantled so totally, and in a few weeks Congress had begun to enact a "little NRA."

Most spectacular of the "little NRA" legislation was the Guffey-Snyder Act of 1935, which, in effect, re-enacted the old bituminous coal code. By threatening a national coal strike, John L. Lewis' United Mine Workers conspired with unionized northern operators to bully Congress into making coal a public utility subject to federal regulation. The Guffey bill guaranteed collective bargaining, stipulated uniform scales of wages and hours, created a national commission which would fix prices and allocate and control production, authorized closing down marginal mines, and levied a production tax to pay for the mines and to rehabilitate displaced miners. When the Court found the act un-

[56] F. A. Bradford, "The Banking Act of 1935," *American Economic Review*, XXV (1935), 661–672; A. D. Gayer, "The Banking Act of 1935," *Quarterly Journal of Economics*, L (1935), 97–116; Eccles, *Beckoning Frontiers*, pp. 222–228; Schlesinger, *Politics of Upheaval*, pp. 291–301; Rexford Tugwell, *The Democratic Roosevelt* (Garden City, N.Y., 1957), p. 373; Elliott Bell, "The Decline of the Money Barons," in Hanson Baldwin and Shepard Stone (eds.), *We Saw It Happen* (New York, 1938), pp. 135–167; Robert Doughton to J. R. Hix, July 22, 1935, Doughton MSS., Drawer 7. Cf. Amos Pinchot to Felix Frankfurter, July 15, 1935, Donald Richberg MSS., Box 1.

[57] James Boylan, "The Daily Newspaper Business in the National Recovery Administration" (unpublished M.A. essay, Columbia University, 1960); *Time*, XXV (June 17, 1935), 15.

constitutional in 1936, Congress in 1937 passed the Guffey-Vinson Act, which re-enacted the original Guffey law, save for the wages and hours provisions, to which the Court had taken exception.[58]

On August 27, the historic first session of the Seventy-fourth Congress came to an end. It had opened on a wintry day in January, finished on a sultry Tuesday in August, a demanding, wearying session that had tried everyone's good temper. "I was so tired," Roosevelt told Morgenthau, "that I would have enjoyed seeing you cry or would have gotten pleasure out of sticking pins into people and hurting them."[59] But no other session of Congress had ever adopted so much legislation of permanent importance: social security, the Wagner Act, holding company, banking and tax measures, a relief law that gave birth to such remarkable progenies as the Federal Theatre and the NYA, and a host of "minor" pieces of legislation like the Guffey Coal Act.[60]

Confronted by another cornucopia of alphabetic agencies, the country tried to see if it could discern a pattern in what the 1935 Congress had wrought. Both contemporary writers and historians have contended there was such a pattern, a break between the legislation of 1935 and that of 1933 so sharp that the acts of 1935, and the ideology that underlay them, constituted nothing less than a "Second New Deal." But writers have differed sharply on what they meant by that

[58] Joseph Guffey, *Seventy Years on the Red-Fire Wagon* (Lebanon, Pa., 1952), pp. 92 ff.; *New Republic*, LXXXIII (1935), 178; *Time*, XXV (June 10, 1935), 13; Sterling Spero, "Trouble in Coal," *New Republic*, LXXXIII (1935), 212. The 1935 Congress also passed the Connally Act, which prohibited the shipment of "hot oil" in interstate commerce. Subsequently, Congress enacted other legislation based on the NIRA. The Walsh-Healey Act of 1936 stipulated that federal government contracts require the observance of minimum labor standards. The Robinson-Patman Act of 1936 forbade wholesalers or manufacturers to give preferential discounts or rebates to large buyers like chain stores, and the Miller-Tydings Act of 1937 buttressed "fair trade" laws.

[59] Blum, *Morgenthau Diaries*, p. 257.

[60] Among its "minor" accomplishments, Congress reorganized federal power regulation; amended the TVA and AAA statutes; framed new farm mortgage and railroad-retirement laws; adopted the Gold Clause and Federal Register Acts; placed interstate buses and trucks under the Interstate Commerce Commission; diminished the hold of bankers on railway reorganization; approved the Air Mail Act, which stipulated ICC rate control and regulation of labor relations; and enacted the Federal Alcohol Act and the Central Statistical Act. Two "major" pieces of legislation adopted at this session, the Emergency Relief Appropriation Act, approved in the spring of 1935, and the Social Security law, enacted in the Second Hundred Days, are discussed in Chapter 6.

term. Contemporaries and early historians of the New Deal believed that Roosevelt and Congress had moved in 1935 in a radical direction. Tom Stokes wrote: "The messiahs gradually shoved President Roosevelt toward the left."[61] More recent writers have agreed that 1935 marked the birth of a Second New Deal, but they have argued that the change ran in a more conservative direction. While the First New Deal sought to "rebuild America through the reconstruction of economic institutions," the Second, they assert, held the "naïve" conviction "that the classical model of the market was somehow recoverable."[62] The Brandeisians, whose conception of the economy was essentially capitalistic, had conquered the planners, who had urged fundamental structural reforms. "The First New Deal characteristically told business what it must do," Schlesinger writes. "The Second New Deal characteristically told business what it must *not* do."[63]

Both views contain helpful insights, as well as some questionable assumptions, but both share the handicap of exaggerating the extent of the shift from 1933 to 1935. Many of the 1935 measures—social security, utility regulation, progressive taxation—had long been in the works, and it was only a question of time when they would be adopted.[64] Nor can the Second Hundred Days be viewed simply as the triumph of the Brandeis faction. It had won little of substance save the Holding Company Act, and many of the NRA emphases persisted. If Roosevelt found more use for Brandeisian lieutenants in 1935, in part because his earlier advisers were now politically vulnerable, he never wholly adopted the viewpoint either of the Brandeisians or the spenders. As late as 1937, newsmen identified Donald Richberg as the "number-one boy of the White House," and as late as 1938 Roosevelt contemplated reviving, in a revised form, the NRA.[65]

[61] Stokes, *Chip Off My Shoulder,* p. 420; Moley, *After Seven Years,* pp. 291, 313.

[62] Schlesinger, *Politics of Upheaval,* pp. 389, 392. Cf. Leon Keyserling's comments, *ibid.,* pp. 690–692.

[63] *Ibid.,* p. 392. Cf. Basil Rauch, *The History of the New Deal 1933–1938* (New York, 1944); Tugwell, *Roosevelt,* p. 454; Joseph Schumpeter, *Business Cycles* (2 vols., New York, 1939), II, 1047.

[64] Harlan Phillips (ed.), *Felix Frankfurter Reminisces* (New York, 1960), p. 244; Irving Bernstein, *The Lean Years* (Boston, 1960), p. 487; Ickes, *Diary,* I, 163. Ironically, it had been the radical Tugwell who had pressed Roosevelt to postpone the social security bill in 1934.

[65] Christopher Lasch, "Donald Richberg and the Idea of a National Interest" (unpublished M.A. essay, Columbia University, 1955), p. 92.

The President had scant patience with the theological discussions that revolved around him. When someone reported to him a conversation in which Tugwell had once remarked to a Brandeisian, "I do not see why your crowd and ours cannot work together," Roosevelt replied: "I always hate the frame of mind which talks about 'your group' and 'my group' among Liberals. . . . Brandeis is one thousand per cent right in principle," the President added, "but in certain fields there must be a guiding or restraining hand of Government because of the very nature of the specific field."[66] Roosevelt no longer had the same hope of converting businessmen, but he still held the wistful belief that he might, perhaps by showing them he could balance his books, yet win their favor. If the atomizers like Brandeis had not won as much as appeared, the planners like Tugwell could claim still less. At no time had Roosevelt seriously considered the creation of a planned economy, and to represent the events of 1935 as the defeat of the planners is to confuse shadow with substance. A planned economy had never been in the cards. Of the President's close advisers, only Tugwell had collectivist ideas; he had little chance to express them, still less to carry them out, and in 1936 he gracefully resigned.

Only the Tennessee Valley Authority advanced any serious claims to being a planning agency, and these did not bear close scrutiny. The TVA's utopian Director Arthur Morgan insisted that the Authority was "not primarily a dam-building program, a fertilizer job or power-transmission job" but a vast scheme for "a designed and planned social and economic order."[67] Yet his meekness toward the power companies and his archaic interest in craft industry seemed far removed from what most people meant by planning. "That the one planning body so far established by the government should meantime be representing planning as a Davy-Crockett-coonskin-cap retreat from life seems a great shame," noted one critic.[68] Frustrated by Directors Lilienthal and

[66] F.D.R. to Norman Hapgood, February 24, 1936, in Elliott Roosevelt (ed.), *F.D.R.: His Personal Letters, 1928–1945* (2 vols., New York, 1950), I, 561–563.

[67] Arthur Morgan, "Bench-Marks in the Tennessee Valley," *Survey Graphic,* XXIII (1934), 43; George Fort Milton to F.D.R., April 21, 1933, Milton MSS., Box 13.

[68] Jonathan Mitchell, "Utopia—Tennessee Valley Style," *New Republic,* LXXVI (1933), 274; Harry Slattery to Gifford Pinchot, February 7, 1933, Slattery MSS., Letters.

Harcourt Morgan, Arthur Morgan let loose some intemperate attacks on his colleagues that led Roosevelt to dismiss him in 1938.[69] After 1938, the TVA squeamishly avoided even using the word "plan."[70] The TVA was the most spectacularly successful of the New Deal agencies, not only because of its achievements in power and flood control, but because of its pioneering in areas from malaria control to library bookmobiles, from recreational lakes to architectural design. No other agency did so much to alter the mores of the region. Yet TVA never fulfilled itself as an experiment in regional planning; it remained chiefly a corporation to produce and sell power and fertilizer.

Even the most precedent-breaking New Deal projects reflected capitalist thinking and deferred to business sensibilities. Social security was modeled, often irrelevantly, on private insurance systems; relief directors were forbidden to approve projects which interfered with private profit-taking. The HOLC gave no relief to homeowners who were unemployed; had a commercial agency investigate each applicant to determine whether he was a sound "moral risk"; and foreclosed more mortgages than the villain of a thousand melodramas. By the middle of 1937, it had acquired enough properties to house a quarter of a million people; by the spring of 1938, when it was dispossessing workers who could not make payments because they had been thrown out of work in the brutal recession that year, the HOLC had foreclosed mortgages on more than a hundred thousand homes.[71]

Roosevelt's program rested on the assumption that a just society could be secured by imposing a welfare state on a capitalist foundation. Without critically challenging the system of private profit, the New Deal reformers were employing the power of government not only to discipline business but to bolster unionization, pension the elderly, succor the crippled, give relief to the needy, and extend a hand to the forgotten men. The achievements of the Second Hundred Days, wrote

[69] Ickes, *Diary*, II, 337; George Fort Milton to Robert La Follette, Jr., November 4, 1936, Milton MSS., Box 20; Judson King to Charles A. Beard, April 11, 1938, King MSS., Box 18.

[70] Norman Wengert, "TVA—Symbol and Reality," *Journal of Politics*, XIII (1951), 383.

[71] Thurman Arnold, *Symbols of Government* (New Haven, 1935), pp. 111, 121; Lowell Harriss, *History and Policies of the Home Owners' Loan Corporation* (Washington, 1951), pp. 1–3, 23–25, 71–101; Miss Lonigan to Henry Morgenthau, Jr., May 23, 1938, FDRL PSF 26.

William Allen White, were "long past due"; they represented a "belated attempt to bring the American people up to the modern standards of English-speaking countries."[72]

It remained to be seen whether such a conception of reform would suffice. Roosevelt, observed the English writer H. N. Brailsford, was doing what Lloyd George had done between 1906 and 1914, but at a quicker tempo. The British reforms, which rested on the conviction that the profit system was compatible with aid to the underdog, had rendered working-class life less precarious and was one of the important reasons that the depression struck Britain less heavily than America. Yet, Brailsford warned, this was all they had done. "America, with boundless faith, has just adopted the Liberal specifics that ceased among us to arouse our extravagant enthusiasm many a long year ago."[73] This was disturbing criticism, but it left Roosevelt undismayed. Three years later, when the period of New Deal reform had come to a halt, the President told Anne O'Hare McCormick: "In five years I think we have caught up twenty years. If liberal government continues over another ten years we ought to be contemporary somewhere in the late Nineteen Forties."[74]

[72] Emporia *Daily Gazette*, August 26, 1935.
[73] H. N. Brailsford, "In Retrospect," *New Republic*, LXXXII (1935), 121. Cf. Ludovic Naudeau, "Le Rooseveltisme ou la troisième solution," *L'Illustration*, XCCV (1936), 374–375.
[74] Anne O'Hare McCormick, "As He Sees Himself," *The New York Times Magazine*, October 16, 1938, p. 2.

CHAPTER 8

The New Deal at High Tide

W HEN Anne O'Hare McCormick returned to the United States in 1936 after interviewing European leaders whose strained, prematurely aged faces revealed the heavy price they had paid for power, she marveled at the serenity of Franklin Roosevelt. "On none of his predecessors has the office left so few marks as on Mr. Roosevelt. He is a little heavier, a shade grayer; otherwise he looks harder and in better health than on the day of his inauguration. His face is so tanned that his eyes appear lighter, a cool Wedgwood blue; after the four grilling years since the last campaign they are as keen, curious, friendly and impenetrable as ever."[1]

Almost everyone remarked on Roosevelt's inscrutability. Few men claimed they truly knew him. "To describe Roosevelt," reflected William Phillips, Roosevelt's Under Secretary of State, "you would have to describe three or four men for he had at least three or four different personalities. He could turn from one personality to another with such speed that you often never knew where you were or to which personality you were talking. . . ."[2] In depicting Roosevelt, cartoonists frequently used this motif of the multiple image. In one cartoon, he looks into a multisurface mirror at four different images of himself; in others, he smiles back at his own image. Through all runs the conception that a political implication underlay his every move, and that the man

[1] Anne O'Hare McCormick, "Still 'A Little Left of Center.'" *The New York Times Magazine*, June 21, 1936, pp. 2–3.
[2] William Phillips, COHC.

could manipulate the different aspects of his own personality for political advantage.[3] Everything about the President—his wife, his children, his dog Fala, his stamp collection, his fishing trips—seemed to have a public quality.

Roosevelt's public face—grinning, confident, but inscrutable—masked the inner man, and no one knew for certain what the man, in moments of reflection, really felt. Some doubted that those moments of personal communion ever came. He seemed the compleat political man, the public self so merged with the private self that the President could no longer divorce them. His relentless cheerfulness, his boyish hobbies, his simple view of God, his distaste for theoretical speculation, vexed critics who thought him no more than a Rover Boy who had never grown up. Others believed that beyond that gay reserve was an infinitely complex man whose detachment served his political ends, but who had an inner serenity that came, in unequal parts, from his secure childhood, his early acquaintance with grief, for he was to be a cripple the rest of his days, and his private understanding with God.[4]

Both friend and foe marveled at his serenity.[5] Burdened by the weightiest of responsibilities, he remained high-spirited and untroubled. "He was like the fairy-story prince who didn't know how to shudder," Moley has written. "Not even the realization that he was playing nine-pins with the skulls and thighbones of economic orthodoxy seemed to worry him."[6] Roosevelt advised Tugwell: "You'll have to learn that public life takes a lot of sweat; but it doesn't need to worry you. You won't always be right, but you mustn't suffer from being wrong. That's what kills people like us." If a truck driver were doing your job, he told Tugwell, he would probably be right 50 per cent of the time. "But you aren't a truck driver. You've had some preparation. Your

---

[3] William McKinley Moore, "FDR's Image—A Study in Pictorial Symbols" (unpublished Ph.D. dissertation, University of Wisconsin, 1946), pp. 135–136.

[4] James Roosevelt and Sidney Shalett, *Affectionately, F.D.R.* (New York, 1959), pp. 236, 315. James MacGregor Burns has written: "No biographer of Roosevelt, I think, feels that he really understands the man, nor, evidently, does any member of his family." *The New York Times Book Review,* October 18, 1959, p. 10.

[5] "More than any other person I have ever met," testified his White House physician, "he had equanimity, poise, and a serenity of temper that kept him on the most even of keels." Ross McIntire, *White House Physician* (New York, 1946), p. 77.

[6] Raymond Moley, *After Seven Years* (New York, 1939), p. 192.

percentage is bound to be higher."[7] A man with little formal religious attachment, Roosevelt believed that he walked with God, that he was the instrument of the Lord, and that the Lord would care for him in moments of trial.[8]

One of the rare breaks in his composure came at the Democratic convention in Philadelphia in 1936. As the President began his stiff-legged march toward the stage in Franklin Field, he reached out to shake the hand of the white-bearded poet Edwin Markham, was thrown off balance, and sprawled to the ground. White-faced and angry, he snapped: "Clean me up."[9] But most of the time he bore his handicap with astonishing good humor. To board a train, he had to be wheeled up special ramps; to go fishing, or to get to the second story of a meeting hall, he had to be carried in men's arms like a help-less child. To walk, he had to be harnessed in leg braces, and endure the strain of dragging, painfully, a cumbersome dead weight of pounds of steel.[10] Yet he carried it all off with a wonderfully nonchalant air, and often made his incapacity the subject of some seemingly carefree jest. He would roar with laughter and say: "Really, it's as funny as a crutch." Or at the end of a conversation, Roosevelt, who had not been able to walk since 1921, would often remark: "Well, I'm sorry, I have to run now!"[11] Indeed, so vigorous did he seem that most Americans never knew he remained a cripple in a wheelchair.[12] Frequently, in fact, writers gave the impression that Roosevelt had fully conquered his infirmity. One "who had summoned from the depths of character the incredible patience to win his battle for health and make himself

[7] Rexford Tugwell, "The Experimental Roosevelt," *Political Quarterly*, XXI (1950), 266–267.

[8] Arthur Schlesinger, Jr., *The Coming of the New Deal* (Boston, 1959), pp. 586–587; Frances Perkins, COHC, VII, 556–557.

[9] Michael Reilly and William Slocum, *Reilly of the White House* (New York, 1947), pp. 98–100; Grace Tully, *F.D.R. My Boss* (New York, 1949), p. 202.

[10] Schlesinger, *Coming of New Deal,* p. 574; Donald Richberg, *My Hero* (New York, 1954), p. 290; Samuel Rosenman, *Working with Roosevelt* (New York, 1952), p. 22; Harold Ickes, *The Secret Diary of Harold Ickes* (3 vols., New York, 1954), I, 449.

[11] Tully, *F.D.R.,* p. 320; Rosenman, *Roosevelt,* p. 37.

[12] News photographers in the twenties voluntarily destroyed their own plates when they showed Roosevelt in poses that revealed his handicap, and by the thirties there was a strong taboo, sometimes enforced by the Secret Service, against depicting the President's paralysis. Moore, "F.D.R.'s Image," pp. 427 ff., 477 ff., 634 ff.; *Editor and Publisher*, LXV (1932), 36.

walk again among men," the President was leading the nation, as he had himself, to full "recovery" from the paralysis that once had afflicted both man and country.[13]

After the lacerating Second Hundred Days, Roosevelt saw 1936 as a time of healing, when he would unite the party by salving the wounds of his opponents and avoid new conflicts with Congress until he had won re-election. On the day the 1935 Congress adjourned, the publisher Roy Howard asked Roosevelt to grant "a breathing spell to industry, and a recess from further experimentation. . . ." A few weeks before, Raymond Moley had written that the public was "developing a terrific thirst for a long, cool swig of political quiescence." Roosevelt asked Moley to frame a response to Howard; his basic program, the President's letter of reply stated, had "now reached substantial completion and the 'breathing spell' of which you speak is here—very decidedly so."[14] When Congress reconvened in 1936, Roosevelt delivered an inflammatory Annual Message in which he spoke of having "earned the hatred of entrenched greed," yet he recommended no new reforms and demonstrated his sensitivity to attacks on his failure to balance the budget by a drive to slash government spending. Throughout the campaign year of 1936, the President stepped up the radicalism of his rhetoric, but accentuated the conservatism of his deeds.[15]

In his plans for a short session, Roosevelt reckoned without the Supreme Court. Early in 1936, in a 6–3 decision, the Court held the AAA's processing tax unconstitutional. Justice Roberts, in a wretchedly argued opinion, found that the levy was not a legitimate use of the taxing power but "the expropriation of money from one group for the benefit of another."[16] In a biting dissent, Justice Stone scolded the majority for this "tortured construction of the Constitution." "Courts,"

[13] Edwin C. Hill, cited in Rita Halle Kleeman, *Gracious Lady* (New York, 1935), p. 300; *Look,* October 25, 1938, p. 16; Moore, "FDR's Image," p. 646.

[14] Samuel Rosenman (ed.), *The Public Papers and Addresses of Franklin D. Roosevelt* (13 vols., New York, 1938–50), IV, 352–353; Moley, "On Reading Ecclesiastes," *Today,* IV (July 13, 1935), 13; *Public Papers,* IV, 357; Stephen Early to Moley, September 3, 1935, FDRL PPF 743.

[15] *Public Papers,* V, 13; Moley, *After Seven Years,* pp. 330–331; Arthur Schlesinger, Jr., *The Politics of Upheaval* (Boston, 1960), pp. 502–504; James Burns, *Roosevelt: The Lion and the Fox* (New York, 1956), pp. 269–270.

[16] United States *v.* Butler et al., 297 U.S. 61. Cf. Charles Collier, "Judicial Bootstraps and the General Welfare Clause," *George Washington Law Review,* IV (1936), 211–242.

he observed, "are not the only agency of government that must be assumed to have capacity to govern."[17] The Butler decision, which was followed by a sharp decline in farm prices, aroused deep resentment. Senator Bankhead denounced Roberts' opinion as "a political stump speech," Governor Earle of Pennsylvania declared the Court had become a "political body" with "six members committed to the policies of the Liberty League," and in Iowa the six justices who handed down the decision were hanged in effigy.[18]

The Butler decision played hob with Roosevelt's desire to balance the budget. It deprived the government of anticipated receipts from processing taxes, and the Court ordered the return to processors of $200 million already collected. Together with the $2 billion bonus which Congress enacted over Roosevelt's veto, it created an immediate need for new revenue. Keynesians within the administration, who had never liked the regressive processing taxes, rejoiced in the opportunity to agitate for an undistributed-profits tax scaled to penalize size. They hoped the new levy would check the growth of huge personal fortunes, serve as an antitrust weapon, bring corporations within the sphere of regulatory agencies, and, most important, stimulate the economy by draining idle pools of capital. But Senate conservatives had no intention of approving a proposal which Raymond Moley warned would throw businessmen into "paroxysms of fright" and reduce the rich to "rags and tatters." The bill, Senator Pittman observed, was "intended to kill some criminals, but to kill them, it appeared to me that the G-men were ordered to shoot into a crowd." The Revenue Act of 1936 did include a modest graduated tax on undistributed earnings, but this marked a victory in principle rather than in substance for the New Dealers.[19]

No sooner had the Butler decision been handed down than Agricul-

[17] U.S. v. Butler, p. 87. Cf. Homer Cummings to Justice Stone, January 8, 1936, Stone to Cummings, January 9, 1936, FDRL PSF 18; Alpheus T. Mason, *Harlan Fiske Stone* (New York, 1956), pp. 405–418.

[18] Baltimore *Sun,* January 7, 1936; "Address of George H. Earle, January 18, 1936," Earle MSS., Speech and News File, No. 73; Schlesinger, *Politics of Upheaval,* p. 488; Clifford Hope to Chester Stevens, April 21, 1936, Hope MSS., Tax (Legis.) folder, Legislative Correspondence, 1935–36.

[19] John Blum, *From the Morgenthau Diaries* (Boston, 1959), pp. 305–319; Key Pittman to F.D.R., October 17, 1936, Pittman MSS., Box 16; Marriner Eccles, *Beckoning Frontiers* (New York, 1951), pp. 256–265; Randolph Paul, *Taxation in the United States* (Boston, 1954), pp. 188–199.

ture officials scurried to throw together a new farm law that would permit the government to disburse funds to the farmer without incurring the wrath of the Court. In a few weeks time Wallace had won the approval of Congress for a measure which was rammed through over Republican protests that, in an election year, the administration was concerned chiefly with keeping surplus Democrats on the government payroll and not letting the farm vote lie fallow.[20] The Soil Conservation and Domestic Allotment Act offered farmers bounties for not planting soil-depleting commercial crops and for sowing instead soil-enriching grasses and legumes such as clover and soybeans.[21]

The new farm law reflected the intense national concern with soil conservation aroused by the terrible dust storms of the early 1930's. The drought, which began in 1932 and continued each year until 1936, converted a huge area from Texas to the Dakotas into a "Dust Bowl." The dry death scorched pastures and cornfields and turned plowed land into sand dunes. Farmers watched helplessly as cattle fell in their tracks and died. In Vinita, Oklahoma, in 1934, the sun topped 100° for thirty-five straight days; on the thirty-sixth, it climbed to 117°. In the spring of 1935, trains arrived hours late. A conductor on the Santa Fe's Navajo reported: "Noon was like night. There was no sun, and, at times, it was impossible to see a yard. The engineer could not see the signal-lights." As far east as Memphis, people covered their faces with handkerchiefs, a dust cloud seven thousand feet thick darkened the city of Cleveland, yellow grit from Nebraska sifted through the White House doors, and bits of western plains came to rest on vessels in the Atlantic three hundred miles at sea. That winter, red snow fell on New England.[22]

While the 1936 farm act was, as Republicans charged, largely a

[20] "It's a mighty poor substitute for the old Triple A," conceded an Oregon Democrat, "but it's the best we could do." Representative Walter Pierce to Charles W. Smith, March 2, 1936, Pierce MSS., File 3.6.

[21] *Congressional Digest*, XV (March, 1936), 68–96; Paul Murphy, "The New Deal Agricultural Problem and the Constitution," *Agricultural History*, XXIX (1955), 160–169; Louis Taber, COHC, pp. 304 ff.; O. C. Stine, COHC, pp. 356–357.

[22] *Literary Digest*, CXIX (April 20, 1935), 10; Lawrence Svobida, *An Empire of Dust* (Caldwell, Ida., 1940); Vance Johnson, *Heaven's Tableland* (New York, 1947); Paul Sears, "The Black Blizzards," in Daniel Aaron (ed.), *America in Crisis* (New York, 1952), pp. 287–302; Margaret Bourke-White, "The Drought," *Fortune*, X (October, 1934), 76–83; *Literary Digest*, CXIX (March 30, 1935), 7.

1. "Hooverville," New York City

2. Franklin Roosevelt at Indianapolis in the 1932 campaign

3. Franklin Roosevelt barnstorming with John Garner at Topeka,
September, 1932

4. Alfred E. Smith and Franklin Roosevelt mend political fences,
November, 1932

Wide World

5. President-elect Roosevelt confers with his advisers on a train from Washington to New York: Admiral Cary Grayson, Norman Davis, Raymond Moley, Rexford Tugwell, and William Woodin

U.P.I.

6. The Bank of the United States fails

Franklin D. Roosevelt Library

7. Relief administrators Harry Hopkins, Harold Ickes, and Frank Walker leaving the White House

Wide World

8. The mammoth NRA parade in New York City, September 13, 1933

9. A New York City breadline

U.P.I.

10. A family of "Okies" stalled on the desert near the California border

11. A sharecropper's shack, Arkansas

12. Migrants from Oklahoma seek work in the California pea fields

13. President Roosevelt signs the Social Security Act. Observers, left to right, Representative Doughton, E. A. Witte, Senator Wagner, co-author of the bill, Senator Barkley, Senator La Follette, Senator Lonergan, Secretary Perkins, Senator King, Representative Lewis, co-author of the bill, and Senator Guffey

14. Unemployed file claims under Social Security Act

15. City Hall, Libby, Montana—before the PWA

16. New City Hall, Libby, Montana, a PWA project

17. A Farm Security Administration project, California

18. Dust storm approaching Spearman, Texas, April, 1935

19. Bonneville Dam, under construction

20. Norris Dam, Tennessee Valley Authority

21. Strikers riot at Fisher Body plant, Cleveland

22. Huey Long leads the Louisiana
     State University band

Wide World

23. The Reverend Charles E. Coughlin

Wide World

24. Federal Art Project, Melrose, New Mexico

25. Federal Art Project, Utah

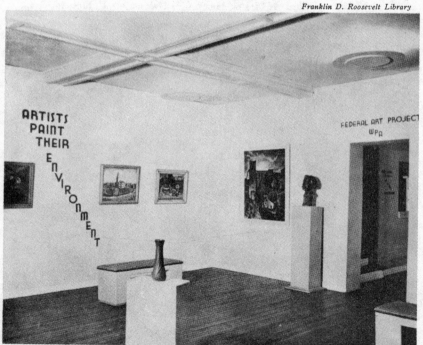

26. Cordell Hull,
Secretary of State,
with Under Secretary
of State Sumner Welles

Franklin D. Roosevelt Library

27. Senator Alben W. Barkley, James A. Farley, and F.D.R., June, 1937

U.P.I.

28. President Roosevelt with Vice-President-elect Henry Wallace, November, 1940

Wide World

29. Eleanor Roosevelt votes at Hyde Park, November, 1936

Franklin D. Roosevelt Library

30. Wendell Willkie in the 1940 campaign, Elkton, Indiana

31. President Roosevelt campaigning in Newburgh, New York, November, 1940

32. Supreme Court Justices Charles Evans Hughes,
Harlan Fiske Stone, Owen Roberts, Hugo Black,
Stanley Reed, Felix Frankfurter, and Frank Murphy

33. Secretary of War Henry L. Stimson draws first number to determine
order of selection for draft registrants, October, 1940

Franklin D. Roosevelt Library

34. Champion campaigner, in his successful bid for an unprecedented
third term, 1940

subterfuge to permit the administration to go on paying money to farmers to restrict production, it represented too the long-range commitment of Wallace and his aides to a greater emphasis on soil conservation. The dust storms were perceived as a penance Man paid for his "sin" in mistreating the "wounded land"; if the process were not reversed, the United States would, like ancient civilizations, disappear into oblivion. In October, 1933, on Tugwell's recommendation, Hugh Hammond Bennett, the "father of soil conservation," was named to head the newly created Soil Erosion Service of the Department of the Interior. In 1934, Congress passed the Taylor Grazing Act, which subjected grazing on the public domain to national regulation. In 1935, as grit from the western plains created a copper pall over Washington like the skies of a smoky steel town, Congress created a permanent Soil Conservation Service in the Department of Agriculture. Under the crusading Bennett, the Service taught farmers the proper way to till their hillsides. The New Deal's conception of the common man, Alistair Cooke has noted wryly, was one who could "take up contour plowing late in life."[23] In its emphasis on soil conservation, the New Deal added a new dimension to the conservation movement. *Fortune* observed: "It is conceivable that when the history of our generation comes to be written in the perspective of a hundred years the saving of the broken lands will stand out as the great and most enduring achievement of the time."[24]

The same sense of working against time to save the nation's resources from "the bastard claims of a bastard acquisitive culture" permeated much of the rest of the New Deal's conservation program.[25] When President Roosevelt released the report of the Mississippi Valley Committee, headed by Morris Cooke, Stuart Chase commented: "Another generation of the strenuous pursuit of the main chance, and it is safe to say that Old Man River would roar in lonely majesty, rid of that dirty animal *homo sapiens,* his waters clean at last." Cooke himself told a reporter that if the country did not adopt the Committee's twenty-year program, "our country as a vital force in human affairs will disappear in, say, three generations."[26] Animated by the

[23] Alistair Cooke, *A Generation on Trial* (New York, 1951), p. 28.

[24] "The Grasslands," *Fortune,* XII (November, 1935), 60. Cf. Wellington Brink. *Big Hugh* (New York, 1951).

[25] Stuart Chase, "Old Man River," *New Republic,* LXXXII (1935), 178.

[26] *Ibid.;* Kenneth Trombley, *The Life and Times of a Happy Liberal* (New

conviction that they were racing doom, the conservationists speeded up their timetable: the government established national parks in the Olympic rain forest and in the Shenandoah, new national monuments like Joshua Tree and White Sands, new big-game refuges like the Hart Mountain Antelope Range in southeastern Oregon; demonstrated the feasibility of multipurpose river basin development in the Tennessee Valley; and outlined bold programs of resources development. In the first three years of the New Deal, the government acquired more than twice the acreage in forest lands as had been purchased in the previous history of national forests.[27]

Indispensable to the work of soil conservation and reforestation was the Civilian Conservation Corps. By September, 1935, over five hundred thousand young men lived in CCC camps. In all, more than two and a half million would enlist: boys from the southern pines and the sequoia country of the West, boys from Bayonne and Cicero who had never seen forests or mountains before. They thinned four million acres of trees, stocked almost a billion fish, built more than 30,000 wildlife shelters, dug diversion ditches and canals, and restored Revolutionary and Civil War battlefields. They fought spruce sawflies and Mormon crickets in western forests, grasshoppers in the Midwest, gypsy moths in the East. They built a network of lookout towers and roads and trails so that fires could be detected and reached more easily; when fires broke out, regiments of Roosevelt's "Tree Army" were rushed to the front—forty-seven firefighters lost their lives. Above all, they planted trees—saplings of cedar and hemlock and poplar—in burned-over districts, on eroded hillsides, on bleak mountain slopes ruthlessly stripped of virgin timber. Of all the forest planting, public and private, in the history of the nation, more than half was done by the CCC.[28]

York, 1954), p. 120; "Minutes, Mississippi Valley Committee," Morris Cooke MSS., Box 264. In addition to the Mississippi Valley Committee, the National Resources Committee, the Great Plains Committee, the President's Committee on Water Flow, and numerous other committees inventoried resources to make recommendations to planning and action agencies.

[27] *Report of the Secretary of Agriculture 1936* (Washington, 1936).

[28] David Coyle, *Conservation* (New Brunswick, 1957), p. 103; Major John Guthrie, "The CCC and American Conservation," *Scientific Monthly*, LVII (1943), 401–412; Bernard Roth, "Remember the CCC," *Soil Conservation*, XVIII (1953), 205; Schlesinger, *Coming of New Deal*, pp. 338–340; Luther Gulick, *American Forest Policy* (New York, 1951), p. 111.

The policies Roosevelt had pursued in his first term had met with so favorable a response that the Republicans faced an uphill battle in their campaign to oust him in 1936. To oppose the President, the G.O.P. named Alf Landon of Kansas, the only Republican governor elected in 1932 who survived the Democratic tidal wave of 1934. As his running mate, the convention chose Colonel Frank Knox, a strongly nationalistic Chicago publisher who had served as a Rough Rider under Teddy Roosevelt and boasted a sombrero pierced by two bullets at San Juan Hill. Landon, dismissed by many as an Old Guard wheelhorse, was, in fact, a good deal more liberal than his sponsors. Like Knox a former Bull Mooser, he had fought the Ku Klux Klan in Kansas, had a superb record on civil liberties, and had demonstrated that he favored the regulation of business. He had endorsed any number of New Deal projects in words that were to come back to plague him during the campaign.[29]

In a race against the dynamic Franklin Roosevelt, Landon seemed badly mismatched. Chester Rowell, the head of the California delegation to the Republican convention, thought Landon "a pretty poor specimen" and even a supporter like Gifford Pinchot conceded he was no "world-beater."[30] Many Landon backers were distressed by the Governor's ineffective delivery on the radio. "For God's sake break your sentences and make them shorter in your speeches," pleaded one of his followers.[31] In contrast to the worldly Roosevelt, Landon appeared to be what Jim Farley, in an indiscreet moment, called him: the governor of a "typical prairie state." Yet it was precisely Landon's quality of provincialism which appealed to many Americans. Republican speakers contrasted the steady-going Landon, with his agreeable smile and comfortable air, with the sophisticated Roosevelt with his unsettling and vaguely dishonest charm. Since Landon spoke so wretchedly, people knew he was sincere. One correspondent, who told Landon not to change his style, observed: "The people of this country want exactly what the Pilgrim Fathers wanted, and exactly what the

[29] Willis Thornton, *The Life of Alfred M. Landon* (New York, 1936), pp. 76–77; William Allen White to Judson King, August 27, 1936, King MSS., Box 15; Landon to F.D.R., January 16, 1934, FDRL OF 444–B.

[30] Chester Rowell to Myrtle Rowell, June 11, 1936, Rowell MSS.; Gifford Pinchot to Judson King, June 6, 1936, King MSS., Box 15.

[31] James R. Schmidt to Alf Landon, *c.* May 27, 1936, Landon MSS. Landon, lamented H. L. Mencken, "simply lacks the power to inflame the boobs." Mencken to Henry Morrow Hyde, September 17, 1936, Hyde MSS.

pioneers who treked to Kansas wanted. . . ."[32] This nostalgia for the old days permeated the Republican campaign. The G.O.P., which sought to glorify the horse-and-buggy days Roosevelt had derided, blanketed the country with sunflowers, and adopted as its campaign song "Oh, Susanna," the ballad of the Oregon Trail.

Landon's modestly liberal record won him few votes. His appeal lay rather in offering opponents of President Roosevelt a way to express their displeasure at New Deal policies. Roosevelt's style of politics deeply offended men of conservative temperament. He refused to be bound by the restraints that confined other leaders in the past. He abandoned the gold standard, operated the government with unbalanced budgets, and moved in a world of "non-Euclidian" economics.[33] The fixed world of the predepression years—the world where parallel lines never met—had been torpedoed by the commandants of the alphabet agencies. Roosevelt was seen as the agent of willful change, a man who expressed the spirit of a decade in which the swing clarinetist Benny Goodman invaded Carnegie Hall, a man whose Surgeon General Thomas Parran dared to speak the forbidden word "syphilis" over the radio, a man whose defiance of tradition would even permit him to change the day of the year on which the old New England holiday of Thanksgiving was celebrated.[34]

Many of the nation's wealthy were almost incoherent with rage at President Roosevelt. They refused to say his name, but referred to him as "that man" or "he." One of Roosevelt's wealthy neighbors, Howland Spencer, hated the President so much that he exiled himself in the Bahamas and returned only after the Republicans won the con-

---

[32] Ickes, *Diary,* I, 603; Otis Peabody Swift to Alf Landon, July 30, 1936, Landon MSS.

[33] Robert and Helen Lynd, *Middletown in Transition* (New York, 1937), p. 125; I. F. Stone, "Company in the Doghouse," *New Republic* XCIII (1938), 251.

[34] Parran defied taboos to conduct an effective campaign of education about venereal disease. Parran, Bernard de Voto sardonically remarked, "as I understand it, holds letters patent on the disease." De Voto to Hans Zinsser, May 25, 1937, De Voto MSS., Box 2. One critic believed the New Deal had fathered swing, for swing was "merely a tonal expression of this age of apparent license, in which the common property rights of man, upon which all enduring civilizations have been built, seem to be purloined by many. Under such a social frenzy no one has any rights." William Roberts Tilford, "Swing! Swing! Swing!" *Etude,* LV (1937), 835. In 1939, Roosevelt decreed that Thanksgiving would fall on November 23 rather than the last Thursday, November 30.

gressional elections in 1946. Terrified by the prospect of confiscatory taxation, the wealthy viewed Roosevelt as a reckless leveler. Accustomed to having their authority unchallenged, businessmen resented the rising empires of government and labor. Many businessmen were openly defiant. Arthur Young, U.S. Steel vice-president in charge of industrial relations, told the American Management Association that rather than obey the Wagner bill he would "go to jail or be convicted as a felon." The Association then gave Young a medal for "outstanding and creative work in the field of industrial relations."[35] In the fall of 1935, the American Bankers Association convention rejected its own nominating committee's recommendation for second vice-president, because he was a business associate of Marriner Eccles, and chose instead a "Main Street banker," Orval Adams of Salt Lake City. Viewing government as a sovereign power would treat an upstart rival, Adams proposed that bankers "declare an embargo" and "decline to make further purchases." "The Bankers of America," Adams asserted, "should resume negotiations with the Federal Government only under a rigid economy, a balanced budget, and a sane tax program."[36]

Landon won the support of a number of Wilsonian Democrats who believed that Roosevelt had jettisoned virtues like character, thrift, and self-reliance, had forsaken goals like the equality of opportunity to compete for a new emphasis on security and social rights, and had given himself over to a brazen cadre of planners. Like the advance guard of a revolutionary army, these *novi homines* had moved into the Walsh-McLean mansion on Massachusetts Avenue in the summer of 1935. Here, where the Prince of Wales and the King of the Belgians had once been entertained, the New Dealers drafted plans for greenbelt towns, some working at desks in the grand ballroom, some in Mrs. McLean's brocaded boudoir.[37] Most of all, old-time Democrats were discountenanced by policies which they thought fomented class animosity and encouraged minority-bloc politics. Newton Baker protested:

[35] *Fortune,* XII (July, 1935), 140.

[36] Eccles, *Beckoning Frontiers,* pp. 232–234; Grant Macfarlane to Elbert Thomas, February 19, 1936, Elbert Thomas MSS., Box 14. Cf. Schlesinger, *Coming of New Deal,* pp. 471–480; Marquis Childs, *They Hate Roosevelt* (New York, 1936); Thomas Paul Jenkin, *Reaction of Major Groups to Positive Government in the United States, 1930–1940* ("University of California Publications in Political Science," I [Berkeley and Los Angeles, 1945]), 298–328.

[37] Marquis Childs, *I Write from Washington* (New York, 1942), pp. 10–11; *Literary Digest,* CXX (August 31, 1935), 10.

"The trouble with this recognition of the class war is that it spreads like a grease stain and every group formed around a special and selfish interest demands the same sort of recognition. As a consequence our Government for the last three years has been the mere tossing of tubs to each whale as it grows bold enough to stick its head out of the water."[38]

For three years, Al Smith had taken a strong stand behind the policies of Grover Cleveland. On January 25, 1936, at a Liberty League banquet at the Mayflower Hotel in Washington before a gathering of two thousand that numbered a dozen du Ponts, business leaders like Winthrop Aldrich, and discontented Democrats like Dean Acheson, Smith charged that the Supreme Court had been forced to work overtime "throwing the alphabet out the window three letters at a time." The Roosevelt administration, he declared, had been turned in a Socialist direction. "It is all right with me if they want to disguise themselves as Norman Thomas or Karl Marx, or Lenin, or any of the rest of that bunch," Smith shouted, "but what I won't stand for is allowing them to march under the banner of Jefferson, Jackson and Cleveland." When the November elections came, Smith warned, he and other Democrats of like mind would probably "take a walk." Together with Smith, men like John W. Davis, another former Democratic presidential nominee, bolted the party, convinced it had been taken over by Republican progressives and hitherto unknown social workers and professors. "Who is Ickes? Who is Wallace? Who is Hopkins, and, in the name of all that is good and holy, who is Tugwell, and where did he blow from?" Smith asked.[39]

Such appeals probably cost Landon more votes than they won him. Infuriated Democrats who had supported Smith in 1928 denounced him as a "Benedict Arnold" who had betrayed his party, and as a turn-

[38] Newton Baker to Walter Lippmann, January 27, 1936, Baker MSS., Box 149; D. F. Houston to W. W. Ball, December 31, 1937, Ball MSS.; George Creel, *Rebel at Large* (New York, 1947), p. 337.

[39] "Dinner, American Liberty League, January 25, 1936, Hotel Mayflower, Washington, D.C.," Jouett Shouse MSS., American Liberty League, 1934–36; Smith, "I Am an American Before I Am a Democrat," *Vital Speeches,* III (1936), 18. The Jeffersonian Democrats of Virginia stated: "Clad in its new and fantastic vestments of the New Deal [the Democratic party] resembles true Democracy about as much as Earl Browder resembles Washington, Frankfurter Jefferson, Tugwell Madison, or Ickes Wilson." Copy in the Thomas Lomax Hunter MSS.

coat who, unable to resist the blandishments of the rich, had turned in
his brown derby for a top hat.[40] "It's a long way from Mamie
O'Rourke to the grand dames of Park Avenue, Al," commented the
New York *Post*.[41] The Liberty League was so patently an alliance of
the wealthy that the Republican National Committee asked it to refrain from endorsing the Landon ticket. Landon himself found it
impossible to disassociate himself from conservative party leaders, including its chairman John Hamilton. After the election, Landon
wrote: "One of the things that was wrong with the campaign was that
I didn't have enough help on the Progressive and Liberal side to pick
up the ball and emphasize our really liberal stand on many questions."
As the campaign progressed, Landon, who from the outset had favored
retrenchment, a balanced budget, and the gold standard, made more
and more conservative statements at the same time that he made extravagant promises to farmers and to the aged. By Election Day, many
of the disaffected Democrats—Newton Baker, Dean Acheson, James
P. Warburg—had returned to Roosevelt, in part no doubt out of disillusionment with Landon's campaign.[42]

It seemed unlikely that any Republican candidate could outpoll
Franklin Roosevelt in 1936; G.O.P. hopes rested on the possibility
that a combination of Coughlin, Townsend, and other dissident elements under the leadership of Huey Long would draw enough votes
away from the President to allow the Republican challenger to slip
through. But all threats from Huey Long in 1936 had ended dramati-
cally at the State House in Baton Rouge on a September night in
1935, when a white-clad figure stepped out of the evening shadows
and shot down Louisiana's senator. Dr. Carl Austin Weiss, who had
witnessed the dictatorship of Dollfuss in Vienna in his student days,
had been outraged by Long's attempt to dislodge an obscure judge,

[40] Leslie Fowden to Lindsay Warren, January 7, 1936, Warren MSS., Box 15;
The New York Times, October 3, 1936. Lillian Wald wrote sadly: "I mar-
velled that bitterness could so blind a man who for so many years has stood for
better things." Lillian Wald to Herbert Lehman, January 29, 1936, Lillian Wald
MSS., Correspondence.

[41] New York Post, *Press Time* (New York, 1936), pp. 95–96. Governor Earle
compared the Liberty League dinner at which Smith spoke to Blaine's "Bel-
shazzar" speech. George Earle, "Federal Relief Work in Pennsylvania," address,
Philadelphia, April 20, 1936, Earle MSS., Speech and News File No. 75.

[42] Landon to Henry Haskell, November 13, 1936, Landon MSS.; Schlesinger,
Politics of Upheaval, p. 634.

Weiss's father-in-law. As he staggered through the corridors, Huey asked: "I wonder what they shot me for?" Thirty-one hours later, he was dead.

Long's Share the Wealth followers in every part of the country mourned the death of their hero. A New York woman wrote Roosevelt: "God have mercy on poor dead good Huey Long and for give the man who kild him he was a good man Huey Long was. . . ." Alongside paid advertisements to St. Rita and St. Joseph, devout Catholics in southern Louisiana inserted notices that read: "Thanks to Huey P. Long for answers to my prayers."[43] Gerald L. K. Smith, who wired Roosevelt protesting "financing potential assassins with government money," laid claim to the Long mantle, but Louisiana politicians brushed him aside, liquidated the organization, and sought an arrangement with the President. In an accommodation that was jeered as the "Second Louisiana Purchase," the administration was happy to oblige.[44]

As the Long organization disintegrated, the main threats to Roosevelt from the messiah groups came from Dr. Townsend and Father Coughlin. Roosevelt, it has been said, nullified their appeal by reform legislation which wiped out the discontent on which they fed. But Townsend's strength reached its apex after the passage of the Social Security Act; in fact, he gained recruits precisely because of the inadequacy of the New Deal pension system, which left millions of elderly Americans unprotected.[45] Townsendism, which, as *Today* remarked, was "easily the outstanding political sensation" of 1935, claimed control of two western governors and the legislatures of seven western states. In November, 1935, it threw a bad scare into politicians east of the Mississippi. Verner Main, the only candidate to support the Townsend Plan in a field of five, won the Republican primary in

[43] Anne Sloane to F.D.R., n.d., FDRL OF 1403; Harnett Kane, *Louisiana Hayride* (New York, 1941), p. 162. "The poor man cannot have represenatives because they, like Lincoln and Huey, are killed and by whom and for what?" asked a Kansan. "We know that big interest—Standard oil and so on—are at the bottom of this." V. J. Parsens to F.D.R., n.d., FDRL OF 1403. Cf. P. L. Gassaway to B. F. Gassaway, September 13, 1935, Gassaway MSS., Correspondence, 1935.

[44] Smith to F.D.R., September 16, 1935, FDRL OF 1403; Kane, *Louisiana Hayride*, p. 183.

[45] Edwin Witte to Merrill Murray, December 11, 1935, Grace Abbott to Edward Costigan, January 30, 1936, Costigan MSS., V.F. 4.

Michigan's Third Congressional District with 13,419 votes; his nearest opponent polled only 4,806 votes. Main went on to defeat his Democratic opponent in the by-election in December. The *Townsend National Weekly* gloated: "As Main Goes, So Goes the Nation."[46]

At the height of its power, the Townsend movement began to collapse. When a former national publicity director of the organization leveled charges of corruption at the doctor and Robert Clements, Republicans and Democrats seized the opportunity to unite to scotch the Townsend menace. A probe by a hostile congressional committee, while unable to prove any wrongdoing by Townsend, revealed that Clements had profited handsomely. In March, 1936, after a political and financial disagreement, Townsend ordered Clements to resign; one week later, the doctor broke with Congressman McGroarty. The Townsend strength had never been as great as western politicians had been scared into believing, and, with the adroit Clements gone, the political organization fell apart. In an Oregon primary, four rival Townsend candidates battled for the same post. Townsend himself headed in a new political direction. After being subjected to severe questioning, Townsend stalked out of the committee room, and the House cited him for contempt. Arm and arm with him as he marched out of the hearings walked Gerald L. K. Smith.[47]

In June, 1936, Smith, Father Coughlin, and some of the Townsendites joined forces to create a new third party. Representative William Lemke of North Dakota, the Union party's presidential candidate, had no hope of victory, but he counted on the battalions of Townsend and Coughlin to roll up a respectable vote. Townsend gave Lemke his personal support, but did not dare try to win official endorsement from his organization. A national Townsendite convention in Cleveland in mid-July shook with dissension. Father Coughlin ripped off his coat and clerical collar, and denounced the President as "the great betrayer

[46] "The Townsend Threat," *Today*, V (December 28, 1935), 12; *Townsend National Weekly*, I (December 2, 1935), 1; *News-Week*, VI (December 28, 1935), 5–7.

[47] *The New York Times*, May 22, 1936; Schlesinger, *Politics of Upheaval*, p. 552; Gerold Frank, "Huey Long the Second," *Nation*, CXLIII (1936), 93–94. In March, 1937, Townsend received a thirty-day jail term. At the last second, as he was about to start serving the sentence, a breathless messenger boy ran up with a pardon from President Roosevelt. *The New York Times*, March 13, 1937; Dr. Francis Townsend, *New Horizons* (Chicago, 1943), p. 211.

and liar" and as "Franklin Double-Crossing Roosevelt." But an Oklahoma Democrat, Gomer Smith, won almost as much applause when he denounced Gerald Smith and defended the President as a "church-going, Bible-reading, God-fearing man," a "golden-hearted patriot who had saved the country from communism."[48]

By the time Coughlin's National Union for Social Justice convened in August, the alliance had been all but shattered. The manager of the National Union convention declared that Townsend and Smith would speak at the meeting "only over his dead body."[49] Coughlin announced he would swing at least nine million votes for Lemke, or he would quit broadcasting. "As I was instrumental in removing Herbert Hoover from the White House so help me God I will be instrumental in taking a Communist from the chair once occupied by Washington," he promised.[50] Yet Coughlin not only scorned Townsend and Smith; he even ignored Lemke. Before the end of the campaign, both Townsend and Lemke had denounced Smith for his fascist sympathies, and Townsend, his own organization in the process of disruption, gave Lemke lukewarm support. Progressives abhorred the Jew-baiting of Coughlin and Smith. Catholic liberals like Frank Walsh, who had once looked with favor on the Radio Priest, now repudiated him.[51]

As Lemke's hopes withered, Coughlin became more violent. On September 25 in Cincinnati, he called the President "anti-God," and advocated using bullets when an "upstart dictator in the United States succeeds in making a one-party government and when the ballot is useless." Increasingly he resorted to the kind of effects Hitler employed

[48] Time, XXVIII (July 27, 1936), 17–19; The New York Times, July 17, 1936; H. L. Mencken to Henry Morrow Hyde, July 28, 1936, Hyde MSS.; H. L. Mencken, Heathen Days (New York, 1943), pp. 295–297.

[49] When a reporter questioned Coughlin about a tour with the other two leaders, he responded: "Why should they tag me around?" The New York Times, August 13, 14, 1936. One writer noted: "He ruthlessly prevented the Reverend Smith from speaking until late on Saturday, squirmed in his seat while Gerald talked, elaborately pretended to go to sleep, grinned slyly at those around him." Jonathan Mitchell, "Father Coughlin's Children," New Republic, LXXXVIII (1936), 74.

[50] Literary Digest, CXXII (Aug. 15, 1936), 4–5.

[51] Frank P. Walsh to John Lapp, October 1, 1936, Walsh MSS. Neither farm nor labor groups gave Lemke the backing which Coughlin had expected. Donald McCoy, Angry Voices (Lawrence, Kan., 1958), pp. 115–155.

in Germany. In Chicago, he walked between an honor guard of 2,500 men who wore armbands and carried flags.[52] When he launched personal diatribes against the duly constituted head of the American government, his own Church intervened to restrain him. The Vatican summoned his bishop, Michael Gallagher, to Rome; *Osservatore Romano* upbraided him for creating disrespect for authority; the Papal Secretary of State, Cardinal Pacelli (later Pope Pius XII), embarked for a "vacation" in the United States; and Monsignor John Ryan, on a national radio network, rebuked Coughlin for calling Roosevelt a Communist.[53] Despite all these tribulations, the Union party ticket gave the Democrats no little concern. Reports that the Union party might win 20 per cent of the Irish-Catholic ballots in Massachusetts and Rhode Island indicated that Roosevelt might be in trouble among Catholic voters in industrial states.[54]

The Franklin Roosevelt who campaigned in 1936 had come a long way in four years. He no longer had the same faith in an all-class alliance. In early May, aboard the yacht *Potomac,* he told Moley angrily that businessmen as a class were stupid, that newspapers were just as bad; nothing would win more votes than to have the press and the business community aligned against him. In his opening fusillade of the campaign, on a sticky June night in Philadelphia before over a hundred thousand at Franklin Field, Roosevelt lashed out at "eco-

[52] *The New York Times,* September 26, 1936; *Social Justice,* September 14, 1936.

[53] *The New York Times,* September 3, 10, 1936; James Shenton, "The Coughlin Movement and the New Deal," *Political Science Quarterly,* LXIII (1958), 364–366. A Coughlinite censured the Monsignor: "Do you realize the enormity of your sin???? You have split the Catholic world in twain." A. Ryan to Msgr. John Ryan, October 25, 1936, Ryan MSS. When Coughlin persisted in branding Roosevelt a "scab President," even the complaisant Bishop Gallagher required him to apologize publicly. *The New York Times,* November 1, 1936.

[54] Thomas Simonton to Maurice P. Davidson and Frank P. Walsh, Progressive National Committee MSS., Box 2. In western Massachusetts, Mrs. Agnes Reavey, a Coughlinite, defeated six male rivals to win the Democratic nomination for Congress in the second district. Representative John McCormack of Boston ran scared when 40,000 voters took the trouble to put stickers of the Coughlinite candidate for senator on the ballot. No less than three Catholic congressmen in the Bay State—John Higgins, Arthur Healey, and Joseph Casey—abandoned Roosevelt to curry Coughlin's support. Springfield (Mass.) *Union,* September 17, 1936; McCormack to F.D.R., September 26, 1936, FDRL PPF 4057; *The New York Times,* October 13, 1936.

nomic royalists" who took "other people's money" to impose a "new industrial dictatorship."[55]

For the next four months, he ignored Landon and campaigned instead against Herbert Hoover and the interests. At the close of the campaign, at Madison Square Garden on October 31, he let out all the stops. The forces of "organized money," he told a roaring crowd, "are unanimous in their hate for me—and I welcome their hatred." "I should like to have it said of my first Administration that in it the forces of selfishness and of lust for power met their match. . . . I should like to have it said—" Another deafening roar from the crowd. "Wait a moment!" the President cried. "I should like to have it said of my second Administration that in it these forces met their master." The Garden was pandemonium.[56]

By 1936, Franklin Roosevelt had forged a new political coalition firmly based on the masses in the great northern cities, and led in Congress by a new political type: the northern urban liberal Democrat typified by New York City's Robert Wagner. Wagner, a true son of the city—he became uneasy whenever the el train near his apartment ceased to rumble—spoke for an industrial liberalism considerably more advanced than that advocated by most traditional Republican farm-state progressives. While old-stock Americans in the small towns clung to the G.O.P., the newer ethnic groups in the cities swung to Roosevelt, mostly out of gratitude for New Deal welfare measures, but partly out of delight with being granted "recognition." Under Harding, Coolidge, and Hoover, one out of every twenty-five judicial appointments went to a Catholic; under Roosevelt, more than one out of every four.[57] At the 1936 convention, northern cities demonstrated their new

[55] Moley, *After Seven Years*, p. 337; *Public Papers*, V, 232–233; Raymond Clapper, *Watching the World* (New York, 1944), pp. 86–87. Roosevelt's reference to "economic royalists" stung. "I can't be for Roosevelt any more," one prominent woman told Frances Perkins. "He called me a dirty name." Frances Perkins, COHC, VII, 7.

[56] *Public Papers*, V, 568–569; Burns, *Roosevelt*, pp. 282–283. Cf. Newton Baker to William Allen White, November 2, 1936, White MSS., Box 182.

[57] Samuel Lubell, *The Future of American Politics* (New York, 1952), pp. 78–79; Oscar Handlin, *The American People in the Twentieth Century* (Cambridge, 1954), pp. 183–184, 203. Cf. John Dingell to F.D.R., June 26, 1937, James Roosevelt MSS., Box 36; M. C. Gonzales to Richard Dillon, October 20, 1934, Dillon MSS.

power in the party when delegates wiped out the century-old two-thirds rule, long one of the chief defenses of southern Democrats.[58] On Election Day, of the cities of one hundred thousand or more, Roosevelt would capture 104, Landon two.[59]

Even the depression did little at first to shake the loyalty of Negroes to the party of Lincoln. In Detroit, Cleveland, Philadelphia, and other cities in 1932, Negroes cast their ballots for Herbert Hoover. Cincinnati's heavily Negro Ward 16 gave Roosevelt less than 29 per cent. While Roosevelt won 59 per cent of the white vote in Chicago, he polled only 23 per cent of the Negro ballots.[60] The early New Deal provoked sharp criticism from Negro leaders. They censured the NRA for displacing Negro industrial workers and raising prices more than wages: the Blue Eagle, contended one writer, was "a predatory bird"; the NRA had but one meaning, "Negroes Ruined Again."[61] They protested AAA policies which drove Negro tenant farmers and sharecroppers from the land, while white landowners pocketed government checks. Negroes could live neither in the TVA's model town of Norris nor in the subsistence homestead community of Arthurdale; the New Deal's showplace communities were Jim Crow towns. The Negro, wrote the N.A.A.C.P. organ Crisis in 1935, "ought to realize by now that the powers-that-be in the Roosevelt administration have nothing for them."[62]

[58] San Francisco *Chronicle,* June 25–26, 1936. The fight to abolish the rule was led by Senator Bennett Champ Clark of Missouri, whose father had had a majority of the delegates at the 1912 convention but had been denied the nomination.

[59] This was the peak year of metropolitan pluralities for Roosevelt. New York City gave the President a whopping 1,367,000 plurality. Samuel Eldersveld, "The Influence of Metropolitan Party Pluralities in Presidential Elections since 1920," *American Political Science Review,* XLIII (1949), 1197.

[60] Ernest Collins, "Cincinnati Negroes and Presidential Politics," *Journal of Negro History,* XLI (1956), 132–133; Elmer Henderson, "Political Changes Among Negroes in Chicago During the Depression," *Social Forces,* XIX (1941), 540.

[61] John P. Davis, "Blue Eagles and Black Workers," *New Republic,* LXXXI (1934), 9, and, "What Price National Recovery?" *Crisis,* XL (1933), 271–272. Cf. Donald Richberg to Eleanor Roosevelt, October 23, 1934, Richberg MSS., Box 2.

[62] John P. Davis, "The Negro and TVA," National Association for the Advancement of Colored People, August, 1935, mimeographed, in Robert Wagner MSS.; Paul Conkin, *Tomorrow a New World* (Ithaca, 1959), pp. 200–203; "Social Security—for White Folk," *Crisis,* XLII (1935), 80.

Negro leaders also rebuked Roosevelt for the failure of New Deal Congresses to enact civil rights legislation. After a recrudescence of lynching in 1933, the N.A.A.C.P. drafted a federal antilynching bill, which was introduced in 1934 by Senators Costigan and Wagner. New lynchings in that year intensified the plea for federal action. A Georgian wrote Senator Wagner: "In Georgia thay Ar Killing the Colored Woman and Saturday Night Oct 27th thay mobed a Negro man he is Not Dead but looking to Die Any Day i Want you to Do your best to Stop this. . . . i wont Rite my Name fer that it Would be Publish and thay Will Kill Me as i A Negro."[63] The President denounced lynching and was willing, after some urging, to support a vote on an antilynching measure, so long as it did not tie up other reform legislation. But he refused to make it "must" legislation, for, if he did, Southern committee chairmen might kill every economic proposal he asked them to advance. Without the President's help, supporters of the bill failed to break a filibuster of southern Democrats abetted by Senator Borah. Not a single piece of civil rights legislation was adopted in Roosevelt's four terms in the White House. Vexed at the President's stand, the N.A.A.C.P. broke with him on April 27, 1935. Vice-Dean Charles Houston of Howard University asked: "Is the Democratic Party determined to make it impossible for self-respecting Negroes to support it in 1936?"[64]

But there were counterforces pulling the Negro toward the Roosevelt coalition. Roosevelt appointed Negroes to more important posts than they had ever held, and New Deal policies frequently broke the pattern of racial discrimination. Negroes were especially impressed by the sincerity of Mrs. Roosevelt and Secretary Ickes. The Democrats took pains to distribute more than a million photographs of Mrs. Roosevelt flanked by two young Negro R.O.T.C. officers at Howard University. Harold Ickes, who had been president of the Chicago N.A.A.C.P., created a new post of director of "Negro economics," made a point of hiring Negroes in the Department of the Interior, and gave them a large share of the new housing in his slum-clearance projects. Most of all, Negroes swung to Roosevelt because they had been

[63] Anon. to Robert Wagner, November 14, 1934, Wagner MSS.
[64] Walter White, *A Man Called White* (New York, 1948), pp. 169–170; Joseph Gavagan, COHC, p. 46; statement in Costigan MSS., V.F. 3.

granted relief. In many areas, Negroes, hit harder than any group by the depression, survived largely because of relief checks. Unlike the CCC, the WPA imposed no quotas. The NYA, through the noted Negro leader Mary McLeod Bethune, funneled funds to thousands of young Negroes. Negro intellectuals might fret at the inequities of the New Deal, but the masses of Negroes began to break party lines in gratitude for government bounties and nondiscriminatory treatment.[65]

As early as 1932, G.O.P. politicians had come back with surprising intelligence from Negro wards: "They're getting tired of Lincoln." Disenchanted with the Republican party, Robert Vann, publisher of the Pittsburgh *Courier,* told Negro voters: "My friends, go turn Lincoln's picture to the wall. That debt has been paid in full." In 1934, Negroes began to switch to the Democrats, and by 1936 they were substantially Democratic. That year every Negro ward in Cleveland voted Democratic; Cincinnati's Ward 16 gave the President over 65 per cent of the ballots; Roosevelt ran better than twice as well among Negro voters in Chicago as he had four years before. The President took Pittsburgh's Negro Third Ward by nearly 10–1. By mid-1938, a *Fortune* poll would reveal that 84.7 per cent of Negro respondents counted themselves pro-Roosevelt.[66]

Roosevelt's urban coalition united city leaders like Pittsburgh's Dave Lawrence not only behind the national party program but in support of "little New Deals" in the states. Under Governor George Earle, Pennsylvania passed a stringent child labor law, ended the notorious coal and iron police, adopted a "little Wagner Act," and stepped up taxes on corporations. Under Herbert Lehman, New York extended workmen's compensation, outlawed yellow-dog contracts, and provided unemployment insurance. Michigan boasted a "little New Deal" under

[65] Gunnar Myrdal, *An American Dilemma* (New York, 1944), pp. 463–464, 503; George Rawick, "The New Deal and Youth" (unpublished Ph.D. dissertation, University of Wisconsin, 1957), Ch. 10; "Minutes of the fourth meeting of the inter-departmental group concerned with the special problems of Negroes," June 1, 1934, Leon Henderson MSS., Box 7; Edward Litchfield, "A Case Study of Negro Political Behavior in Detroit," *Public Opinion Quarterly,* V (1941), 267–274.

[66] W. L. White, *What People Said* (New York, 1938), p. 335. Joseph Alsop and Robert Kintner, "The Guffey: Biography of a Boss, New Style," *Saturday Evening Post,* CCX (March 26, 1938), 6; Joseph Guffey, *Seventy Years on the Red-Fire Wagon* (Lebanon, Pa., 1952), pp. 75, 170–171; Collins, "Cincinnati," p. 133; Henderson, "Chicago," p. 540; *Fortune,* XVIII (July, 1938), 37.

Frank Murphy; California waited until 1939 to develop its unique version under Culbert Olson.[67]

In the 1936 campaign, John L. Lewis, a lifelong Republican who had backed Hoover four years before, massed the battalions of the C.I.O. behind Roosevelt. C.I.O. unions contributed the enormous sum of $770,000 to the President's campaign chest; $469,000 of it came from Lewis' United Mine Workers.[68] These contributions marked a historic shift in the financial base of the Democratic party; as business sources dried up, the U.M.W.'s donation made the union the party's largest single benefactor. In 1932, bankers and brokers had subscribed 24 per cent of all sums of $1,000 or more; in 1936, they gave less than 4 per cent.[69] Lewis and Sidney Hillman also persuaded George Berry, leader of the A.F. of L. Printing Pressmen's Union, to join with them to create Labor's Non-Partisan League to help re-elect Roosevelt. Hillman and Dubinsky led a mass exodus of Socialists from the garment unions into the Roosevelt coalition. The defection of Socialist union leaders, and the immense appeal of Franklin Roosevelt's New Deal, all but destroyed the Socialist party.[70]

The creation of Labor's Non-Partisan League appeared to indicate that American unions had decided to adopt the remedy British Labour theoreticians had long prescribed: the creation of a labor party in the United States. In fact, Lewis was doing no more than modifying the Gompers tradition, although modifying it in a significant way. No A.F. of L. walking delegate ever had more concern with protecting the

[67] Richard Keller, "Pennsylvania's Little New Deal" (unpublished Ph.D. dissertation, Columbia University, 1960); Guffey, *Red-Fire Wagon*, pp. 99–100; Message to General Assembly, January 28, 1935, Speech and News File No. 67, George Earle MSS.; David Ellis, James Frost, Harold Syrett, and Harry Carman, *A Short History of New York State* (Ithaca, 1957), pp. 420–431; *Knickerbocker Press*, October 24, 1936; Robert Burke, *Olson's New Deal for California* (Berkeley and Los Angeles, 1953.)

[68] These figures include a loan of $50,000 from the U.M.W. to the Democratic National Committee and of $35,000 from the Amalgamated Bank to the American Labor Party.

[69] Louise Overacker, "Labor's Political Contributions," *Political Science Quarterly*, LIV (1939), 60.

[70] In 1936, Thomas received only 187,720 votes, compared to the 884,781 he had polled four years before. By 1940, the Socialist vote had sunk to 99,557. David Shannon, *The Socialist Party of America* (New York, 1955), pp. 227–249; Norman Thomas to Maynard Kreuger, November 1, 1936, Socialist Party MSS.

job interest of union members than Lewis, and he regarded his huge contribution to Roosevelt as an offering for which he expected a measurable return. When Lewis sat with the President at a rally in Harrisburg in 1936, he had, one reporter noted, "a slightly possessive air as though this had been a *fête de Versailles* that he had ordered and paid for, as indeed he had." Lewis told a writer later: "The United Mine Workers and the CIO have paid cash on the barrel for every piece of legislation that we have gotten. . . . Is anyone fool enough to believe for one instant that we gave this money to Roosevelt because we were spellbound by his voice?"[71]

In the 1936 elections, Labor's Non-Partisan League was credited with helping swing Ohio, Illinois, and Indiana to Roosevelt and with turning Pennsylvania Democratic. Roosevelt carried traditionally Republican steel towns like Homestead 4–1 and the Jones and Laughlin ward in Pittsburgh 5,870 to 925. In New York, the newly formed American Labor party won 300,000 votes under its emblem. Never before had union leaders done so effective a job of mobilizing the labor vote in a national campaign.[72] Yet the final outcome was less a testimony to their efficiency than of the sense many workers had that the President was their spokesman and, more than that, their friend. As one workingman put it: "Mr. Roosevelt is the only man we ever had in the White House who would understand that my boss is a sonofabitch."[73]

The labor vote indicated the sharpening class cleavage over political issues that marked the Roosevelt years. As early as 1932, the Roosevelt vote divided on class lines. Yet political scientists who analyzed the 1936 returns found to their astonishment that the class division was not as wide as anticipated, in part because Roosevelt received a surprising percentage of votes from the more prosperous.[74] He had no

[71] Marquis Childs, *I Write from Washington* (New York, 1942), p. 123; Saul Alinsky, *John L. Lewis* (New York, 1949), p. 177.

[72] Alf Landon fretted: "The labor leaders are all tied up with this administration." Landon to Raymond Gram Swing, August 11, 1936, Landon MSS.

[73] Eric Goldman, *Rendezvous with Destiny* (New York, 1952), p. 345. Cf. *United Mine Workers Journal*, XLVII (November 1, 1936), 5–6.

[74] W. F. Ogburn and Estelle Hill, "Income Classes and the Roosevelt Vote in 1932," *Political Science Quarterly*, L (1935), 186–193; Alfred Winslow Jones, *Life, Liberty and Property* (Philadelphia, 1941), pp. 316–317; William F. Ogburn and Lolagene C. Coombs, "The Economic Factor in the Roosevelt

difficulty winning the backing of reformer-businessmen like the Boston merchant Edward Filene. "Why shouldn't the American people take half my money from me?" Filene asked. "I took all of it from them." He held the support of thoughtful financiers like Russell Leffingwell of the House of Morgan, although Leffingwell no longer had his old enthusiasm. "It hurts our feelings to have you go on calling us money changers and economic royalists," he wrote the President. Still more significant was the allegiance of businessmen on the make in the South and West, many of whom had feuded with Wall Street. A. P. Giannini, the powerful California chain banker, urged Roosevelt's re-election and warmly supported the New Deal. Finally, the President drew strength from businessmen in newer industries, typified by Thomas Watson, whose International Business Machines flourished in the 1930's.[75]

Roosevelt campaigned in 1936 as the leader of a liberal crusade which knew no party lines. He mentioned the Democratic party by name no more than three times in the whole campaign. He insisted on an alliance of Democrats and Farmer-Laborites in Minnesota and got the Democratic candidates for governor and senator to withdraw; he encouraged the La Follette Progressives in Wisconsin; he worked with the American Labor party in New York; and he repudiated his own party's nominee in Nebraska when he asked Cornhusker voters to return the independent George Norris, "one of our major prophets," to the Senate.[76]

By 1936, almost all of the progressives had drifted into the Roosevelt camp, in part because they believed they had no choice save to

Elections," *American Political Science Review,* XXXIV (1940), 719–727; Burns, *Roosevelt,* pp. 287–288.

[75] Schlesinger, *Coming of New Deal,* p. 494; Leffingwell to F.D.R., September 1, 1936, FDRL PPF 866; San Francisco *Chronicle,* June 26, 1936; Watson to Marvin McIntyre, June 19, 1936, FDRL PPF 2489. Cf. Joseph P. Kennedy, *I'm for Roosevelt* (New York, 1936).

[76] Stanley High, "Whose Party Is It?" *Saturday Evening Post,* CCIX (February 6, 1937), 37; Stokes, *Chip Off My Shoulder,* pp. 456–457; *Knickerbocker Press,* October 22, 1936; *Public Papers,* IV, 458–459; memorandum from Leo Crowley to Marvin McIntyre, September 8, 1936, FDRL PPF 6659. In *Roosevelt: The Lion and the Fox,* James Burns advances skilfully the thesis that Roosevelt failed to weld an effective farmer-labor coalition. I am impressed rather by how far Roosevelt did go in breaking with his party to build a new liberal alliance.

support the President in order to stave off a return to Hooverism, in part because the New Deal, and especially the Second Hundred Days, seemed to be a fulfillment of the hopes of progressive reformers like Jane Addams and Florence Kelley.[77] When Congress adopted the Social Security bill, Lillian Wald wrote Frances Perkins: "I wish that sister Kelley had been saved to know that. Her life would have been more happily ended."[78] PWA housing projects in Chicago bore the names of Jane Addams and Julia Lathrop. In the Tennessee Valley, Roosevelt was bringing to a brilliant conclusion the battle over Muscle Shoals which had bloodied the progressives for so many years. Some progressive Republican Senators like Charles McNary, Gerald Nye, and William Borah sat out the campaign; others like Hiram Johnson left no doubt they favored the President. Still other progressives like George Norris, James Couzens, and the La Follettes campaigned actively for the President. Norris, who thought Roosevelt's defeat would be "a national calamity," commented: "Not within my recollection has there been a President who has taken the advanced ground which President Roosevelt has taken to free the common man from the domination of monopoly and human greed." In one of the last letters of his life, Lincoln Steffens wrote of Roosevelt: "My trust has grown in him. No one else in sight."[79] By the fall of 1936, the *New Republic* got few takers when it offered five dollars "for the name of an American citizen of either sex, of recognized intellectual distinction and progressive outlook, who is willing to admit publicly that he intends to vote for Landon. . . ."[80]

[77] In the midst of the Second Hundred Days, Felix Frankfurter reported: "The Progressives—Ickes, Norris, Hi Johnson, Gruening, Bob LaF.—feel much better about F.D. than they did three or four months ago." Frankfurter to Louis Brandeis, July 21, 1935, Brandeis MSS., G9.

[78] Lillian Wald to Frances Perkins, August 12, 1935, Lillian Wald MSS., Correspondence.

[79] William Borah to James J. Davis, October 6, 1936, Davis MSS., Box 24; James A. Farley to Francis Heney, September 26, 1936, Heney MSS.; George Norris to Frank P. Walsh, August 31, 1936, Frank Walsh MSS., Progressive National Committee Correspondence; Donald McCoy, "The Progressive National Committee of 1936," *Western Political Quarterly*, IX (1956.), 468; Harry Barnard, *Independent Man* (New York, 1958), pp. 312–322; Steffens to Jo Davidson, July 24, 1936, Steffens MSS.

[80] *New Republic*, LXXXVIII (1936), 270. Among the names suggested were William Allen White, Charles Taft, H. L. Mencken, Walter Lippmann, Henry Ford, and Gifford and Amos Pinchot. Shortly after the election, in a letter to Ezra Pound, Amos Pinchot referred to "our democratic administration, for which

In less than four years, Eleanor Roosevelt had established herself as the conscience of the administration. In her first year as First Lady, the indefatigable Mrs. Roosevelt traveled forty thousand miles. In one two-week period in the spring of 1933, she unveiled a memorial at Bear Mountain Park, urged the graduating class of Todhunter School to see as many kinds of life as possible, received an honorary degree from the Washington College of Law, told the graduating class of the Malcolm Gordon School to put joy in their lives, received six hundred disabled veterans on the South Grounds of the White House, and flew to Los Angeles. That spring, the *New Yorker* printed its famous cartoon of a bemused coal miner crying: "For gosh sakes, here comes Mrs. Roosevelt!"[81] By June, 1935, the Washington *Star* was running a headline on its society page: MRS. ROOSEVELT SPENDS NIGHT AT WHITE HOUSE.[82]

Sometimes naïve, she had one virtue that more than compensated for her gullibility: a warm, unaffected sympathy with other people. When a second bonus army descended on Washington in 1933, she waded through ankle-deep mire to lead the marchers in singing "There's a long, long trail a-winding."[83] An ardent exponent of Negro rights, the First Lady kept Walter White of the N.A.A.C.P. posted on the President's views, and on occasion secured him an unusual one hour and twenty minutes' conference with Roosevelt.[84] She helped persuade Congress to eliminate the noisome alley dwellings in the District of Columbia; she won White House audiences for reformers; she was the Good Fairy who saw to it that in a world of pressure groups and partisan decisions, the President did not neglect people and causes that had no other voice in places of power.[85]

I have no more use on the whole than you have for the bloody bishops." Amos Pinchot to Ezra Pound, January 26, 1937, Amos Pinchot MSS., Box 60.

[81] *Time*, XXI (June 12, 1933), 16; *The New Yorker*, IX (June 3, 1933), 13. A short while later, she did go down in a coal mine. Her patrician mother-in-law wrote the family: "I hope Eleanor is with you this morning. . . . I see she has emerged from the mine. . . . That is something to be thankful for." Roosevelt and Shalett, *Affectionately, F.D.R.*, pp. 272–273.

[82] *Time*, XXV (June 24, 1935), 9.

[83] Schlesinger, *Coming of New Deal*, p. 15; Josephus Daniels to Eleanor Roosevelt, Daniels MSS., Box 1.

[84] Eleanor Roosevelt to Walter White, May 2, 1934, White to Edward Costigan, May 8, 1934, White to Eleanor Roosevelt, May 29, 1934, Costigan to Joseph Robinson, June 11, 1934, Costigan MSS., Vertical File 3.

[85] Alfred Steinberg, *Mrs. R* (New York, 1958).

When Roosevelt toured the country in 1936, thousands of men and women pressed up to railroad tracks for a glimpse of the President. There was something terrible about their response, he told Ickes. He could hear people crying out: "He saved my home," "He gave me a job." In Detroit's Cadillac Square, the cheering was frenzied, almost out of control. In New England, people lined the streets not only in town but well out into the country. A crowd of 150,000 took over Boston Common. In Connecticut, campaign cars could barely maneuver through city streets. In New York City, Roosevelt's party drove for more than thirty miles without passing a block that was not jammed with people.[86]

Roosevelt won loyalty not simply to the New Deal as an abstraction but to particular agencies which acted directly on people as government rarely had before. On an average day, the Farm Credit Administration saved three hundred farms from foreclosure. "I would be without a roof over my head if it hadn't been for the government loan," wrote one farmer. "God bless Mr. Roosevelt and the Democratic party who saved thousands of poor people all over this country from starvation." The HOLC and FHA performed a similar service for distressed urban homeowners. An FHA loan of $110.12 to Chief Albert Looking Elk Martinez of the Pueblo tribe in Taos, New Mexico, enabled him to remodel his ancient house; he replaced the adobe hearth with a modern gas stove and covered the earthen floor with bright lineoleum. A California real-estate broker, who had been a lifelong Republican, wrote the President: "Your work saved our humble little home from the Trust Deed sharks and are we a happy couple in our little home, and listen too, the Real Estate Business is now over 100% better than in 1932, life is 1000% better since you took Charge of our United States."[87]

The chief theme of Roosevelt's campaign was "Four Years Ago and Now." In packed city squares, in jammed open-air stadiums, to throngs from the rear of train platforms, he would say: "You look happier today than you did four years ago." His speeches, noted one

[86] Ickes, *Diary*, I, 695; Breckinridge Long MS. Diary, October 22, 1936; Burns, *Roosevelt*, pp. 281–282; Childs, *I Write from Washington*, pp. 117–118.

[87] Herman Kahn, "The First New Deal," paper delivered to the Mississippi Valley Historical Association, St. Louis, April 28, 1955; Blum, *Morgenthau Diaries*, p. 50; *Fortune*, XI (April, 1935), 186; H. F. and Genevieve Heard to F.D.R., July 9, 1936, FDRL PPF 1135.

reporter, were less campaign addresses than "friendly sermons of a bishop come to make his quadrennial diocesan call. Bishop Roosevelt reported on the excellent state of health enjoyed throughout his vast diocese, particularly as compared with the miserable state that had prevailed before he took high office."[88] He could point out that at least six million jobs had been created in three years, and that national income was half again as high in 1936 as in 1933. Industrial output had almost doubled since he took office; by the fall of 1936, Detroit was rolling out more automobiles than in any year save 1929 and three times more than in 1922, and the electrical industry was selling more current than at any time in the past. In May, 1936, *The New York Times* Index of Business Activity climbed to 100 for the first time since 1930. Corporation profit sheets, which showed a $2 billion deficit in 1933, ran $5 billion in the black in 1936. Los Angeles department and specialty stores reported rises in fur sales of 5 to 50 per cent over an excellent season the year before, and Montgomery Ward rejoiced in the largest August sales in its history.[89] From the first quarter of 1933 to the third quarter of 1936, net income of farm operators almost quadrupled.[90] When Alf Landon got word of the latest business indexes and a prognosis of further gains, he gave up the election as lost.[91]

To be sure, there were weak spots. Construction was notoriously sluggish, and some of the recent gains seemed more speculative than sound. Eight million Americans in the seventh year of the depression still had no jobs. Yet there was a smell of prosperity in the air. In the people milling through the exhibits at the Texas Centennial, at ring-

[88] Childs, *I Write from Washington*, pp. 119–120; Stokes, *Chip Off My Shoulder*, p. 449; *Public Papers*, V, 407, 450.

[89] Simon Kuznets, *Commodity Flow and Capital Formation in the Recent Recovery and Decline, 1932–1938*, National Bureau of Economic Research, Bulletin 74 (New York, 1939), pp. 9–10; Thomas Watson to F.D.R., September 24, 1936, FDRL PPF 2489: *Women's Wear Daily*, September 1, 9, 1936. During 1936, which saw a succession of three-million-share days, a long list of stocks, including Westinghouse, Wright Aeronautical, and U.S. Rubber preferred, jumped fifty or more points. Denver *Post*, January 1, 1937; A. Wilfred May, "Why the Great Bull Market?" *New Republic*, LXXXVII (1936), 336–339.

[90] Harold Barger, *Outlay and Income in the United States, 1921–1938* (New York, 1942), pp. 148–152; "Senator Capper, The Situation, Sept. 23, 1935," Arthur Capper MSS.

[91] Interview with Alf Landon, Topeka, June 29, 1961.

side at the heavyweight championship fights of the unsmiling, lethal
Joe Louis, in the newspaper headlines reporting "U. S. Steel Shows
First Net Since '31," there was a heady sense of a nation once more in
the money. Not only had gains been made, but Roosevelt's measures
had the curious result of driving poverty out of sight. The applesellers
disappeared from the streets, the breadlines from the cities. If you
lived in a comfortable neighborhood, you had to look hard to find
people on relief.[92]

On November 2, Jim Farley sent the President a detailed report on
election prospects. "After looking them all over carefully and discount-
ing everything that has been given in these reports," Farley wrote the
President, "I am still definitely of the opinion that you will carry every
state but two—Maine and Vermont."[93] On election night, the Roosevelt family and intimates gathered at Hyde Park to sing to Tommy
Corcoran's accordion and await the returns. As the night wore on, it
became clear that Farley's prediction was being borne out precisely.
Roosevelt had been re-elected by the most sweeping electoral margin of
any candidate since James Monroe. The stanchly Republican state of
Pennsylvania had gone Democratic for the first time since Buchanan's
triumph in 1856. Lemke had polled only 882,000 votes in the entire
country. Immediately after the election, Father Coughlin, likening his
fate to that of the persecuted Christ, announced his retirement from
politics.[94]

[92] Childs, *I Write from Washington*, pp. 19, 26; *Today*, V, (February 8,
1936), 24.

[93] Farley to F.D.R., November 2, 1936, FDRL PSF 20. *The New York Times*
commented on Farley's forecast: "Nobody believes this." "If there can be any
political certainty in the confused situation," it observed, "it is that LANDON
will carry more States than HOOVER did four years ago." *The New York Times,*
October 23, 1936. See too Key Pittman to William Pittman, January 30, 1936,
Pittman MSS., Box 20; Jay Hayden to Alf Landon, November 6, 1936, Landon
MSS.

[94] Samuel Lubell has argued that Lemke's vote was the product of Irish-
American Anglophobia and German-American concern that Roosevelt would
lead the country into another war against the Reich, but it seems unlikely that
foreign affairs were that compelling in 1936. Moreover, the Lemke vote does
not appear to correlate on ethnic grounds as neatly as Lubell suggests. Samuel
Lubell, *The Future of American Politics* (New York, 1952), pp. 143–144; Ralph
Smuckler, "Region of Isolationism," *American Political Science Review,* XLVII
(1953), 401; Edward Blackorby, "William Lemke: Agrarian Radical and Union
Party Presidential Candidate," *Mississippi Valley Historical Review,* XLIX
(1962), 80–81.

The President had carried along with him so many Democratic candidates that when the new Congress met in January, 1937, it would be impossible to squeeze all seventy-five Democrats in the customary left side of the Senate chamber, and twelve freshmen would have to sit with the Republicans. The Grand Old Party had become the butt of the jokes of politicians and radio comedians. "As Maine goes," Farley gibed, "so goes Vermont." "The Republican Party," lamented a Pennsylvania judge, "is reduced to the impotency of the Federalist Party after Thomas Jefferson's sweeping victory in 1800." Roosevelt no longer had any political liabilities, observed a writer for *The New York Times*. "If he were to say a kind word for the man-eating shark, people would look thoughtful and say perhaps there *are* two sides to the question."[95] Franklin Roosevelt's fortunes were at high tide.

[95] George Maxey to James J. Davis, January 9, 1937, Davis MSS., Box 24; "Topics of the Times," *The New York Times*, November 22, 1936.

CHAPTER 9

A Farewell to Arms

ON MARCH 5, 1933, one day after Roosevelt entered the White House, the German Reichstag placed absolute power in the hands of Adolf Hitler. Little more than a week before, Yosuke Matsuoka had marched the Japanese delegation out of the League of Nations in Geneva. During the Hundred Days, while Congress debated farm subsidies and banking legislation, Jews were being beaten on the streets of Germany. While Roosevelt's "forest army" planted trees on western hillsides, Hitler was rebuilding the *Reichswehr*. All of the New Deal was to be carried on under the shadow of the menace of fascism.

This was a peril most Americans chose to ignore. Absorbed in meeting the endless personal crises posed by the depression, they had little patience with admonitions to direct their attention abroad.[1] Some of the resistance to involvement in world affairs came from nationalists like Senator William Borah, long convinced the League of Nations was "nothing more than a cog in the military machine of Europe."[2] Perhaps even more intense opposition arose from those who believed that the horrors of World War I made participation in another war unthinkable. Opinion leaders publicly avowed their guilt for leading the country into war in 1917, and resolved they would never again

[1] "Let us turn our eyes inward," urged Pennsylvania's Governor George Earle. "If the world is to become a wilderness of waste, hatred and bitterness, let us all the more earnestly protect and preserve our own oasis of liberty." "Armistice Day Address of George H. Earle, Nov. 11, 1935," Earle MSS., Speech and News File No. 73.

[2] Borah to James Nelson, December 7, 1935, Borah MSS., Box 391.

so abuse the trust people had placed in them. Rabbi Stephen S. Wise announced: "I would as little support a war to crush Hitlerism as a war for the strengthening of Jewish claims in Palestine."[3] Historians as well as journalists claimed that the United States had been tricked into war in 1917 by the twin forces of profiteering and Allied propaganda.[4] The war, historians argued, had been as tragic a blunder as the Civil War; in their appreciation of politics as the art of compromise, revisionists made Bryan the hero of pre-World War I diplomacy, Stephen Douglas the cynosure of the Civil War crisis.[5]

Many argued that if war were to be avoided, Hitler should be accommodated. His plea for "living space" appeared not unreasonable and his desire to unite all Germans under one flag seemed to some a more faithful application of the principle of self-determination than Wilson had achieved at Versailles. In the Albert Shaw lectures in 1935, the highly respected newspaperman Frank Simonds said of the Versailles Treaty: "We, therefore, offered Japan, Germany, and Italy partnership in a peace association in which we were assured lasting prosperity and they were condemned to continuing poverty." By making themselves out to be "have not" nations, Germany and Italy persuaded some Americans that the fascist powers were the sharecroppers of world politics. William Allen White asserted that it was the duty of the United States to call a peace conference at which the demands of "these underprivileged nations—Germany, Japan, Italy" could be considered. "Are men really Christian sufficiently in their heart of hearts," White asked, "to bring justice to those who are underprivileged?"[6]

Buffeted by the gales of isolationism, Franklin Roosevelt trimmed sail. In the 1932 campaign, he dismayed internationalists by bowing to the demand of the militantly chauvinist publisher, William Randolph Hearst, that he disavow his Wilsonian past and announce publicly that he opposed American entrance into the League of Nations. Josephus

[3] *Literary Digest,* CXIII (May 14, 1932), 19.

[4] Probably the most influential book, and one written with more restraint and plausibility than many earlier works, was Walter Millis' *Road to War* (Boston and New York, 1935).

[5] James Randall, "The Blundering Generation," *Mississippi Valley Historical Review,* XXVII (1940), 27.

[6] Frank Simonds, *American Foreign Policy in the Post-War Years* (Baltimore, 1935), p. 62; Walter Johnson, *The Battle Against Isolation* (Chicago, 1944), p. 38; *Today,* V (December 21, 1935), 24.

Daniels complained that Roosevelt was trying "to out-Reed Reed in opposition to the League," while Wilsonians like the Ohio Democrat Wendell Willkie worked for the nomination of the more internationalist Newton Baker. After Roosevelt's nomination, Baker confided: "I have no particular difficulty in supporting Frank if I can convince myself it does not mean that I am supporting Hearst. Unless I can rid myself of that suspicion I shall take to the woods and support nobody."[7] Once in office, Roosevelt continued to voice the rhetoric of isolationism. At San Diego in October, 1935, the President declared: "Despite what happens in continents overseas, the United States of America shall and must remain, as long ago the Father of our Country prayed that it might remain—unentangled and free."[8]

Although Roosevelt's Secretary of State, Cordell Hull, sought to lower trade barriers, the President's closest advisers—men like Moley and Tugwell—were economic nationalists. They argued that the causes of the depression lay within national borders and had to be solved there. Roosevelt himself never accepted wholly the view of these autarchists—he resisted strenuously the advocates of "dumping"—but, as he said in his Inaugural Address, he favored "the putting of first things first."[9]

For a few weeks in the spring of 1933, Roosevelt gave the internationalists hope. He sent a stirring message to the world stressing the need for international economic co-operation, and made plans to participate in an international conference in London to deal with the paralysis of world trade. Yet even before Roosevelt took office so many subjects had been excluded from the agenda that the conference had been reduced to yet another conclave where the most critical questions could not be discussed. Roosevelt's own advisers derided the expectations of the internationalists. Tugwell noted: "The argonauts leave

[7] Josephus Daniels to Newton Baker, March 10, 1932, Baker MSS., Box 84; Wendell Willkie, "Democratic Party's Share in Isolationism," *United States News*, XVII (September 8, 1944), 28–29; Newton Baker to Josephus Daniels, July 6, 1932, Baker MSS., Box 84. Cf. Daniel Roper to William Dodd, August 29, 1932, Dodd MSS., Box 41; Francis Sayre, *Glad Adventure* (New York, 1957), p. 153; David Reay to John W. Davis, January 4, 1932, Reay MSS.

[8] Samuel Rosenman (ed.), *The Public Papers and Addresses of Franklin D. Roosevelt* (13 vols., New York, 1938–50), IV, 410.

[9] "I shall spare no effort to restore world trade by international economic readjustment, but the emergency at home cannot wait on that accomplishment," the President stated. *Public Papers*, II, 14; Raymond Moley, *After Seven Years* (New York, 1939), p. 70.

this morning. Moley pronounces the word sardonically and I think we feel alike that the results of the Conference are sure to be pretty slim —in the vague, hazy sense that Hull means. His rambling, lisping speeches on the evils of protection, and his extolling of the old *laissez-faire* liberal internationalism have become harder and harder to bear."[10] Hull held a trump card: Roosevelt's promise to ask Congress for authority to engage in reciprocal tariff agreements. When Hull sailed from New York on May 31, he had a copy of the bill in his pocket. On the eve of the conference, the President cabled him in London that he would not submit the tariff bill because, if Congress did not adjourn immediately, the paper money wildmen might win the day. Hull, who had hoped to show the bill to other delegates as proof of America's determination to reverse its high-tariff policies, now had nothing to offer. "I left for London with the highest of hopes," he wrote later, "but arrived with empty hands."[11]

No sooner had the International Economic Conference convened in London on June 12, 1933, than a wrangle broke out between the gold-bloc countries headed by France and those, in particular the United States and Great Britain, which had gone off gold. Premier Édouard Daladier called permanent stabilization of currencies "the indispensable condition of a useful economic conference."[12] Roosevelt, too, had favored some kind of stabilization, but as the weeks went by he came to fear that stabilization would jeopardize his attempts to raise domestic prices. As rumors came out of London of an imminent stabilization agreement, security prices broke in New York, and some commodity prices began to fall. Roosevelt quickly vetoed the proposal. One of the delegation's financial advisers, George Harrison, governor of the New York Federal Reserve Bank, sailed for home, and Delegate James Cox cabled the President: "If you love us at all, don't give us another week like this one."[13] The United States delegation, as ill-chosen a

[10] "Notes from a New Deal Diary," May 31, 1933. Cf. Louis Fischer, "Le 'Brain Trust' et la Conférence mondiale," *L'Europe Nouvelle,* XVI (1933), 543–545; Ray Atherton to Jay Pierrepont Moffat, March 28, 1933, Moffat MSS.

[11] Cordell Hull, *The Memoirs of Cordell Hull* (2 vols., New York, 1948), I, 249–255; Robert Ferrell, *American Diplomacy in the Great Depression* (New Haven, 1957), pp. 259–266; Bernard Flexner to Louis Brandeis, September 17, 1933, Brandeis MSS., G5.

[12] Rene Pinon, "La Conférence de Londres," *Revue des Deux Mondes,* XVI (1933), 232.

[13] Arthur Schlesinger, Jr., *The Coming of the New Deal* (Boston, 1959), p.

body as had ever represented this country at an international parley, seemed to have lost what little cohesion it had had. It could not agree on its own policies, let alone offer leadership to the world. The Americans at London, noted one European writer, seemed, like Pirandello's characters, to be "Six Delegates in Search of a Chief."[14]

At this juncture, with the conference foundering, Roosevelt dispatched Raymond Moley to London. What Roosevelt had in mind was then, and is today, an enigma. In the week Moley was on the high seas, the conference did little business. His trip was followed with an attention to detail that had marked no transatlantic journey since Lindbergh's flight—his dash by Navy plane and destroyer to the President's schooner *Amberjack* off Nantucket, his voyage on the *Manhattan*, his dramatic stop at Cobh, and his course from Plymouth to London. There was something both comical and hauntingly sad about the eagerness with which the world's statesmen awaited the message it assumed Roosevelt's Man Friday was carrying to an anxious world. Hull was pushed aside, and Moley was the hero of the hour. The newspapers sang: "Moley, Moley, Moley, Lord God Almighty." Moley quickly gave his approval to a harmless statement, to be issued by the United States, Britain, and the gold-bloc countries, avowing they would do their best to stabilize currencies.[15]

On July 1, Roosevelt pulled the rug out from under Moley. He not only rejected the new compact, but sent Moley a stinging cable which objected to the attempts of other nations to hamstring New Deal price-raising by forcing a commitment to premature stabilization. Two days later, from the cruiser *Indianapolis,* he fired a "bombshell message" which blasted all hopes for immediate economic co-operation. He not only announced to the world that he had repudiated Moley's agreement, but he scolded the other nations as though they were back-

---

216; Ferrell, *Diplomacy in Great Depression,* p. 266; J. F. T. O'Connor MS. Diary, June 16, 1933.

[14] Roger Nathan, "Histoire désabusée de la Conférence de Londres," *L'Europe Nouvelle,* XVI (1933), 592. When Georges Bonnet made a speech which undercut Roosevelt's policies, Cox astonished him by coming up to say: "Your words deserve to be carved in marble." *L'Echo de Paris,* July 4, 1933, cited in Frances Pingeon, "French Opinion of Roosevelt and the New Deal" (unpublished M. A. essay, Columbia University, 1962).

[15] Ferrell, *Diplomacy in Great Depression,* pp. 266–269; Moley, *After Seven Years,* pp. 224–254. For Hull's bitterness at Moley, see "Diary of Trip to Great Britain in 1933," Henry Stimson MSS.

ward schoolchildren. He rejoiced that the "old fetishes of so-called international bankers" were "being replaced by efforts to plan national currencies" and added testily that currency matters could be discussed when the majority of nations learned to balance their budgets and live within their means.[16]

The President's mischievous message destroyed the conference. The American press hailed Roosevelt's dispatch, published on the Fourth of July, as a "new Declaration of Independence," but most European observers were outraged by both the substance and the tone of the remarks. Hull had to struggle to keep the delegates from formally denouncing the United States; Ramsay MacDonald was so incensed that he called Roosevelt "that person." Few could square the President's invective with his message to the nations of the world of May 16, when he had called, among other things, for "stabilization of currencies." It was not the world that had changed but Roosevelt.[17]

There is probably no act of Roosevelt's White House years for which he has been more universally censured. Much of the criticism has been extreme. He was accused of opposing the internationalists led by France, yet France, with its *idée fixe* on gold, was at least as self-centered as the United States. If the United States boggled at tariff reduction, neither the British, the French, nor the Commonwealth nations evidenced any more eagerness to lower tariff walls. Moreover, there is much to be said for Roosevelt's view that stabilization at a low level of prices would have been an error. President Roosevelt, concluded John Maynard Keynes, had been "magnificently right."[18]

Yet the failure of the London Economic Conference was deplorable. The United States may have been no more blameworthy than other countries, but, as the most powerful nation in the world, and as the principal creditor, it had a special obligation to lead. The meeting

[16] *Public Papers,* II, 264–265; Ferrell, *Diplomacy of Great Depression,* pp. 269–272; Schlesinger, *Coming of New Deal,* pp. 218–220; Raymond Moley to the editor of *The New York Times,* February 2, 1948.

[17] Ernest Lindley, *The Roosevelt Revolution: First Phase* (New York, 1933), p. 211; J. Ramsay MacDonald to Cordell Hull, July 5, 1933, FDRL PSF 32; *Public Papers,* II, 186.

[18] Schlesinger, *Coming of New Deal,* p. 223; H. V. Hodson, *Slump and Recovery 1929–1937* (London, 1938), pp. 197–198, 205–206; memoranda of July 4, 6, 8, Samuel Davis McReynolds MSS.; George Peel, "The World Economic Conference: The Result," *Contemporary Review,* CXLIV (1933), 129–137.

marked the last opportunity of democratic statesmen to work out a co-operative solution to common economic problems, and it ended in total failure. Henceforth, international trade would be directed by national governments as a form of bloodless warfare. In the very year that Hitler assumed power, the collapse of the conference sapped the morale of the democratic opponents of fascism. Roosevelt's policies strengthened those forces in Britain and France which argued the futility of international parleys and celebrated the virtues of economic self-interest.[19] The Nazis hailed the President's message as a "bomb" which ended a futile meeting that was only a "remnant of an anti-quated parliamentary order." Roosevelt, declared Hjalmar Schacht, had the same idea as Hitler and Mussolini: "Take your economic fate in your own hands."[20]

By 1934, the nation had come to look to the white-haired, saintly looking Cordell Hull as the Galahad of internationalism in a nation-alist administration. This view of Hull hit well wide of the mark. No saint, Judge Hull was a hard-bitten, hot-tempered Tennessee moun-taineer. He spoke with a slight impediment, and when someone like Moley seemed to undercut him, he would get in a " 'Cwist' mood." His commitment to collective security proved strong on words and weak on deeds. On some crucial issues, he was the most powerful in-fluence for appeasement within the administration. But on the single question of the tariff, Hull's reputation for internationalism was well earned. His economic thinking a blend of Adam Smith and the cotton South, Hull regarded himself as a disciple of "Locke, Milton, Pitt, Burke, Gladstone, and the Lloyd George school."[21]

Hull knew that in the autarchic world of the 1930's no nation would adopt free trade. He advanced instead the idea, once favored by Blaine and McKinley, of reciprocity—of empowering the President,

[19] A British journal commented: "The worst feature about the whole episode is that it will intensify the impression all too prevalent in Europe that America is a country it is impossible to work with." *The Spectator*, CLI (July 7, 1933), 6.

[20] *Die Bank*, August 2, 1933; *Der Angriff*, July 4, 1933; *Völkischer Beobachter*, Norddeutsche Ausgabe, July 13, 1933. Cf. Beniamino de Ritis, "Il Rubicone americano," *Nuova Antologia*, CCCLXVIII (1933), 227–233; *Le Temps*, July 8, 1933.

[21] Grace Tully, *F.D.R. My Boss* (New York, 1949), p. 174; Hull, *Memoirs*, I, 21, 196; William Allen, "The International Trade Philosophy of Cordell Hull, 1907–1933," *American Economic Review*, XLIII (1953), 101–116.

free of the pressures of congressional logrolling, to negotiate agreements with other nations to lower duties. In 1933, Roosevelt had quashed the proposal as impolitic; but by 1934, Hull had won the backing of Henry Wallace, whose enormously influential pamphlet, *America Must Choose*, charted a middle way between total self-sufficiency and internationalism. Wallace's tract won over Republican internationalists like Henry Stimson, who was impressed by Wallace's argument that the only alternative to burgeoning foreign trade was a controlled economy. Roosevelt, disturbed by the extreme economic nationalist course the nation was pursuing and encouraged by new support, asked Congress on March 2, 1934, for authority to negotiate tariff agreements.[22]

The President's request met strenuous opposition from business interests and from Republican congressmen, who objected both to lowering duties and to surrendering yet another congressional prerogative. On March 29, when the House adopted the bill 274–111, only two Republicans voted for it. But the Democratic strength proved great enough to win Senate approval in early June. The Trade Agreements Act of 1934 authorized the President to raise or lower existing tariff rates up to 50 per cent for countries which would reciprocate with similar concessions for American products. In the next four years, Hull concluded eighteen such treaties.

The passage of the act, instead of resolving conflicts over trade policies within the administration, merely shifted the scene of battle. The crux of the difficulty lay in Hull's use of the "most-favored-nation clause"; if he negotiated a treaty with any one country lowering the tariff on particular items, any other nation could do business with the United States on the same terms, so long as it did not discriminate against us. Peek and Moley protested that the United States should make no concessions unless it got specific privileges in return. Moley argued, not unreasonably, that Roosevelt was guilty of hypocrisy in backing Hull's most-favored-nation approach, and then nullifying most of its effect by imposing quotas. Peek damned the most-favored-nation principle as "unilateral economic disarmament." In November, 1935, Peek delivered a caustic philippic against internationalistic policies which so exasperated Roosevelt that he wrote him a tart letter

[22] Schlesinger, *Coming of New Deal*, pp. 83–84; Henry Morrow Hyde MS. Diary, May 11, 1933; Henry Stimson MS. Diary, May 18, 1934.

saying his speech was "rather silly" and that it sounded "like a Hearst paper." Four days later, Peek resigned.[23]

Internationalists rejoiced in the triumph of their principles, and, throughout the decade, the public was given the image of Hull working miracles in breaking the fetters of international commerce. In fact, the reciprocal trade agreements were of little economic consequence. Roosevelt never seemed sure whether they were supposed to be a way to revive world trade by lowering America's tariff wall or a shrewd device to expand America's foreign markets. Certainly they did not redress the imbalance caused by the position of the United States as a creditor nation. The most that can be said for them is that they stopped a bad situation from getting worse. They constituted a more liberal trade policy than that pursued by many other powers in an era of quotas and barter agreements, and they helped remove some discrimination against us. They probably won some political good will, especially in Latin America, and they served as an instrument of economic warfare against the fascist powers.[24]

As a solution to the problem of recovering America's lost foreign markets, reciprocity was a less popular panacea than the recognition of Soviet Russia. Businessmen like James Mooney, vice-president of General Motors, and Thomas Morgan, president of Curtiss-Wright and Sperry Gyroscope, argued that recognition of the U.S.S.R. would lead to a revival of trade with the world's largest buyer of American industrial and agricultural equipment. While the A.F. of L., patriotic societies like the D.A.R., and prominent Catholic leaders opposed any action that might appear to sanction bolshevism, Roy Howard, head of the Scripps-Howard newspaper chain, observed: "I think the men-

[23] Moley to F.D.R., November 30, 1935, FDRL PPF 743; Peek to F.D.R., November 12, 1934, F.D.R. to Peek, November 22, 1935, Peek MSS.; Gilbert Fite, *George N. Peek and the Fight for Farm Parity* (Norman, Okla., 1954), pp. 281–285; George N. Peek, with Samuel Crowther, *Why Quit Our Own* (New York, 1936), pp. 35–37. Peek told his old boss, Bernard Baruch: "His letter was a kind of mean one. He called it silly and put the lie on it and various other things." Telephone conversation of George Peek with Bernard Baruch, November 26, 1935, Peek MSS.

[24] Grace Beckett, *The Reciprocal Trade Agreements Program* (New York, 1951); Power Yung-Chao Chu, "A History of the Hull Trade Program, 1934–1939" (unpublished Ph.D. dissertation, Columbia University, 1957). Cf. Richard Snyder, "Commercial Policy as Reflected in Treaties from 1931 to 1939," *American Economic Review*, XXX (1940), 793; Cordell Hull to Carter Glass, July 28, 1936, Glass MSS., Box 4.

ace of Bolshevism in the United States is about as great as the menace of sunstroke in Greenland or chilblains in the Sahara." The conservative North Carolina Senator Josiah Bailey reflected: "There is no reason to have horror about the Russian movement, if it does prove to be a success." He even thought it might be welcome when contrasted to "the Kreuger, Insull, and Mitchell behavior."[25] Heartened by the shift in sentiment, Roosevelt began to explore gingerly the possibilities of renewed relations with Russia, less, perhaps, out of illusions about the nature of the system or the lure of a trade boom than from the hope that the Soviet Union might serve as a buffer to Japanese and Nazi expansion.[26]

After a series of covert negotiations, Maxim Litvinov arrived in Washington on November 7, 1933, at the invitation of the President. When State Department officials greeted him at Union Station, none of the Americans wore a top hat, since Litvinov did not represent a government officially recognized by the United States.[27] For nine days, the Soviet envoy conferred with the President and State Department officers. Late in the night of November 16, Roosevelt and Litvinov formally exchanged eleven letters and one memorandum which signified the restoration of diplomatic relations between the two powers. The Russians agreed to guarantee religious liberty to Americans in the U.S.S.R. and to curb communist propaganda in this country; the vexing debt question was left for future settlement.

Years later, critics charged that the success of the worldwide communist conspiracy dated from the decision by Roosevelt to defy the wishes of the American people and grant recognition to the Soviet regime. In fact, the President's action won wide support even in conservative quarters. At a farewell dinner for Litvinov at the

[25] James Mooney, "Soviet Trade and the United States," *Economic Review of the Soviet Union,* VI (1931), 223–226; Smith W. Brookhart to Carter Glass, September 27, 1933, Glass MSS., Box 9; Paul Boller, Jr., "The 'Great Conspiracy' of 1933," *Southwest Review,* XXXIX (1954), 102; Frederick L. Schuman to Samuel Harper, February 4, 1933, Harper MSS.; Bailey to Mrs. Josiah Bailey, February 23, 1933, Bailey MSS., Personal File. Cf. Hugh Gibson to E. M. House, February 17, 1933, House MSS.

[26] Robert Paul Browder, *The Origins of Soviet-American Diplomacy* (Princeton, 1953), pp. 99–112; John Blum, *From the Morgenthau Diaries* (Boston, 1959), pp. 54–56; Travis Jacobs, "America Recognizes Russia: A Conspiracy?" (unpublished M.A. essay, Columbia University, 1960).

[27] William Phillips, *Ventures in Diplomacy* (Boston, 1952), p. 157; Phillips, COHC, p. 104.

Waldorf-Astoria, executives of the House of Morgan, the Pennsylvania Railroad, the Chase National Bank, and other firms feted the Soviet emissary. Thomas Watson, president of International Business Machines, asked every American, in the interest of good relations, to "refrain from making any criticism of the present form of Government adopted by Russia."[28] Subsequent events disappointed both friends and foes of recognition. The benefits anticipated did not materialize—there was no substantial increase in trade, the Russians did not noticeably curb propaganda, and they proved a mercurial agency in international politics. They had made the ritualistic promises of good behavior we had asked of them, but that was all. On the other hand, no ill resulted from recognition; the United States had merely acknowledged what all other great powers had long since granted. In short, American recognition of the Soviet Union was an event of monumental unimportance.[29]

At first glance, Roosevelt's "Good Neighbor" policy toward Latin America seemed, like the recognition of Russia, to represent a liquidation of policies of the old regime: the right of intervention and the right to pass judgment on the governments of other nations. Yet it was the old regime, starting with Dwight Morrow's mission to Mexico in 1927, which began the new departure, and Roosevelt's first months in office actually marked a retreat toward imperialism.[30] A short while before Roosevelt took office, he received a memorandum from his friend Sumner Welles, soon to become Assistant Secretary of State in charge of Latin American affairs, which advised that our policies "should never again result in armed intervention by the United States in a sister republic."[31] But when the Cuban government was over-

[28] Meno Lovenstein, *American Opinion of Soviet Russia* (Washington, 1941), pp. 139–146; Bernard Baruch to Louis Howe, November 18, 1933, FDRL PPF 88; Boller, "Great Conspiracy," pp. 111–112; "American-Russian Chamber of Commerce Dinner," *Soviet Union Review,* XI (1933), 257–259; Browder, *Origins,* p. 171.

[29] Cf. Browder, *Origins,* pp. 176–222; William Appleman Williams, *American-Russian Relations, 1781–1947* (New York, 1952), pp. 231–255.

[30] Alexander De Conde, *Herbert Hoover's Latin-American Policy* (Stanford, Calif., 1951); Ferrell, *Diplomacy of Great Depression,* pp. 221–222. Cf. E. David Cronon's contrast of Morrow and Josephus Daniels in *Josephus Daniels in Mexico* (Madison, Wis., 1960), p. 287.

[31] Charles Griffin (ed.), "Welles to Roosevelt: A Memorandum on Inter-American Relations, 1933," *Hispanic American Historical Review,* XXXIV (1954), 190–192.

thrown in 1933, Welles, who took over the U.S. embassy in Havana, asked the President to send a naval armada and even suggested landing troops. To support Welles's adventure as kingmaker, Roosevelt dispatched no less than thirty ships, but neither the President nor his associates would approve even a limited military intervention. According to one apocryphal tale, when Roosevelt sounded out Garner, the Vice-President told him not to intervene. "But suppose they kill Americans?" the President persisted. "Wait and see who the Americans are," Garner replied.[32] In 1933, neither evangelical diplomacy nor the rights of investors aroused much enthusiasm. Yet Roosevelt did abide by Welles's decision to withhold recognition from the regime of Dr. Ramón Grau San Martín, a university professor put in power by a revolutionary junta. Denied recognition, Grau's government was doomed. When he resigned in January, 1934, Grau announced: "I fell because Washington willed it."[33]

Roosevelt's Cuban policy heightened distrust of a President who had been a jingoist in the Mexican crises of the Wilson era and who had once boasted that he had written the Haitian Constitution. Yet by the time Grau stepped down, Roosevelt was already headed in another direction. At Montevideo in December, 1933, Cordell Hull voted to support the proposition that no state had the right to intervene in the affairs of another. A few days later, Roosevelt even more unequivocally renounced the policy of armed intervention.[34] Delighted by the President's statement, which they hailed as *"Un Sensacional Discurso,"* Latin Americans were even more pleased by his deeds. In May, 1934, the United States and Cuba signed a treaty which abrogated the Platt Amendment; in August of that year, the last Marines left Haiti, although the United States continued to supervise Haitian finances; in 1936, America relinquished its right to intervene in Panama. Reciprocal trade agreements brought modest benefits both to the United States and to Latin America. When Roosevelt traveled to South Amer-

[32] Henry Morrow Hyde MS. Diary, Nov. 1, 1933.

[33] E. David Cronon, "Interpreting the New Good Neighbor Policy: the Cuban Crisis of 1933," *Hispanic American Historical Review*, XXXIX (1959), 538–567; Bryce Wood, *The Making of the Good Neighbor Policy* (New York, 1961), pp. 48–117; Robert Smith, *The United States and Cuba* (New York, 1960), pp. 144–157; Francisco Cuevas Cancino, *Roosevelt y la Buena Vecindad* (Mexico City, 1954), pp. 192–194.

[34] Frank Freidel, *Franklin D. Roosevelt: The Apprenticeship* (Boston, 1952), pp. 223–235, 270–285; Hull, *Memoirs*, I, 333–335; *Public Papers*, II, 545.

ica in 1936, he found himself idolized as *"el gran democrata"* whose New Deal served as a model of the kind of reform Latin America needed. Aristocratic, literate, strong-willed, a leader of the people against the plutocracy, Roosevelt was lauded by the spokesman of the Brazilian Congress as "the *Man*—the fearless and generous man who is accomplishing and living the most thrilling political experience of modern times."[35]

The Good Neighbor policy met harsher criticism at home. Catholic leaders threatened Roosevelt with political reprisals if he did not intervene to check anticlerical activities in Mexico. American nationalists were outraged when President Lázaro Cárdenas expropriated almost all the foreign-owned oil industry in Mexico in 1938. Cordell Hull, as deeply attached to vested rights as to international law, believed he was dealing with communist brigands who were flouting the State Department. He persuaded the reluctant Morgenthau not only to boycott Mexican silver but to attack the Mexican peso on the world market. Representative Hamilton Fish suggested that the United States retaliate by running its border west from Nogales, thereby giving Arizona a port on the Gulf of California. "Poor Mexico!" cried the weekly magazine *Hoy*. "So far from God, and so close to the United States."[36]

Happily, the United States ambassador in Mexico City, Josephus Daniels, a black-string-tie country editor who had drunk deeply at the well of Bryanism, sympathized with the land and oil reforms and refused to serve as the agent of American oil interests. At times he carried his opposition to the State Department's stringent line to the point of insubordination, no doubt in part because he knew that when the chips were down he could count on the backing of his former aide, Franklin Roosevelt. In 1941, an amicable settlement was reached.[37] Yet not even this evidence of Roosevelt's good intentions appeased his liberal critics. For years they had called for a posture of nonintervention in Latin America. Now they complained that the

[35] *El Universal* (Mexico City), December 29, 30, 1933; *La Prensa* (Buenos Aires), November 21, 1936; Donald Dozer, *Are We Good Neighbors?* (Gainesville, 1959), p. 31.

[36] E. David Cronon, "American Catholics and Mexican Anticlericalism, 1933–1936," *Mississippi Valley Historical Review*, XLV (1958), 201–230; Patrick O'Brien to Joseph O'Mahoney, February 5, 1935, O'Mahoney MSS.; Cronon, *Daniels*, pp. 185–248; Dozer, *Good Neighbors*, pp. 96–97.

[37] Cronon, *Daniels*, pp. 57–59, 93–94, 115–129, 134–135, 150–152, 248–271;

hands-off formula they had sought permitted vile dictatorships like that of the Dominican Republic's Rafael Trujillo to flourish. They shared *El Universal*'s uneasiness that the Good Neighbor doctrine was "transforming itself into a league of 'mestizo' dictators, with the United States destined to guarantee the slavery of Latin American peoples."[38]

Roosevelt's performance and, even more, his rhetoric seemed to prove that he shared the assumptions of the isolationists. In fact, he was deeply troubled by reports from Europe and Asia. The American consul general at Berlin, George Messersmith, informed Washington that some of the Nazis were psychopaths and others knew no reason.[39] William Dodd, the American ambassador to Germany, said of the Führer: "He is such a horror to me, I cannot endure his presence." Dodd warned that Nazism meant war, that Hitler aimed "to annex part of the Corridor, part of Czechoslovakia and all Austria."[40] News correspondents like Edgar Ansel Mowrer were horrified by what they saw. Frederick L. Schuman wrote from Germany in 1933: "Mowrer tells me he feels like a man who is supposed to go to sleep and say nothing in a room in which another man is murdering a third in the next bed."[41]

American political leaders did not attempt to conceal their distaste for Hitler's persecution of the Jews. When the German ambassador complained about an antifascist incident in Chicago, Hull replied sternly that until the Germans gave up their indefensible policy toward the Jews, they would have to expect such outbreaks all over the world.

Josephus Daniels to Claude Bowers, August 22, 1933, Daniels MSS., Box 698; Jesús Silva Herzog, *Petróleo Mexicano* (México, 1941); Wood, *Good Neighbor*, pp. 203–259.

[38] Dozer, *Good Neighbor*, p. 51. Cf. David I. Walsh to Rafael Trujillo, November 30, 1937, Walsh MSS.

[39] Hull, *Memoirs*, I, 236. Cf. Henry Stimson MS. Diary, January 25, 1934; Edgar Ansel Mowrer to Frank Knox, n.d. [c. November 1933], Knox MSS., Box 1; Leland Stowe, *Nazi Means War* (New York, 1934.)

[40] William Dodd, Jr., and Martha Dodd (eds.), *Ambassador Dodd's Diary, 1933–1938* (New York, 1941), p. 126; Franklin L. Ford, "Three Observers in Berlin: Rumbold, Dodd, and François-Poncet," in Gordon Craig and Felix Gilbert (eds.), *The Diplomats, 1919–1939* (Princeton, 1953), pp. 447–460; Dodd to William Phillips, May 29, 1935, Dodd MSS., Box 44; Dodd to R. Walton Moore, November 5, 1934, Moore MSS., Box 5; Harold Ickes, *The Secret Diary of Harold Ickes* (3 vols., New York, 1954), I, 494.

[41] Schuman to Samuel Harper, May 17, 1933, Harper MSS.

At a huge Madison Square Garden rally attended by Mayor Fiorello La Guardia, Al Smith, and others, twenty thousand people judged Hitler guilty of a crime against civilization. Even some of the pacifists, disturbed by the unanticipated consequences of their beliefs, found themselves supporting collective-security measures. "I believe in being neutral in the sense of non-belligerent," explained Emily Balch, "but not in the sense of treating the aggressor and his victim alike."[42]

Roosevelt, frankly baffled by what to do, sought some way to choke fascist expansion that would not involve a head-on fight with the isolationists. In May, 1933, he promised that the United States would not interfere with League sanctions against an aggressor, if the European powers agreed to a substantial reduction of armaments. John Bassett Moore, the first American to sit on the World Court, denounced this timid offer, hedged with numerous safeguards, as "the gravest danger to which the country had ever been exposed, a danger involving our very independence."[43] Nothing came of the proposal, in part because Hiram Johnson mobilized the Senate isolationists, in part because Roosevelt himself had not yet determined to follow a bold course.[44]

Yet he continued to seek a method both to preserve peace and to bolster those elements in Britain and France which were attempting to curb the fascists. In the spring of 1935, he toyed with the possibility of using his executive powers to support sanctions against Germany, if they were applied. On different occasions, he asked for the right to impose a discriminatory embargo. Alternatively, he explored the idea of some kind of personal intervention, perhaps through a meeting of the chiefs of state. He wrote Dodd: "If five or six heads of the important governments could meet together for a week with complete inaccessibility to press or cables or radio, a definite, useful agreement

[42] Jay Pierrepont Moffat MS. Diary, September 21, 1933; F.D.R. to George Earle, December 22, 1933, in Elliott Roosevelt, *F.D.R.: His Personal Letters, 1928–1945* (2 vols., New York, 1950), I, 379–380; Emily Balch to Eleanor Moore, January 17, 1936, Balch MSS.

[43] Hull, *Memoirs,* I, 228; Cordell Hull to Norman Davis, May 29, 1933, Davis MSS., Box 21; Malcolm Davis, COHC, pp. 159 ff.; "An Appeal to Reason," *Foreign Affairs,* XI (1933), 571.

[44] Robert Divine, "Franklin D. Roosevelt and Collective Security, 1933," *Mississippi Valley Historical Review,* XLVIII (1961), 42–59; Walter Lippmann to Newton Baker, September 4, 1933, Baker MSS., Box 149; Hiram Johnson to Charles K. McClatchy, May 19, 1933, Johnson MSS.

might result or else one or two of them would be murdered by the others! In any case it would be worthwhile from the point of view of civilization."[45] If internationalists, largely unaware of the drift of the President's thinking, despaired because he would not make firm commitments abroad, nationalists feared that the President would not long refrain from using his personal influence in world affairs; that he might even choose foreign war as a way out of domestic crisis. As early as 1935, Charles Beard sounded the alarm: "The Jeffersonian party gave the nation the War of 1812, the Mexican War, and its participation in the World War. The Pacific War awaits."[46]

Beard singled out the Far East as the breeding ground of war, for it had been here that Japan had raised the first serious challenge to the treaty structure of the postwar years when it began military operations in Manchuria in September, 1931. The Japanese seizure of Chinese territory seemed a patent violation of the League Covenant, the Nine-Power Treaty, and the Kellogg-Briand Pact; Japan's actions, as our Ambassador to China informed the State Department with almost comic solemnity, "could not be interpreted as a pacific method of settling a dispute between Japan and China."[47] Convinced that if the world permitted Japan to go unanswered the system of collective security would be destroyed, Hoover's Secretary of State Henry Stimson had sought some way to reweave the fabric of peace. As the evangel of peace, Stimson seemed curiously miscast. An ardent admirer of Teddy Roosevelt's doctrine of the strenuous life, Stimson had written of the "joy of war," bore himself like a soldier, thought even of his civilian career as being "on active service," and followed a code of the warrior not unlike that of the Japanese Samurai.[48] Yet in this instance his instincts were right; peace could be won, to use the main

[45] F.D.R. to Edward House, April 10, 1935, FDRL PPF 222; FDR to Dodd, January 9, 1937, FDRL PSF (Dodd).

[46] Beard, "National Politics and War," *Scribner's*, XCVII (1935), 70. "As I have often said to you," Beard wrote Raymond Moley that spring, "I consider the foreign implications of our domestic policy and the hazards of a futile and idiotic war in the Far Pacific more important than old age pensions and all the rest of it." Beard to Moley, May 18, 1935, FDRL PPF 743.

[47] Conversation with Sir Miles Lampson, September 23, 1931, Nelson Trusler Johnson MSS., "Memoranda of Conversations 1931."

[48] Henry Stimson MS. Diary, October 9, November 19, 1931; Raymond

epithet directed against him, only through war—or rather at the risk of war.

Stimson toyed with the notion of some kind of sanctions against Japan, yet he knew the United States would not go to war over Manchuria, and he was not certain himself that this was a wise course. Instead, he turned to that last refuge of diplomats, the power of public opinion. On January 7, 1932, he dispatched a note which embodied what became known as the "Stimson doctrine," although it had been Hoover who had first suggested applying nonrecognition to Manchuria. Stimson's note stated that the United States would not admit the legality of any situation, treaty, or agreement which might impair the treaty rights of the United States or which was secured in violation of the Pact of Paris.

Although internationalists rallied to Stimson, much of the world was bewildered. The British found it hard to believe that he attached such importance to nonrecognition, a slender reed compared to the powers Geneva already possessed.[49] Neither did any provision of the Kellogg Pact empower any one state to enforce it; Salmon Levinson, the American who had played such an important role in the development of the Pact, was distressed by Stimson's interpretation of it. Few doubted that Japan had violated China's territorial integrity, but many questioned how real that integrity was, especially with respect to Manchuria, over which China had only nominal sovereignty. Even the most ardent collective-security advocates conceded the government was corrupt and its control of its own territory problematic. China, Sir John Simon stated, was but a "geographical expression."[50]

If the Open Door and the integrity of China were venerable American precepts, so too was the tradition that the United States would

---

Clapper MS. Diary, September 23, November 5, 12, 1931; Richard Current, *Secretary Stimson* (New Brunswick, 1954), pp. 10–11, 26, 247–248; Henry L. Stimson and McGeorge Bundy, *On Active Service in Peace and War* (New York, 1948), pp. 89–100.

[49] On the effectiveness of the doctrine, see Robert Langer, *Seizure of Territory* (Princeton, 1947), pp. 59–66, 287–290. Cf. Arnold McNair, "The Stimson Doctrine of Non-Recognition," *British Yearbook of International Law*, XIV (1933), 65–74.

[50] Salmon Levinson to Jane Addams, December 22, 1932, Levinson MSS., Vol. 1; Ferrell, *Diplomacy of Great Depression*, p. 181; Chester Rowell to Ray Lyman Wilbur, January 1, 1932, Rowell MSS.

not go to war to uphold them.[51] Hoover not only firmly opposed sanctions but would not even permit Stimson to use the threat of force. The President, Stimson lamented, had not "the slightest element of even the fairest kind of bluff."[52] Stimson felt, too, often quite unfairly, that the western powers had also let him down; by February, 1932, he had wearied of the thought of drafting yet another note invoking the Nine-Power Treaty, for he feared he would only receive "yellow-bellied responses." Without support in Washington for a large policy, unable to forge an effective coalition of western powers, Stimson complained that he was armed with "spears of straw and swords of ice."[53]

On January 9, 1933, Stimson conferred at Hyde Park for six hours with President-elect Franklin Roosevelt, who a week later issued a statement supporting Stimson's Far Eastern line. Stimson "is going out of office a big man," noted one diplomat, "since he was able to 'sell' his Far Eastern policy to the new Administration."[54] Revisionist historians, echoing this judgment, have seen in the Hyde Park meeting a kind of diabolic succession whereby Stimson transmitted to Roosevelt a "strong" stance in the Orient that led inevitably to World War II.[55] In fact, Roosevelt needed no conversion to a policy of resisting Japanese expansion. He had long had a big-Navy man's preoccupation with the war that would someday come with Japan, and he told Tugwell a few days after his meeting with Stimson that such a war might better come now than later.[56] In his very first week in office, Roosevelt

[51] Paul Schroeder, *The Axis Alliance and Japanese-American Relations 1941* (Ithaca, 1958), p. 181; Stimson and Bundy, *On Active Service*, pp. 243–244.

[52] Stimson MS. Diary, January 26, 1932. Richard Current has contended that Hoover viewed nonrecognition as an alternative to sanctions, while Stimson thought of it as preliminary to sanctions. "The Stimson Doctrine and the Hoover Doctrine," *American Historical Review*, LIX (1954), 513–542. Ferrell denies that there were two distinct doctrines, and stresses the area of agreement between Hoover and Stimson. *Diplomacy of Great Depression*, p. 169.

[53] Current, *Stimson*, p. 98; Stimson and Bundy, *On Active Service*, p. 256. Cf. Elting Morison, *Turmoil and Tradition* (Boston, 1960), pp. 392–402.

[54] *The New York Times*, January 17, 1933; Jay Pierrepont Moffat MS. Diary, January 30, 1933.

[55] Charles A. Beard, *American Foreign Policy in the Making* (New Haven, 1946), pp. 136–143; Charles Tansill, *Back Door to War* (Chicago, 1952), p. viii.

[56] "Memorandum of Conversation with Franklin D. Roosevelt, Monday, January 9, at Hyde Park, New York," Henry Stimson MS. Diary, January 9, 1933; Henry Stimson, COHC, pp. 16 ff.; Ferrell, *Diplomacy of Great Depression*, pp.

discussed with his cabinet the possibility of war with Japan, and Secretary of the Navy Claude Swanson announced plans to build the fleet to treaty strength.[57] Three months later, the President allocated $238 million from NIRA funds for new warships, a move which helped touch off a new naval armaments race in the Pacific.

Despite this ominous beginning, Roosevelt's first term was marked by a moralistic isolationism with respect to the Far East even more pronounced than under Hoover, despite the fact that Japan gave evidence that the success of the Manchurian adventure had whetted appetites for new conquests. In April, 1934, Eiji Amau, spokesman of the Tokyo Foreign Office, claimed special rights for Japan in East Asia; in December of that year, the Japanese government denounced the Washington Naval Treaty; in January, 1936, Japan, unable to win parity, walked out of the London Naval Conference; and in November, 1936, Tokyo signed an Anti-Comintern Pact with Nazi Germany. Nettled by Japanese truculence, the Roosevelt administration responded with a series of sermons by Parson Hull. Even these verbal condemnations proved too strong for our ambassador in Tokyo, Joseph Grew, who believed that they weakened the position of the Japanese moderates, on whose influence he, as well as many other officials, placed inordinate reliance.[58] When Congress passed the Tydings-McDuffie Act in March, 1934, granting the Philippines independence at the end of ten years, its action was taken as fresh evidence of the desire of the United States to withdraw from the Orient.

Midway through Roosevelt's first term, congressional isolationists taught him he could not win approval for even the mildest of internationalist ventures. Early in 1935, the President urged entry into the

---

238–245; Tugwell, "Notes from a New Deal Diary," January 17, 1933; Sumner Welles, *Seven Decisions That Shaped History* (New York, 1951), p. 68.

[57] James A. Farley, *Jim Farley's Story* (New York, 1948), p. 39; William Neumann, "Franklin D. Roosevelt and Japan, 1913–33," *Pacific Historical Review*, XXII (1953), 143–153. Cf. Ferrell, *Diplomacy of Great Depression*, p. 244.

[58] Schroeder, *Axis Alliance*, p. 6; Toshikazu Kase, *Eclipse of the Rising Sun* (London, 1951), p. 32; Norman Davis to F.D.R., October 31, 1934, F.D.R. to Davis, November 9, 1934, FDRL PSF 33; Joseph Grew, *Turbulent Era*, ed. Walter Johnson (2 vols., Boston, 1952), II, 953; "Minutes of the Annual Meeting, American Council, Institute of Pacific Relations," December 18, 1936, Elbert Thomas MSS., Box 21; Hull, *Memoirs*, I, 276–277.

World Court. It was a proposal of little more than symbolic importance at a time when the West placed too much reliance on the value
of symbols in international affairs, but isolationist senators were no
more willing to concede Roosevelt a symbolic than a real victory. "We
are dealing today with one simple proposition—shall we go into foreign politics?" Senator Hiram Johnson responded. "Once we are in, it
does not make any difference whether we are in a little way or
whether we are in a long way . . . once we are in, we are in. . . ."
Huey Long protested: "We are being rushed pell-mell to get into
this World Court so that Señor Ab Jap or some other something from
Japan can pass upon our controversies." In the Senate debate, Minnesota's Thomas Schall shouted: "To hell with Europe and the rest of
those nations!"[59]

With sixty-eight Democrats in the Senate, and the knowledge that
both Coolidge and Hoover had supported the proposition in the past,
Roosevelt felt reasonably confident he would win approval. The isolationist Senator Borah later observed: "We could not count when the
fight began over ten votes. . . ." But the President reckoned without
the kind of outcry that could be stirred up by William Randolph
Hearst, Will Rogers, and, most of all, Father Coughlin. So many
thousands of wires flooded Washington in response to a radio appeal
by Coughlin that messengers carted them by wheelbarrow to the
Senate Office Building.[60] When the vote was taken, fifty-two Senators
approved the Court protocols, thirty-six opposed; the proposal fell
seven votes short of the two-thirds needed. Twenty Democrats had
bolted the President.[61] Roosevelt knew he had not sustained a meaningful defeat, but he was disturbed by its implications. "As to the
thirty-six Senators who placed themselves on record against the principle of a World Court," he wrote Senator Joseph Robinson, "I am

[59] *Time*, XXV (February 11, 1935), 13–15.

[60] Borah to F. F. Johnson, February 8, 1935, Borah MSS., Box 391; Peter
Morris, "Father Coughlin and the New Deal" (unpublished M.A. essay, Columbia University, 1958), pp. 71–73; Charles Sumner Hamlin MS. Diary, February
8, 1935; National Council for the Prevention of War files, Drawer 75. For the
important role of Midwestern opposition, see Ray Billington, "The Origins of
Middle Western Isolationism," *Political Science Quarterly*, LX (1945), 60.

[61] "The action of the Senate was positively disgraceful," reflected the Wilsonian Carter Glass, "but the combination of Hearst, Father Coughlin and Huey
Long was too much for a few Senators." Glass to Mrs. Joseph Ruebush, February 18, 1935, Glass MSS. See Denna Frank Fleming, *The United States and
the World Court* (Garden City, N.Y., 1945), pp. 117–137.

inclined to think that if they ever get to Heaven they will be doing a great deal of apologizing for a very long time—that is if God is against war—and I think He is."[62]

The defeat of the World Court proposal can be explained in part by the alarm aroused by the findings of the Nye Committee. On February 8, 1934, Senator Gerald Nye of North Dakota, a progressive Republican, had urged an investigation of the armaments industry. Nye was aided by the timely publication of a striking indictment of the munitions makers in the March, 1934, issue of the business magazine *Fortune*. In "Arms and the Men," *Fortune* stated the "axioms" of the armament titans were "(a) prolong war, (b) disturb peace," and charged that in World War I English and French industries had maintained the Kaiser's war drive by selling Germany vital supplies for armaments. On April 12, the Senate, indignant at the arms traffic, approved the resolution and Vice-President Garner named Nye to head the inquiry.[63]

The Nye investigation turned up some startling evidence to buttress *Fortune*'s contentions. It made public a letter from an officer of Electric Boat to his agent in Peru: "It is too bad that the pernicious activities of our State Department have put the brake on armament orders from Peru by forcing the resumption of formal diplomatic relations with Chile."[64] But by the end of 1934 it had switched its emphasis from a probe of arms makers to an exposé of the manner in which economic interests had influenced American intervention on the side of the Allies in 1917. After haling J. P. Morgan and the four du Pont brothers to the stand, the committee charged that Morgan had profited handsomely as agent for the Allies, and that bankers and munitions makers had reaped unconscionable profits. The United States had entered the war, the committee concluded, neither to save the world for democracy nor to defend its own interests, but as the result of the intrigues of profiteers. Nye declared bluntly: "When Americans went into the fray, they little thought that they were there

[62] F.D.R. to Robinson, January 30, 1935, *Personal Letters*, I, 450. Cf. F.D.R. to Elihu Root, February 9, 1935, FDRL PPF 2201.

[63] "Arms and the Men," *Fortune*, IX (March, 1934), 53–57; Hull, *Memoirs* I, 398, and "Memorandum Cooperation with the Nye Committee," FDRL PSF 33; Dorothy Detzer, *Appointment on the Hill* (New York, 1948), pp. 155–157.

[64] U.S. Congress, Senate, *Munitions Industry*, Hearings before the Special Committee Investigating the Munitions Industry, U.S. Senate, 73d Cong., Pursuant to S. Res. 206, Part I, Sept. 4–6, 1934 (Washington, 1934), p. 199.

and fighting to save the skins of American bankers who had bet too boldly on the outcome of the war and had two billions of dollars of loans to the Allies in jeopardy."[65]

The Nye Committee reinforced the image of Woodrow Wilson as the evil American. In Dos Passos' *U.S.A.*, Wilson is Satan. Savage says of him: "A terrifying face, I swear it's a reptile's face, not warm-blooded. . . ."[66] The Nye Committee rubbed old wounds when it declared that Wilson had "falsified" when he had denied knowledge of the secret treaties. Old Wilsonians like Carter Glass were outraged by Nye's reflection on Wilson's integrity; Democrats queried whether the Republican chairman had not turned the inquiry into a partisan vendetta; and Senator Tom Connally sputtered: "Some checker-playing, beer-drinking back room of some low house is the only place fit for the kind of language which the Senator from North Dakota . . . puts into the *Record* about a dead man, a great man, a good man. . . ."[67] But despite the uneasiness of the Democrats, and in the teeth of attempts by the administration to undermine the probe, the Nye Committee made its mark.

Under the stimulus of the Nye Committee revelations and the threat of war posed by Hitler's announcement of rearmament plans in 1935 and the mounting crisis in Ethiopia, congressional isolationists pushed for neutrality legislation which would define American conduct in the event of a European war. Roosevelt, impressed by an article in *Foreign Affairs* written by Charles Warren, former Assistant Attorney General under Wilson, had already directed the State Department to look into the question.[68] But Roosevelt, like Hoover before him, wanted dis-

[65] Charles Seymour, *American Neutrality 1914–1917* (New Haven, 1935), p. 85. Even Franklin Roosevelt, who viewed the Nye investigation with disfavor, seems to have been impressed by some of its reasoning. In October, 1934, he wrote Josephus Daniels: "Would that W.J.B. had stayed on as Secretary of State—the country would have been better off." Freidel, *Roosevelt: The Apprenticeship*, p. 250. Cf. John Edward Wiltz, "The Nye Committee Revisited," *Historian*, XXIII (1961), 211–233.

[66] John Dos Passos, *1919* (New York, 1932), p. 375; Melvin Landsberg, "A Study of the Political Development of John Dos Passos from 1912 to 1936" (unpublished Ph.D. dissertation, Columbia University, 1959).

[67] *Congressional Record*, 74th Cong., 2d Sess., p. 502. Cf. Josephus Daniels to Newton Baker, February 3, 1936, Baker MSS., Box 84; F.D.R. to Edward House, September 17, 1935, House MSS.; letters in Carter Glass MSS., Box 8.

[68] Charles Warren, "Troubles of a Neutral," *Foreign Affairs*, XII (1934), 377–394; "A Memorandum on Some Problems Arising in the Maintenance and

cretionary power to decide whether an embargo should be applied and, if applied, to which nation. Congress opposed giving the President this prerogative, for it smacked too much of the distinction between good and bad nations, which it thought had led the United States into the World War. Even more, it seemed a second cousin to League of Nations sanctions.

Roosevelt and Hull preferred no neutrality legislation at all to a law that would shackle the President; but when week after week passed in the Congress of the Second Hundred Days without any sign of White House action, many of the President's supporters became mutinous. Representative Fred Sisson of New York wrote the President's aide, Stephen Early, angrily: "Thousands and thousands of women's votes have been lost by this stalling." "I tell you, Steve," Senator Pittman told Early on the phone, "the President is riding for a fall if he insists on designating the aggressor in accordance with the wishes of the League of Nations. . . . He will be licked as sure as hell."[69] Roosevelt pleaded, but he got nowhere. Congress not only would not grant discretionary powers but, with headlines black with imminence of war in Ethiopia, it refused to adjourn until it had tied the President's hands.[70]

With Borah's help, Senator Pittman drafted a compromise which provided for a mandatory embargo to be applied not to all war materials but only to implements of war to belligerents. It would expire February 29, 1936. The measure prohibited American vessels from carrying munitions to nations at war, and empowered the President at his discretion to withhold protection from U.S. citizens traveling on belligerent vessels. It took the Senate only twenty-five minutes to rush the resolution through. On August 31, the President signed it. A resolution which satisfied neither side, the Neutrality Act of 1935 served chiefly to manacle the President while Congress was recessed. Hiram

---

Enforcement of the Neutrality of the United States," by Charles Warren, R. Walton Moore MSS., Box 13.

[69] Early memorandum of August 16, 1935, FDRL OF 1561; Pittman to Early, August 19, 1935, FDRL PPF 745.

[70] "Of what avail is it to settle the utility issue, as close as that is to my heart, to settle monetary policies, to settle strike issues, or anything else, when the peace of the whole world is involved in what we may do here and our civilization may be at stake?" asked Washington's Senator Homer Bone. *Congressional Record,* 74th Cong., 1st Sess., p. 13776. See National Council for the Prevention of War files, Drawer 80.

Johnson rejoiced: "So today is the triumph of the so-called isolationists and today marks the downfall . . . of the internationalist who has been devoting his gigantic energy in the last seventeen years to involving us in marching abroad and who would take us into Europe's troubles and into Europe's difficulties. . . ."[71]

When Italian troops invaded Ethiopia early in October, 1935, Roosevelt, who had dragged his feet during the neutrality debates, acted with unanticipated swiftness in invoking the Neutrality Act. Although technically no state of war yet existed, he issued a proclamation of neutrality immediately. "They are dropping bombs on Ethiopia—and that is war," he told Harry Hopkins. "Why wait for Mussolini to say so."[72] He not only imposed an arms embargo but cautioned Americans not to travel on ships of either belligerent, and warned businessmen that any trade whatsoever with the belligerents, even in raw materials, would be conducted at their own risk. When, despite this warning, exports to Italy mounted, Hull invoked a "moral embargo" on shipments in abnormal quantities to belligerents; Hull's list of commodities included oil, which the League sanctions list omitted. Roosevelt won commendation from the isolationists for his vigorous enforcement of the Neutrality Act, but the President knew well that the more he enforced the act, the more he assisted sanctions against Italy. By warning Americans not to sail on transatlantic vessels of belligerents, for example, he aimed to cripple Italian passenger trade by removing Americans from the danger of mythical Ethiopian submarines. While Roosevelt cleverly kept his actions divorced from those of Geneva, he moved in the direction of collective security.[73]

Beyond this policy of "malevolent neutrality," Roosevelt would not go. He would not commit the United States to a concert of powers against Italy, and, given the state of American opinion, he hesitated to recommend an embargo on oil.[74] When the move to impose effec-

---

[71] *Congressional Record,* 74th Cong., 1st Sess., p. 14432. The nationalist Johnson did not like the resolution itself. See Robert Divine, *The Illusion of Neutrality* (Chicago, 1962), pp. 81–121.

[72] Robert Sherwood, *Roosevelt and Hopkins* (New York, 1948), p. 79.

[73] Hull, *Memoirs,* I, pp. 428–443; Joseph Green, "Memorandum," September 14, 1935, R. Walton Moore MSS., Box 13; William Phillips MS. Diary, VI, October 7, 1935, 1130; Henderson Braddick, "A New Look at American Policy during the Italo-Ethiopian Crisis, 1935–36," *Journal of Modern History,* XXXIV (1962), 64–73.

[74] Philip Marshall Brown, "Malevolent Neutrality," *American Journal of*

tive economic sanctions on Italy collapsed, the League powers blamed the United States. Exporters had disregarded Hull's moral embargo; in the last three months of 1935, American shipments of oil to Italy tripled. Yet the failure of the United States to act, while it no doubt inhibited the western European powers, seems to have served mainly as a convenient rationalization for their own indecisiveness.[75] Neville Chamberlain noted in his diary: "U.S.A. has already gone a good deal further than usual. . . . If we backed out now because of Mussolini's threats, we should leave the Americans in the air. . . ."[76] The United States, in turn, was shocked by the revelation on December 11, 1935, of the cynical agreement between Sir Samuel Hoare and Pierre Laval, the British and French foreign ministers, to partition Ethiopia. The Ethiopian war, which might have aroused the West to the dangers of fascism—Vittorio Mussolini reported that bombing and burning Abyssinian natives had been "exceptionally good fun"—served rather to reconfirm isolationists in their suspicion of European democracies and to demonstrate to fascist leaders they had little to fear from a divided West.

By 1936, Roosevelt had almost given up the fight for control of foreign policy. The administration's neutrality proposal that year conceded much to the isolationists. Yet since it authorized the President to limit exports of war materials to belligerents to "normal quantities," Roosevelt's critics charged he was trying to secure not neutrality but co-operation with League sanctions against Italy.[77] Politicians of both parties feared to offend Italian-Americans who kicked up a storm against the bill; in Faneuil Hall, Governor James Michael Curley

---

International Law, XXX (1936), 88–90; F.D.R. to J. David Stern, December 11, 1935, FDRL PPF 1039.

[75] M. J. Bonn, "How Sanctions Failed," *Foreign Affairs*, XV (1937), 350–361; Gaetano Salvemini, *Prelude to World War II* (Garden City, N.Y., 1954), pp. 331–338, 372–373, 379–397; James T. Shotwell and Marina Salvin, *Lessons on Security and Disarmament from the History of the League of Nations* (New York, 1949), pp. 92–109; Herbert Feis, *Seen from E.A.* (New York, 1947), pp. 195–275, 297–308.

[76] Keith Feiling, *The Life of Neville Chamberlain* (London, 1946), p. 272. Cf. Viscount Simon, *Retrospect* (London, 1952), p. 212.

[77] R. Walton Moore, "The Proposal to Make the Provisions of Any Neutrality Legislation that May be Passed Inapplicable to the Italian-Ethiopian War," January 31, 1936, Moore MSS., Box 14; Telephone Conversation of George Peek with Raymond Moley, November 27, 1935, Peek MSS.

praised Mussolini as the defender of peace and Christianity.[78] The influential international lawyer Edwin Borchard accused Roosevelt of seeking to "commit this nation to the idiosyncrasies of Wilsonian ideology," especially the malignant doctrine of "denouncing any revolter as an aggressor and inciting the *posse comitatus* to strangle and crush him. . . . These sanctionists," wrote Borchard, "have destruction and murder in their hearts." Isolationists held the whip hand, yet they were so divided amongst themselves that no real revision of the 1935 resolution was possible. Some, like Senator Arthur Vandenberg of Michigan, favored cutting off almost all trade with the outside world if war broke out abroad, while nationalists like Borah insisted that this was craven and the trader in wartime should have the "protection of his Flag."[79] After six weeks of dispute, Congress simply extended the embargo and limited the extension of loans to belligerents.[80]

When on July 17, 1936, the Spanish Army revolted in Morocco, crossed to the mainland, and plunged Spain into civil war, Roosevelt took an even more isolationist stand than some of the neutrality legislation advocates. He accepted the British and French view that only strict nonintervention held any hopes for localizing the Spanish conflict and avoiding a dreadful world war. He treated the situation in Spain as though it were a conflict of two foreign states with equal rights, not as an uprising against a recognized government, and a democratic government at that. He imposed a moral embargo on the sale of arms and munitions to Spain, denounced a Jersey City exporter who shipped some planes and munitions to the Loyalists as "thoroughly unpatriotic," and asked Congress to tighten the neutrality statute by embargoing shipments of military material to countries engaged in civil wars. In January, 1937, Congress swiftly complied, with but a single dissenting vote.[81]

[78] *Il Progresso,* January 28, 1936; John Norman, "Influence of Pro-Fascist Propaganda on American Neutrality, 1935–1936," in Dwight Lee and George McReynolds (eds.), *Essays in Honor of George Hubbard Blakeslee* (Worcester, Mass., 1949), pp. 193–214. Italian-Americans engaged in street brawls with Negroes, who identified with Ethiopia and who rejoiced in Joe Louis' victory over Primo Carnera.

[79] Borchard to William Borah, February 22, 1936; Borah to Charles Jervey, January 14, 1936, Borah MSS., Box 393.

[80] Robert Divine, *Illusion of Neutrality,* pp. 122–161. In 1934, Congress had passed the Johnson Act forbidding citizens to lend money to, or buy securities from, foreign governments in default to the United States.

[81] F. Jay Taylor, *The United States and the Spanish Civil War* (New York,

In a few months, some who had voted for the resolution had second thoughts. The Spanish "civil war," they feared, was being used by Hitler and Mussolini as a testing ground for the great war that lay ahead. Tens of thousands of Italian soldiers fought for Franco, and the Fascist press celebrated the fall of Malaga as a national victory. The Nazis made no attempt to disguise the airplanes that bombed Bilbao.[82] A number of isolationist progressives came to believe the threat to democratic government more compelling than the doctrines of neutrality. Senator Nye, to the discomfiture of the pacifists, introduced a resolution to repeal the embargo on arms to the Loyalists. Both Secretaries Ickes and Morgenthau supported Nye, and fifteen prominent scientists, including Arthur Compton and Harold Urey, pleaded with the President to lift the embargo to "save the world from a fascist gulf." In January, 1938, sixty members of Congress ostentatiously sent greetings to the Spanish Cortes.[83] For many American intellectuals, the Spanish war was the crucial event of the decade, for it signified an apocalyptic struggle between the forces of democracy and fascism. Placards on college bulletin boards announced: "We dance that Spain may live." A few did more than that. Two or three thousand American volunteers fought in the Abraham Lincoln Brigade and other Loyalist units; most who went died there, including Ring Lardner's son James, who lost his life in the Ebro campaign.[84]

For a brief time in the spring of 1938, it appeared that Roosevelt would lift the embargo, but the American ambassador to Great Britain, Joseph Kennedy, warned him that such a move might spread the Spanish war to the rest of the world. Hull insisted that such action would destroy the work of the Nonintervention Committee.[85] The President also had to reckon with the state of American opinion. Most

---

1956), pp. 56–85; *Public Papers,* V, 620–623; Hull, *Memoirs,* I, 475; R. Walton Moore to F.D.R., January 5, 1937, Moore MSS., Box 14.

[82] The Loyalists received Russian and, for a time, French aid, but not nearly as much as Franco obtained from the Fascist powers.

[83] Taylor, *Spanish Civil War,* pp. 169, 172–173; Frederick Libby to Gerald Nye, May 21, 1938, National Council for the Prevention of War files, Drawer 62; *Science News Letter,* XXXIII (1938), 298; Harold Ickes to F.D.R., January 26, 1939, FDRL PSF 17; *The New York Times,* January 31, 1938; Lewis Schwellenbach to George Flood, March 19, 1938, Schwellenbach MSS., Box 1, LC.

[84] Allen Guttmann, *The Wound in the Heart* (New York, 1962).

[85] Taylor, *Spanish Civil War,* pp. 174–175; Henry Morgenthau, Jr., "Morgenthau Diaries: III," *Collier's,* CXX (October 11, 1947), 79.

of the country was indifferent; one poll found the remarkably high figure of 66 per cent of respondents neutral or with no opinion. Pro-Loyalist intellectuals were a negligible political force compared to the large bloc of pro-Franco Catholics. Although Catholic laymen like Kathleen Norris and George Schuster were anti-Franco, and most Catholics did not regard themselves as Franco supporters, the Catholic press and hierarchy were almost uniformly pro-Franco, and polls revealed that the proportion of Catholics who backed Franco was more than four times as great as the proportion of Protestants.[86] Ickes has recorded that Roosevelt told him that to raise the embargo "would mean the loss of every Catholic vote next fall and that the Democratic Members of Congress were jittery about it and didn't want it done." So the cat was out of the bag, Ickes complained, the "mangiest, scabbiest cat ever."[87]

Although it has been argued that the neutrality laws handcuffed Roosevelt in his efforts to check the aggressors, Congress can scarcely be assigned sole responsibility for American policy toward Spain. Ironically, it was Senator Nye, the symbol of congressional isolationism, who led the move to lift the embargo, while Roosevelt, who had originally opposed such legislation, upheld it. The President's Spanish policy had unfortunate consequences. It helped sustain Neville Chamberlain's disastrous policy of appeasement which permitted Germany and Italy to supply Franco while the democracies enforced "nonintervention" against themselves. "My own impression," wrote Ambassador Claude Bowers in July, 1937, "is that with every surrender beginning long ago with China, followed by Abyssinia and then Spain, the fascist powers, with vanity inflamed, will turn without delay to some other country—such as Czechoslovakia—and that with every surrender the prospects of a European war grow darker."[88]

In 1937, Congress moved toward the creation of "permanent" neutrality legislation. When Senator Pittman learned that the president of

[86] George Gallup and Claude Robinson, "American Institute of Public Opinion Surveys, 1935–1938," *Public Opinion Quarterly,* II (1938), 389; Taylor, *Spanish Civil War,* pp. 152–155; Theodore Newcomb, "The Influence of Attitude Climate Upon Some Determinants of Information," *Journal of Abnormal and Social Psychology,* XLI (1946), 291–302.

[87] Ickes, *Diary,* II, 390. The influence of Catholic opinion on Roosevelt has not been established with any precision, however.

[88] Taylor, *Spanish Civil War,* p. 15.

the British Board of Trade was to pay a visit to the United States, he filed a new neutrality proposal without consulting the Department of State in order to escape the imputation that America was working in concert with Great Britain.[89] The debate that followed revealed both this same skittish concern over avoiding even the appearance of commitment to collective security and the continuing division among isolationists over how to achieve their ends. As a compromise between the desire to avoid war and the fear that a total embargo would cripple the economy, legislators devised the procedure of "cash and carry": belligerents who bought goods from this country had to pay for them on delivery and cart them away in their own vessels, or in the bottoms of some country other than the United States. This proposal provoked the wrath not only of internationalists but of nationalists like Borah, who denounced "the whole sordid cowardly cash and carry proposition" as an abandonment of national honor and a sacrifice of freedom of the seas.[90] But most of the Senate was convinced that this was precisely what had led to American intervention in World War I. The danger that the country might be "dragged in again because the huckster finds himself in trouble [rises] like Banquo's ghost to disturb all of us," Senator Bone explained. Persuaded by the Bone-Nye thesis, Congress adopted the Pittman resolution. In addition to cash and carry, it extended the arms embargo and the ban on loans, forbade American citizens to sail on belligerent ships, and authorized the President, at his discretion, to prohibit the export of certain other materials to belligerent powers. The law, jeered the New York *Herald-Tribune*, should have been entitled: ". . . an act to preserve the United States from intervention in the War of 1914–'18."[91]

[89] Breckinridge Long MS. Diary, January 24, 1937; Edwin Borchard to William Borah, March 1, 1937, Borah MSS., Box 405.

[90] William Borah to Archibald MacLeish, March 15, 1937, Borah MSS., Box 405. Borah announced: "I would sell and ship and I would fight for the right to do so." *Congressional Record,* 75th Cong., 1st Sess., p. 1681. Borah opposed selling munitions, but he favored open trade in other goods in part because cash and carry gave the advantage to the nation that controlled the seas, Great Britain, in any European war. Orde Pinckney, "William E. Borah: Critic of American Foreign Policy," *Studies on the Left,* I (1960), 57. While most business journals espoused strict neutrality, foreign traders wanted their right to seek profits in wartime upheld. Roland Stromberg, "American Business and the Approach of War 1935–1941," *Journal of Economic History,* XIII (1953), 65.

[91] *Congressional Record,* 75th Cong., 1st Sess., p. 1675; New York *Herald-Tribune,* May 1, 1937. Cf. R. Walton Moore to F.D.R., March 4, 1937, Moore

Within a few months after the passage of the Neutrality Act of 1937, the Western democracies confronted new challenges. In July, after a clash at night between Chinese and Japanese troops at the Marco Polo Bridge, just west of Peiping, the Japanese pushed into North China and inflicted heavy losses on civilians in Shanghai. In September, Mussolini harangued eight hundred thousand Germans on the Maifeld in Berlin. Disturbed by the new Japanese assault and the growing Rome-Berlin *rapprochement,* the President pondered making some kind of public statement which would alert the country to the dangers and the world to the need to search out new roads to peace.[92]

On October 5 in Chicago, capital of the isolationist heartland, Roosevelt warned that if aggression triumphed elsewhere in the world America could expect no mercy and, in a striking phrase which immediately caught the attention of the world, proposed a "quarantine" of the war contagion. Isolationist congressmen responded to Roosevelt's address by threatening him with impeachment, the *Wall Street Journal* insisted: "Stop Foreign Meddling; America Wants Peace," and the Chicago *Tribune* charged that he had turned the city into "the center of a world-hurricane of war fright." "It's a terrible thing," he later said, "to look over your shoulder when you are trying to lead— and to find no one there."[93] It has become a convention of historians to explain that Roosevelt in his quarantine speech had sought to lead the country in an internationalist direction, only to find that the people

MSS., Box 14; Raymond Leslie Buell, "The Neutrality Act of 1937," *Foreign Policy Reports,* XIII (1937), 166–180; Robert Divine, *Illusion of Neutrality,* pp. 162–199.

[92] Dorothy Borg, "The Meaning of the Quarantine Speech," microfilm, FDRL. The President deliberately chose not to recognize a state of war in China, so that he would not have to impose an arms embargo. For the debate over this decision, see Blum, *Morgenthau Diaries,* p. 481; Hull, *Memoirs,* I, 557–558; Gerald Nye to Stephen Raushenbush, September 20, 1937, National Council for the Prevention of War files, Drawer 80; R. Walton Moore, "Reasons for Not Invoking the Neutrality Act of May 1, 1937, in Connection with the Far Eastern Situation," October 19, 1937, Moore MSS., Box 14.

[93] *Public Papers,* VI, 406–411; James Burns, *Roosevelt: The Lion and the Fox* (New York, 1956), p. 318; Philadelphia *Inquirer,* October 9, 1937; *Wall Street Journal,* October 8, 1937; Chicago *Tribune,* October 6, 1937; Dorothy Detzer to Gertrud Baer, October 27, 1937, Emily Balch MSS., Box 3; Samuel Rosenman, *Working with Roosevelt* (New York, 1952), p. 167.

were not ready. In fact, much, perhaps most, of the country responded favorably to the address. Even the Chicago *Tribune* conceded that the response had been "gratifying" to the internationalists; press surveys showed, as *Time* concluded, "more words of approval . . . than have greeted any Roosevelt step in many a month"; and most of the letters he received strongly endorsed his talk.[94]

If Roosevelt ever intended his speech to indicate American commitment to a system of collective security, he lost no time in backing away from these implications. The very next day, he told reporters: "Look, 'sanctions' is a terrible word to use. They are out of the window." He refused to clarify how "quarantine" differed from sanctions, or to concede that, if quarantine did not imply coercion, he intended only "moral" condemnation. Privately, he confided to Norman Davis that he regarded "quarantine" as a much milder term than ostracism or similar synonyms.[95] In short, the quarantine speech, often described as a turning point in American foreign policy, changed nothing at all.

The degree to which the United States still pursued an isolationist line quickly became apparent. The day after Roosevelt's address, the League of Nations Assembly adopted a report censuring Japanese aggression and proposed a conference of the Nine-Power Treaty signatories, including two nonmembers of the League, the United States and Japan. The Nine-Power meeting at Brussels was doomed to failure. The American diplomat Jay Pierrepont Moffat commented: "I have never known a conference where even before we meet people are discussing ways to end it."[96] Japan would not attend the meeting, and the other nations could barely bring themselves to go through the perfunctory routines. When Norman Davis, for a time, considered taking a firm position, Hull sternly ordered him to limit himself to "a strong re-affirmation of the principles which should underlie international re-

[94] Chicago *Tribune,* October 9, 1937; *Time,* XXX (October 18, 1937), 19; Frank Knox to George Messersmith, October 6, 1937, Knox MSS., Box 1; Dorothy Borg, "Reaction to the Quarantine Speech," microfilm, FDRL; Public Reaction Letters, Public Addresses, October 5, 1937, FDRL PPF 200B, Boxes 82–84; F.D.R. to Edward House, October 19, 1937, FDRL PPF 222; Travis Jacobs, "Roosevelt's 'Quarantine Speech,' " *The Historian,* XXIV (1962), 483–502.

[95] Press Conference No. 400, October 6, 1937, FDRL; Jay Pierrepont Moffat MS. Diary, October 28, 1937.

[96] Moffat MS. Diary, November 1, 1937.

lationships." The conference, observed the correspondent of *The New York Times,* was "bathetic" and would soon be "pathetic." Amidst British allegations that Roosevelt, who had aroused great hopes, had extended nothing more than "platonic support," the Brussels parley collapsed. It had demonstrated that the West would respond to aggression with nothing but words, and it had more than justified the expectation of some delegates that the United States, confronted by the need to act, would "curl up."[97]

Three weeks later, the Japanese military revealed how far its contempt for the United States could go. On December 12, 1937, Japanese war planes bombed the gunboat U.S.S. *Panay,* lying at anchor in the Yangtse, flying an American flag and with two large American flags spread out on her awnings. The planes went on to bomb three Standard Oil tankers, setting two of them afire and requiring the third to be beached. Japanese aircraft, spitting machine-gun fire, riddled the deck of the *Panay* and strafed one of its small boats carrying the wounded to shore. Two crew members and an Italian journalist died in the bombing; eleven were seriously injured. Two hours after the attack, the *Panay* rolled over and sank to the bottom of the muddy Yangtse.[98]

Roosevelt and some members of his cabinet responded angrily to the *Panay* attack. The President asked Secretary Morgenthau to explore whether he could seize Japanese property in the United States to hold against payment for damages, and dispatched Captain R. E. Ingersoll of the Navy to London to discuss with the Admiralty what to do if the United States and England found themselves at war with Japan. Hull asked the Japanese ambassador bluntly when "these wild, runaway, half-insane Army and Navy officials were going to be properly dealt with." Secretary of the Navy Swanson and the admirals wanted war, Ickes noted. Ickes added: "I confess that Swanson's point of view cannot be lightly dismissed. Certainly war with Japan is in-

[97] U.S. State Department, *Foreign Relations of the United States, 1937* (5 vols., Washington, 1954), IV, 183–188; Memorandum, October 20, 1937, Norman Davis MSS., Box 4; *The New York Times,* November 21, 1937; *The Times* (London), November 25, 1937; Jay Pierrepont Moffat MS. Diary, November 21, 1937.

[98] *The New York Times,* December 13–31, 1937; U.S. State Department, *Papers Relating to the Foreign Relations of the United States, Japan: 1931–41* (2 vols., Washington, 1943), I, 532–547; Joseph Grew MS. Diary, December 13, 1937 *et seq.* The master of a Standard Oil tanker was also killed in the bombing.

evitable sooner or later, and if we have to fight her, isn't this the best possible time?"[99]

Any thought of war was ruled out not only by Tokyo's willingness to apologize but even more by the mood of the American people, who refused to treat the deliberate sinking of an American vessel by Japanese warplanes as a *casus belli.* "The gunboat Panay is not the battleship Maine," commented the *Christian Science Monitor.* Congressional isolationists insisted that the *Panay* affair demonstrated not the need for a more forthright position in foreign affairs but that American ships in China waters should be recalled. "What are they doing there anyway?" Senator Shipstead asked. "How long are you going to sit there and let these fellows kill American soldiers and sailors and sink our battleships?" Secretary Morgenthau stormed at his Assistant Secretary, Wayne Taylor. "A helluva while," Taylor retorted.[100] On Christmas Eve, the Japanese promised to pay every penny the Americans asked in indemnity, noted that the commander of naval forces in Japan had been recalled, and pledged to avoid a recurrence. Secretary Hull, while refusing to grant Tokyo's contention that the bombing had been accidental, accepted the note as "responsive" and the incident was closed.

The *Panay* crisis sped congressional action on a resolution introduced by Louis Ludlow to amend the constitution to provide that, save in case of invasion, the United States could engage in war only when a majority so voted in a national referendum. The Ludlow amendment had wide support, notably from *Good Housekeeping,* which encouraged mothers to urge congressmen to back the referendum.[101]

---

[99] Blum, *Morgenthau Diaries,* p. 486; U.S. State Department, *Papers Relating to the Foreign Relations of the United States, Japan: 1931–1941,* p. 530; Ickes, *Diary,* II, 274. Moffat wrote: "Feeling here is running dangerously high." Moffat to Joseph Grew, December 18, 1937, Moffat MSS., Correspondence, 1937.

[100] *Christian Science Monitor,* December 13, 1937; *The New York Times,* December 14, 1937; Blum, *Morgenthau Diaries,* p. 488. Poll data indicated a sharp increase in popular desire to have American forces withdrawn from China, rather than taking steps "to make them respect our rights." T. A. Bisson, *American Policy in the Far East, 1931–1941* (New York, 1941), pp. 68–69; John Masland, "American Attitudes Toward Japan," *Annals of the American Academy of Political and Social Science,* CCXV (1941), 162.

[101] William Bigelow, "Let's Vote Before We Fight," *Good Housekeeping,* CIV (February, 1937), 4; Lydia Wentworth to Ludlow, Wentworth MSS., Box 6.

Still it had been bottled up by the House Judiciary Committee through most of 1937, and Ludlow could not win enough names for a discharge petition to bring it to the House floor. Two days after the sinking of the *Panay*, the discharge petition had all the required signatures. A cross section of antiwar congressmen, from the liberal Texas Democrat Maury Maverick to the reactionary Republican Hamilton Fish, supported the proposal. If it was fair to hold AAA plebiscites to ask farmers "whether or not they should lead little pigs to slaughter," declared Representative Gerald Boileau of Wisconsin, then "all of the people should be permitted by a referendum vote to determine whether or not the sons and the daughters of these same farmers should be led to slaughter upon the battlefields of foreign countries."[102]

In the midst of the debate over the Ludlow proposal, Speaker Bankhead left the rostrum to read a strongly worded letter from President Roosevelt which objected that the referendum "would cripple any President in his conduct of our foreign relations and . . . would encourage other nations to believe that they could violate American rights with impunity." In 1898, the people would have led the country into war, "in all probability an unnecessary war" which a strong President could have prevented, Roosevelt pointed out to his son.[103] Under the most severe kind of White House pressure, forty-four Democrats who had signed the discharge petition voted against bringing the referendum to the House floor, and another seven abstained. Despite this shift, the House came within twenty-one votes of approving the Ludlow amendment, which was defeated, 209–188, in January, 1938. The 188 votes are a measure both of the President's tenuous control of foreign policy and, as late as 1938, of the hardrock strength of isolationist sentiment in America.

[102] *Congressional Record*, 75th Cong., 3d Sess., p. 282; National Committee for the War Referendum files, National Council for the Prevention of War files, Drawer 82.

[103] *Congressional Record*, 75th Cong., 3d Sess., p. 277; F.D.R. memo to James Roosevelt, January 20, 1938, FDRL PPF 1820, Box 4.

CHAPTER 10

A Sea of Troubles

ROOSEVELT'S Second Inaugural Address on January 20, 1937, indicated he was ready for a more radical turn. "I see one-third of a nation ill-housed, ill-clad, ill-nourished," the President declared.[1] For the next two weeks, the country waited to see what specific legislation the President would demand to improve the lot of this "one-third of a nation." When Roosevelt did act, he caught the country by surprise. In a message which electrified the nation, he asked not for new social legislation but for reform of the Supreme Court.

In the spring of 1936, the Court, which had already wiped out the NRA and the AAA, had gone out of its way to find the Guffey Coal Conservation Act unconstitutional.[2] A month later, in the Morehead Case, the Court shocked even conservatives by ruling, once more by a 5–4 vote, that a New York state minimum wage law was invalid.[3] In the field of labor relations, the Court seemed to have created, as President Roosevelt protested, a " 'no-man's-land,' where no Government—State or Federal" could function.[4] The decisions in these two cases suggested that the Wagner Act would not survive a test in the courts, and that a wages and hours law was out of the question. Roo-

[1] Samuel Rosenman (ed.), *The Public Papers and Addresses of Franklin D. Roosevelt* (13 vols., New York, 1938–50), VI, 4.

[2] Carter *v.* Carter Coal Company, 298 U.S. 238.

[3] Morehead *v.* New York ex rel. Tipaldo, 298 U.S. 587; Stanley High to F.D.R., n.d., FDRL PSF 36; Josephine Goldmark, *Impatient Crusader* (Urbana, Ill., 1953), pp. 177–178; items in Mary Dewson MSS., Box 6.

[4] *Public Papers*, V, 191–192.

sevelt, it seemed, not only faced the prospect of seeing past accomplishments like the Social Security Act destroyed but had been forbidden to attempt anything new.[5]

For at least two years, Roosevelt and his aides had explored proposals to make the Court more amenable to New Deal legislation. The President originally favored amending the Constitution, but he came to feel that this strategy bristled with difficulties, for a modest expenditure could forestall ratification in thirteen legislatures. Moreover, any law passed under such an amendment would still be subject to Court review. The real problem, Roosevelt came to feel, lay not with the Constitution but with the Court. Since so many decisions had been handed down by a divided bench, any reform should be aimed at swelling the Court majority. Captivated by the precedent set by Asquith in threatening to pack the House of Lords by creating several hundred new peers—Roosevelt remembered the episode as an attempt by Lloyd George to secure Irish home rule—the President decided that the only feasible solution was Court packing, a proposition he had earlier viewed with distaste. Yet he moved warily, because he knew Court packing violated taboos, and he sought a formula that both had historical precedent and would divert attention from his desire for a more liberal Court by raising a new issue of judicial reform. The President took puckish delight in the fact that the plan Attorney General Homer Cummings finally hit upon was based upon a recommendation made in 1913, albeit in a different form, by none other than Justice McReynolds.[6]

On February 5, 1937, the President sent Congress his design for reorganizing the judiciary. Roosevelt declared that a deficiency of personnel had resulted in overcrowded federal court dockets; in a single year, he asserted, the Supreme Court had denied 87 per cent of petitions for hearings on appeal, without citing its reasons. In part, this

[5] In fact, the situation was not quite this serious, for Roberts had joined the majority in the Morehead case on technical grounds. Unhappily, by his failure to explain himself, Roberts contributed to the deepening sense of desperation about the Court.

[6] George Creel, *Rebel at Large* (New York, 1947), pp. 291–294; Harold Ickes, *The Secret Diary of Harold Ickes* (3 vols., New York, 1954), I, 274, 467–468, 494–495, 529–530; Joseph Alsop and Turner Catledge, *The 168 Days* (Garden City, N.Y., 1938), pp. 32–37; Carl Brent Swisher (ed.), *Selected Papers of Homer Cummings* (New York, 1939), pp. 147–149; Fred Sims to Pat McCarran, May 18, 1937, McCarran MSS., File 672.

could be attributed to "the capacity of the judges themselves," a problem which raised "the question of aged or infirm judges—a subject of delicacy and yet one which requires frank discussion. . . ." To "vitalize the courts," Roosevelt recommended that when a federal judge who had served at least ten years waited more than six months after his seventieth birthday to resign or retire, the President might add a new judge to the bench. He could appoint as many but no more than six new justices to the Supreme Court and forty-four new judges to the lower federal tribunals.[7] The President's proposal presented Court packing not as a political ruse but as devotion to a principle: the retirement of aged justices in the interest of efficiency.

In this oblique approach to a fundamental issue of government, Roosevelt was too clever by half. Opponents of his plan had no difficulty in proving that Court inefficiency was a bogus issue and showing that what the President really wanted was a more responsive Court. The over-age argument made little sense since one of the President's most consistent supporters had been the eighty-year-old Louis Brandeis, and the President's message offended many of the Justice's admirers. Septuagenarian senators put little stock in the argument that the faculties of septuagenarian judges might be impaired. The seventy-nine-year-old Carter Glass pointed out that Littleton had written his great treatise on property law at the age of seventy-eight and that Lord Coke had been eighty-one when he produced his masterpiece "Coke upon Littleton." Some thought that the President had lent his prestige to the contemporary assault on the Court as "nine old men" and even that he had denied to age the veneration it should command. "I am one of those who have reached the 'senile age,' " a New Jerseyan wrote Senator Bailey, "but hope and pray I may be spared long enough to see our beloved country restored to the people unimpaired."[8]

Roosevelt could scarcely have bungled the presentation of the Court plan more. The proposition had never been mentioned during the recent campaign, and he had handed it to Congress without a word

[7] *Public Papers,* VI, 51–66. The bill also embraced various proposals for reform of judicial procedure, chiefly designed to meet the President's contention that the courts were behind in their business.

[8] James Palmer, Jr., *Carter Glass* (Roanoke, 1938), pp. 300–301; Drew Pearson and Robert Allen, *The Nine Old Men* (Garden City, N.Y., 1936); Herbert Hoover to Charles Evans Hughes, February 19, 1937, Hughes MSS., Box 6; Albert Marr to Bailey, February 14, 1937, Bailey MSS., Supreme Court 1937.

of warning to congressional leaders. Speaker Bankhead protested to Representative Lindsay Warren, who had served with him as one of Roosevelt's assistant floor managers at the 1932 convention: "Lindsay, wouldn't you have thought that the President would have told his own party leaders what he was going to do[?] He didn't because he knew that hell would break loose."[9] Save for Cummings, not even Roosevelt's cabinet officers knew of the proposal, nor had the President taken into his confidence most of his political intimates. The Supreme Court bill reflected less the thinking of New Deal intellectuals than the narrow shrewdness of an Attorney General who had been a Democratic national chairman.[10] Yet too much can be made of Roosevelt's tactical failings. He had little choice save to hand the opposition the one issue around which it could rally.

Men who had feared to oppose his economic policies, because they anticipated popular disapproval, now had the perfect justification for breaking with the President and going to the people. Representative Hatton Sumners, chairman of the House Judiciary Committee, announced: "Boys, here's where I cash in my chips." The administration turned to the Senate only to meet the opposition not just of conservatives like Glass but of usually reliable Roosevelt men like Joseph O'Mahoney of Wyoming, Tom Connally of Texas, and Burton Wheeler of Montana.[11] O'Mahoney protested that nothing in the bill limited the Court's powers of judicial review, nor remedied either the defects of old age or of divided decisions. New judges, he wrote, "would be just as likely to disappoint him as, in similar circumstances, Oliver Wendell Holmes disappointed Theodore Roosevelt when shortly after his appointment, he dissented in the Northern Securities Case."[12]

Many liberals were disquieted by the Court plan. Roosevelt's

[9] Memorandum of Conference with Speaker W. B. Bankhead, February 7, 1937, Warren MSS., Box 17.

[10] Frances Perkins, COHC, VII, 128; Alsop and Catledge, *168 Days,* pp. 36, 46–49; Breckinridge Long MS. Diary, January 8, 1937.

[11] Alsop and Catledge, *168 Days,* p. 67; Tom Connally, *My Name is Tom Connally* (New York, 1954), p. 189; items in Carter Glass MSS., Box 3; "Suggested Memo on Senator Wheeler," n.d., Mary Dewson MSS., Box 6; Joseph O'Mahoney to William Harman Black, April 6, 1937, O'Mahoney MSS.

[12] O'Mahoney to T. S. Taliaferro, Jr., April 6, 1937, O'Mahoney MSS. The Republicans adopted the strategy of silence during the Court fight, and allowed the Democrats to quarrel among themselves. Karl Lamb, "The Opposition Party as Secret Agent: Republicans and the Court Fight, 1937," *Papers of the Michigan Academy of Science, Arts, and Letters,* XLVI (1961), 539–550.

method, and the language he used, suggested he not only wished to reform the Court but to humiliate it, in reprisal for past denials of his own prerogatives. Yet many who agreed with his objections to recent decisions still wished to preserve the dignity of the bench.[13] Moreover, they were disturbed lest the Court be weakened as a protector of civil liberties. Oswald Garrison Villard wrote: "If I were a Negro I would be raging and tearing my hair over this proposal. Woodrow Wilson introduced segregation in the departments in Washington; . . . a future Woodrow Wilson could pack the Supreme Court so that no Negro could get within a thousand miles of justice. . . ."[14] William Allen White protested: "Assuming, which is not at all impossible, a reactionary president, as charming, as eloquent and as irresistible as Roosevelt, with power to change the court, and we should be in the devil's own fix if he decided to abridge the bill of rights by legislation which he could easily call emergency legislation." Roosevelt cajoled Senator Norris into going along, but Norris was disgusted with the President's opportunism. He asked himself how he would have stood if "Harding had offered this bill."[15]

In attempting to alter the Court, Roosevelt had attacked one of the symbols which many believed the nation needed for its sense of unity as a body politic. The Court fight evoked a strong feeling of nostalgia for the days of the Founding Fathers, when, it seemed, life was simpler and principles fixed.[16] One constituent urged Senator McCarran to take a stand "against what Cornwallis and Howe fought for in 1776." The greater the insecurity of the times, the more people clung to the

---

[13] Rexford Tugwell, *The Democratic Roosevelt* (Garden City, N.Y., 1957), p. 392; Roy Howard to Nelson Trusler Johnson, June 1, 1937, Johnson MSS.

[14] Villard to Maury Maverick, May 7, 1937, Villard MSS. Maverick replied: "About some other guy in the White House. So what? . . . England depends on its Parliament, and that eventually is what we must do in this country. If we have Fascism, or if we are going to have Fascism, the Supreme Court certainly cannot stop it." Maverick to Villard, May 8, 1937, Villard MSS.

[15] White to Felix Frankfurter, March 23, 1937, William Allen White MSS., Box 187; Alsop and Catledge, *168 Days*, pp. 94–95; George Norris to Francis Heney, April 25, May 6, 1936, Heney MSS.; Edward Dickson to Hiram Johnson, August 17, 1937, Dickson MSS., Box 8, Folder 9; Donald Richberg to Raymond Clapper, February 26, 1937, Richberg MSS., Box 2.

[16] John Carter, "Nostalgia for the Nursery," *Harper's*, CLXIX (1934), 502–505; Richard Neuberger, "America Talks Court," *Current History*, XLVI (June, 1937), 35. Business interests had both fostered and exploited this attitude toward the Court. Edward S. Corwin, "The Constitution as Instrument and as Symbol," *American Political Science Review*, XXX (1936), 1071–1085.

few institutions which seemed changeless. Congressmen were deluged
with letters expressing intense anxiety over the fate of the Court.
When Senator Bailey announced his opposition to the plan, a South
Carolina lady wrote: "Bully for you! Oh bully for you! *Don't,* don't
let that wild man in the White House do this dreadful thing to our
country."[17]

Roosevelt had counted on conservative opposition, and had probably
anticipated some of the liberal defections, but he failed to reckon with
the power of the Court itself. On March 29, in a 5–4 decision in which
Roberts joined the majority, the Court upheld a Washington minimum-wage law similar to the New York statute it had erased in the
Morehead Case.[18] Two weeks later, Roberts joined once more in a
series of 5–4 rulings which found the Wagner Act constitutional.[19]
These decisions marked a turning point in the history of the Supreme
Court. They upset the historic verdict in the Adkins Case and appeared quite contrary to the Court's rulings in the Carter and the
Schechter cases. The next day Hugh Johnson wrote the President
wryly: "I was taken for a ride on a chicken truck in Brooklyn two
years ago and dumped out on a deserted highway and left for dead.
It seems this was all a mistake." The conservative minority, which had
dominated the Court for so many years, knew it had been eclipsed.
A newspaperman confided to Maury Maverick: "Old McReynolds
was sore as hell, speaking like he seldom speaks, very loud, gyrating
like you Congressmen and Senators do on the floor, and poking his
pencil angrily at the crowd as he shouted his opinion, without reading
it, and his speech was a good deal different from the written one."[20]

[17] John Burrows to Pat McCarran, February 12, 1937, McCarran MSS., File
672; Rose Goode McCullough to Bailey, February 14, 1937, Bailey MSS.,
Supreme Court 1937; Martha Young *et al.* to George Sutherland, March, 1937,
Sutherland MSS., Box 6.

[18] West Coast Hotel Company *v.* Parrish, 300 U.S. 379.

[19] See especially National Labor Relations Board *v.* Jones & Laughlin Steel
Corporation, 301 U.S. 1; Edward S. Corwin, review of *Charles Evans Hughes*
by Merlo Pusey, *American Political Science Review,* XLVI (1952), 1171–1173.

[20] Johnson to F.D.R., April 13, 1937, FDRL PPF 702; Maury Maverick, *A
Maverick American* (New York, 1937), p. 336. There has been a good deal of
speculation about why Roberts joined the liberal majority. It is clear that
Roosevelt's Court message was not responsible, for the minimum-wage decision
was reached before the President sent his message, although it was not handed
down until afterwards. Merlo Pusey, *Charles Evans Hughes* (2 vols., New York,
1951), II, 757, 766–772.

On May 18, prodded by Senator Borah, Justice Van Devanter announced his retirement from the bench. Since Roberts' "conversion" had given Roosevelt a 5–4 court willing to approve New Deal legislation, Van Devanter's departure made possible a 6–3 division. The need for drastic reform no longer seemed pressing. Chief Justice Hughes struck another effective blow at the Court bill. In response to a letter from Senator Wheeler, Hughes composed a devastating reply to Roosevelt's charge of inefficiency. He pointed out that the Court had heard argument on cases it had accepted for review only four weeks earlier, and claimed that an increase in justices would only produce delays: "more judges to hear, more judges to confer, more judges to discuss, more judges to be convinced and to decide."[21] On May 24, the Court gave fresh evidence that the crisis had ended, when in two 5–4 decisions it validated the unemployment insurance provisions of the Social Security Act, and in a third opinion it ruled 7–2 that old-age pensions were constitutional.[22]

The cluster of decisions, Van Devanter's resignation, and Hughes's letter killed the Court reform bill. Before these events, the Senate was probably evenly divided. After them, wavering members joined the opposition, and the proponents lost heart. Senate leaders pressed for a compromise. Senator Byrnes queried: "Why run for a train after you've caught it?" The Senate majority leader, Joe Robinson, thought he could get the President "a couple of extra justices tomorrow," even though men like Borah regarded "two as bad in principle as six."[23] But the President refused to yield. The margin represented by Roberts' switch seemed too thin, especially since Roosevelt had had to promise the next vacancy on the Court to Robinson, who was a conservative. Besides, he had his Dutch up.

Begun in winter, the court fight was consuming all of spring, and now threatened to run its weary way through summer. By mid-June, when Vice-President John Garner, who loathed the Court plan,

[21] *Senate Report 711,* 75th Cong., 1st Sess. (Washington, 1937), pp. 38–40; Alpheus T. Mason, *Harlan Fiske Stone* (New York, 1956), pp. 450–453.

[22] Steward Machine Company *v.* Davis, 301 U.S. 548; Helvering *v.* Davis, 301 U.S. 619; Robert Jackson, *The Struggle for Judicial Supremacy* (New York, 1941), pp. 221–235.

[23] Alsop and Catledge, *168 Days,* pp. 152–153; Frank Cantwell, "Public Opinion and the Legislative Process," *American Political Science Review,* XL (1946), 933; Borah to James Truslow Adams, May 22, 1937, J. T. Adams MSS.

packed his bags and went home to Texas, the Democratic party was tearing itself apart. Still Roosevelt pushed on with the hopeless struggle. In mid-July, in his apartment in the Methodist Building across from the Capitol, the loyal Senator Robinson collapsed and died. His death destroyed what little hope remained for a favorable compromise. Five days later, Roosevelt's close political ally, Governor Herbert Lehman of New York, gratuitously announced his opposition to the proposal.[24] The President had no choice but surrender; on July 22, Senator Logan moved that the bill be recommitted. "Glory be to God!" Hiram Johnson shouted. After 168 days the historic Court battle was over. Vice-President Garner designed a substitute measure in order "not to bloody Mr. Roosevelt's nose"; enacted in August, it embodied procedural reforms, but abandoned altogether the President's proposal to appoint new judges.[25]

In later years, Roosevelt claimed he had lost the battle but won the war. In one sense, this is true. The very month of the defeat of the Court bill, August, 1937, he named to the Van Devanter vacancy Alabama's Senator Hugo Black, the *bête noire* of southern conservatives. Black's appointment, Herbert Hoover protested, meant that the Court was "one ninth packed." Black survived an ugly episode when an enterprising reporter revealed that he had once been a member of the Ku Klux Klan and in his service on the bench more than vindicated those who believed him neither a racist nor a bigot but a champion of minority rights.[26] Within two and a half years after the rejection of the Court measure, Roosevelt had named five of his own appointees to the nine-man bench: Black, Stanley Reed, Felix Frankfurter, William O. Douglas, and Frank Murphy. The new Court— the "Roosevelt court" as it was called—greatly extended the area of permissible national regulation of the economy while at the same time safeguarding the civil liberties of even the most bothersome minority groups.[27] Yet, in another sense, Roosevelt lost the war. The Court

[24] Lehman to Robert Wagner, July 19, 1937, Lehman MSS.; Ickes, *Diary,* II, 171; F.D.R. to Alben Barkley, July 15, 1937, FDRL PSF 36; Representative James Wadsworth to James Truslow Adams, July 16, 1937, J. T. Adams MSS.

[25] Alsop and Catledge, *168 Days,* p. 294; T.R.B., "Washington Notes," *New Republic,* XCII (1937), 18–19.

[26] John Frank, *Mr. Justice Black* (New York, 1949), pp. 99–106; Chicago *Tribune,* October 3, 1937; Rev. Edward Lodge Curran to Louis Brandeis, October 6, 1937, Brandeis MSS., S.C. 20.

[27] Robert Cushman (ed.), "Ten Years of the Supreme Court: 1937–1947,"

fracas destroyed the unity of the Democratic party and greatly strengthened the bipartisan anti-New Deal coalition. The new Court might be willing to uphold new laws, but an angry and divided Congress would pass few of them for the justices to consider.

Alarmed by Roosevelt's attempt to alter the Court in order to win approval for reform legislation, conservatives in Congress were even more distressed by the Jacobin spirit of the C.I.O.'s bid for national power, a development for which they held the President directly responsible. Senators Burke and George viewed the industrial warfare of 1936–37 as yet more evidence of the disrespect for law fostered by those "in high places," while Representative Clare Hoffman of Michigan charged that the strikes were intended to abet the President's assault on the Court.[28]

In June, 1936, the nervy Lewis had decided to mount his first great offensive against organized labor's mightiest enemy, U.S. Steel. He named his Mine Worker lieutenant, Philip Murray, director of the Steel Workers Organizing Committee, and brazenly rented offices in downtown Pittsburgh on the thirty-sixth floor of the Grant Building, an edifice which housed more steel barons than any other in the nation. By the end of the year, as Murray's organizers filtered into Youngstown, scurried up and down the Monongahela, and held furtive meetings in the shadows of mills in company towns, Steel braced for the big strike that would test the defenses of the citadel of the open shop.[29]

But the Big Strike was not to come in steel. On December 28, 1936, after a series of quick victories in Flint, South Bend, and Detroit, General Motors employees in Cleveland struck. Two nights later, workers at Fisher Body No. 1 in Flint "sat down"; instead of walking out, they occupied the factory. For the next six weeks, Flint's workers

---

American Political Science Review, XLI (1947), 1142–1181, XLII (1948), 32–67; Edward S. Corwin, *Constitutional Revolution, Ltd.* (Claremont, Calif., 1941).

[28] J. Woodford Howard, Jr., "Frank Murphy and the Sit-Down Strikes of 1937," *Labor History,* I (1960), 120.

[29] Walter Galenson, "The Unionization of the American Steel Industry," *International Review of Social History,* I (1956), 8–40, and *The CIO Challenge to the AFL* (Cambridge, 1960), pp. 75–91; Robert R. R. Brooks, *As Steel Goes . . .* (New Haven, 1940), pp. 70–107; Herbert Harris, "How the CIO Works," *Current History,* XLVI (May, 1937), 61–66; John Brophy, COHC.

lived in the plant. G.M. hands elsewhere adopted the "sit-down" stratagem or struck in more conventional ways. Guerrilla forces of wrench-wielding unionists seized crucial plants, turned back counter-attacks, and jeered entreaties to abandon the factories. On February 11, 1937, General Motors surrendered. It granted the United Automobile Workers terms which seemed ambiguous but which amounted to a rousing union victory. The bearded workers who filed out of the plant returned to the assembly line flaunting their union buttons.[30]

Less than a month later, Benjamin Fairless announced the stunning news that U.S. Steel had folded without a fight. The corporation which, as Lewis said, had through the years been "the crouching lion in the pathway of labor" had signed a contract granting the S.W.O.C. recognition for its members, a wage boost, and a forty-hour week.[31] Once the C.I.O. pierced the fortifications of G.M. and U.S. Steel, it was able to overrun much of the rest of industry. By the end of 1937, the U.A.W. had brought every car manufacturer save Ford into line. The pace of organization moved so swiftly that workers outdistanced union leaders. "Many of us," reflected a Dodge unionist, "had the feeling we were like the kid that gets a little fire started in the hay-field."[32] In a year, the U.A.W.'s membership jumped from 30,000 to 400,000. By early May, 1937, S.W.O.C., formed less than a year before, claimed 325,000 members; by the end of the month, it had signed up all the subsidiaries of U.S. Steel and such important independents as Jones and Laughlin, Wheeling Steel, and Caterpillar Tractor. As steel and auto went, so went much of the nation. After an eight-week strike in March and April, 1937, Firestone Tire & Rubber capitulated to the United Rubber Workers. By the end of 1937, General Electric, RCA-Victor, and Philco had recognized the United Electrical and Radio Workers. The Textile Workers Organizing Committee, run by Sidney Hillman, brought into line industry titans like American Woolen in Lawrence; by early October, it boasted 450,000 members.[33]

[30] Galenson, CIO Challenge, pp. 134–143; Edward Levinson, Labor on the March (New York, 1938), pp. 151–158; Henry Kraus, The Many and the Few (Los Angeles, 1947); Murray Kempton, Part of Our Time (New York, 1955), pp. 284–287; "Strikes and Lockouts: General Motors Corporation, 1937," Joe Brown Collection, Wayne State University.

[31] Galenson, CIO Challenge, pp. 91–96; "It Happened in Steel," Fortune, XV (May, 1937), 91–94, 176–180.

[32] Richard Harris, Wayne University Oral History Interview.

[33] Despite these enormous gains, the C.I.O. was a million members smaller

In a year's time, the blue-shirted workers had succeeded in doing what their fathers and grandfathers had failed to do or had not even dared to do—unionize much of industrial America. It would be a few more years before some of the giants—Ford, with his ugly crew of strikebreaking hoodlums; Little Steel, under Tom Girdler; the Big Four meat packers—fell into line, but the main task had been done. There had been some violence, notoriously near Girdler's Republic Steel plant in Chicago where police killed ten strikers on Memorial Day.[34] But what was truly noteworthy was the relatively little bloodshed that marked this momentous transfer of social power. In part, the peacefulness resulted from the orderliness of the workers; when Woolworth strikers took an item from the stock, they were careful to leave a dime on the register.[35] Even more important was the reluctance of employers to risk a pitched battle, which might destroy their machinery; they had a respect bordering on awe for the latent violence of the strikers. When Flint police attempted to dislodge G.M. sit-downers, workers responded with a fire of nuts and bolts, pop bottles, steel hinges, metal pipes, and coffee mugs; after three hours of fighting, the police retreated. "The Battle of Bulls' Run" made clear that sit-downers could be ousted forcibly only at a heavy cost of life and property.[36]

Perhaps the most important reason for the success of the sit-downs was that neither state nor federal governments would employ force as they had so often in the past to break the strikes. In the summer of 1936 at Homestead—the bloody ground of the 1890's and forbidden

than the A.F. of L. in 1939. In the thirties, the Teamsters rose from 95,500 to 350,000 to become the biggest union in the Federation; the Hotel and Restaurant Workers grew from a tiny 38,000 in 1929 to an impressive 185,000 a decade later.

[34] U.S. Congress, Senate, 77th Cong., 1st Sess., Senate Committee on Education and Labor, *Report No. 46*, Part 2 (Washington, 1937); Galenson, "Unionization of Steel," pp. 29–37; items in Elbert Thomas MSS., Box 19; Tom Girdler, *Boot Straps* (New York, 1943), pp. 226–373.

[35] Kempton, *Part of Our Time*, p. 289.

[36] Levinson, *Labor on March*, pp. 155–158; George Blackwood, "The Sit-Down Strike in the Thirties," *South Atlantic Quarterly*, LV (1955), 438–448; Ted Laduke, Wayne OHI. Cf. Wilbert Moore, *Industrial Relations and the Social Order* (New York, 1947), pp. 399–449. Some employers proved astonishingly good-natured; the owner of a St. Louis laundry sent his daughter as hostess to the sit-downers, and provided them with coffee, warm bedding, ping-pong, and lotto. John T. Flynn, "Other People's Money" *New Republic*, XC (1937), 111.

land to union organizers for nearly two decades—Pennsylvania's Lieutenant Governor Thomas Kennedy, an official of Lewis' Mine
Workers, told workers that union organizers were free to enter any
steel town in the state, and promised relief funds to strikers if
S.W.O.C. called a walkout. In Michigan, Governor Frank Murphy
refused to send troops to dislodge the sit-downers.[37] A new federal law
prohibited the transport of strikebreakers across state lines, a committee on civil liberties headed by Senator La Follette exposed the
pattern of violence employers had used against their workers, the
Walsh-Healey Act threatened to deprive corporations of government
contracts, and the Wagner law threw the weight of the federal government behind unionization. Although employers were outraged,
President Roosevelt would not use force to oust the sit-downers; the
unions would have to learn for themselves that they could not retain
a "damned unpopular" tactic. "It will take some time, perhaps two
years," Roosevelt reflected, "but that is a short time in the life of a
nation and the education of a nation."[38]

While initially sympathetic to the cause of labor, much of middle-
class America took a less tolerant view than Roosevelt of the sit-downs.
Property-minded citizens were scared by the seizure of factories, incensed when strikers interfered with the mails, vexed by the intimidation of nonunionists, and alarmed by flying squadrons of workers who
marched, or threatened to march, from city to city. When the Flint
sheriff attempted to enforce a circuit-court edict ordering the strikers
to leave the Fisher plants, thousands of union members from distant
towns filed into Flint. Walter Reuther led five hundred from Detroit's
West Side; Akron sent a brigade of rubber workers; Dodge unionists
paraded in military order; and a red-bereted Women's Emergency
Brigade marched outside the plant. Conservatives claimed, mistakenly,
that the sit-down contrivance had been imported from Léon Blum's
France and warned that they manifested the same spirit as the radical

[37] Levinson, *Labor on March,* p. 188; *Steel Labor,* August 1, 1936; John
Fitch, "A Man Can Talk in Homestead," *Survey Graphic,* XXV (1936), 71–
76, 118–120; Howard, "Frank Murphy," pp. 103–140. One governor who did
use force to oust the sit-downers was Connecticut's Wilbur Cross. Wilbur Cross,
Connecticut Yankee (New Haven, 1943), pp. 354–355; Cross to William Kelly,
April 12, 1937, Cross MSS., Box 27.
[38] Thomas Greer, *What Roosevelt Thought* (East Lansing, Mich., 1958), p.
73.

outbreaks which menaced property in Italy in 1919.[39] Only by clever maneuvering did Roosevelt frustrate Senator Byrnes's attempt to get the Senate to condemn the sit-downs.

Roosevelt's sensible attitude toward labor cost him dearly. When Lewis arrogantly demanded the President's co-operation in return for the heavy contribution the U.M.W. had made to his 1936 campaign, Roosevelt turned him down. In June, in a moment of pique, the President pronounced "a plague o' both your houses" on the C.I.O. and "Little Steel." Lewis retorted: "It ill behooves one who has supped at labor's table . . . to curse with equal fervor and fine impartiality both labor and its adversaries when they become locked in deadly embrace."[40] In 1940, John L. would back the Republican presidential candidate. At the same time, foes of the sit-down strikes believed that Roosevelt, in refusing to employ force, was condoning an assault on property rights by lower-class rebels at the very moment he was attacking the sacred institution of the Supreme Court. Throughout 1937, these two issues—Court packing and the sit-downs —were repeatedly linked; that summer an advertisement in the Boston *Traveler* urged: "Come on down to Cape Cod for a real vacation where the CIO is unknown and over 90 percent are Republicans who respect the Supreme Court."[41] Joined together, the President's policies on the Court and the sit-downs threatened to destroy the middle-class base of the Roosevelt coalition.

Many in the middle class had flocked to Roosevelt's standard not because they had been converted to a liberal ideology but because his years in office had been marked by slow but impressive economic gains. By the spring of 1937, the country had finally pulled above 1929 levels of output. In August, disaster struck. With the nation apparently well on the road to recovery, the economy suddenly cracked. Industrial activity fell off with the most brutal drop in the country's history. By December, *The New York Times* Business Index had plunged from 110 to 85, wiping out all the gains made since 1935. In three months,

[39] Alfred Winslow Jones, *Life, Liberty and Property* (Philadelphia, 1941), pp. 127–128; Clarence Niblack memorandum, October 10–19, 1937, W. B. Pine MSS.; Levinson, *Labor on March,* pp. 160–162; Daniel Roper to Claude Bowers, March 31, 1937, Edward House MSS.; Herbert Harris, *American Labor* (New Haven, 1938), p. 294.
[40] James Wechsler, *Labor Baron* (New York, 1944), pp. 98–100.
[41] "The Bandwagon," *New Republic,* XCI (1937), 364.

steel fell from 80 per cent of capacity to 19 per cent. Waves of selling hit the market; the ticker tape fell far behind as more than seven million shares changed hands in a single day; Dow-Jones stock averages lurched from 190 in August to 115 in October. Night clubs failed in Manhattan; new autos piled up in showrooms; gold fled the country. Between Labor Day and the end of the year, two million people were thrown out of work.[42]

The "prosperity" of early 1937, which rested on an insecure base of mass unemployment and a sluggish construction industry, had been achieved largely by the government's deficit spending. In June, 1937, Roosevelt, inordinately worried about the danger of inflation, slashed spending sharply. He cut WPA rolls drastically and turned off PWA pump-priming. At the very same time, Washington collected two billions in new social-security taxes. The government had not only stopped priming the pump but was even "taking some water out of the spout."[43] If business had been ready to take over, none of this would have mattered, but business still lacked the confidence to undertake new investment.[44]

The "recession" touched off a hot debate within the administration over economic policy. Secretary Morgenthau believed that the failure to achieve recovery was caused by the reluctance of business to invest, because it feared federal spending would lead to inflation and heavy taxation. Since the New Deal had failed to bring the country out of the depression, the administration, he argued, should balance the budget and give business a chance to see what it could do. As early as 1936, he had reasoned that although "the patient might scream a

---

[42] Kenneth Roose, "The Recession of 1937–38," *Journal of Political Economy*, LVI (1948), 241; *Time*, XXXI (February 7, 1938), 51–52.

[43] George Soule, "The Present Industrial Depression," *New Republic*, XCIII (1937), 62. Cf. Arthur Smithies, "The American Economy in the Thirties," *American Economic Review*, XXXVI (1946), 26.

[44] Increased costs, especially labor costs, the tight-money policy adopted by the Federal Reserve early in 1937, and worry over the undistributed profits tax made it still less likely that business would take the reins. Kenneth Roose, *The Economics of Recession and Revival* (New Haven, 1954); Sumner Slichter, "The Downturn of 1937," *Review of Economic Statistics*, XX (1938), 97–110; John Blum, *From the Morgenthau Diaries* (Boston, 1959), pp. 279–280, 295–296, 355, 367–375; Marriner Eccles, *Beckoning Frontiers* (New York, 1951), pp. 287–301; Joseph Schumpeter, *Business Cycles* (2 vols., New York, 1939), II, 1018–1019; Douglas Hayes, *Business Confidence and Business Activity* ("Michigan Business Studies," X [Ann Arbor, 1951]).

bit when he was taken off narcotics," the time had come "to strip off the bandages, throw away the crutches," and let the economy see if "it could stand on its own feet."[45]

Strong opposition to Morgenthau's views came from a group of liberal New Dealers headed by Harry Hopkins, Harold Ickes, and Marriner Eccles and including William Douglas and Jerome Frank of the SEC and Robert Jackson in the Justice Department. They were supported by a battery of economists and lawyers: Leon Henderson, Hopkins' economic adviser; Isador Lubin, Commissioner of Labor Statistics; Aubrey Williams, director of the NYA; Lauchlin Currie at Federal Reserve; and Mordecai Ezekiel and Louis Bean in Agriculture. They pointed out that it had been precisely when the budget had been brought into balance in 1937 that the recession had ensued. Since a cutback in spending had triggered the downturn, a sharp increase in spending was the route out. They reasoned that when private investment declined, the government had to spend more. To reduce the peaks of booms and raise the valleys of depressions, the government should spend lavishly in hard times and tax heavily in prosperous periods. "The Government," declared Marriner Eccles, "must be the compensatory agent in this economy; it must unbalance its budget during deflation and create surpluses in periods of great business activity."[46]

Eccles' language suggests the influence of John Maynard Keynes, and some of the New Deal spenders, notably Lauchlin Currie, owed a clear debt to Keynes, while others were undoubtedly influenced by him in a more indirect manner. Yet Eccles, regarded as the most influential Keynesian, recalled he had never heard of Keynes when he first advanced "Keynesian" views and confessed that, even years later, he had

[45] "The Morgenthau Diaries," *Collier's,* CXX (September 27, 1947), 82. Cf. Blum, *Morgenthau Diaries,* pp. 387–388; Henry Morgenthau, Jr., to F.D.R., October 5, 1937, FDRL PSF 26; Melvin Brockie, "The Rally, Crisis, and Depression, 1935–38" (unpublished Ph.D. dissertation, University of California at Los Angeles, 1948); C. O. Hardy, "An Appraisal of the Factors ('Natural' and 'Artificial') Which Stopped Short the Recovery Development in the United States," *American Economic Review,* XXIX, Supplement (March, 1939), 170–182.

[46] Eccles, *Beckoning Frontiers,* p. 252, and "How to Prevent Another 1929 in 1940," FDRL PSF 32; Lauchlin Currie, "A Tentative Program to Meet the Business Recession," Emanuel Goldenweiser MSS., Box 2; Joseph Alsop and Robert Kintner, "We Shall Make America Over," *Saturday Evening Post,* CCXI (November 19, 1938), 15.

not read him save in small extracts. For men like Hopkins, Keynes served as a convenient rationalization for what they were already doing. Moreover, quite apart from considerations of economic policy, the New Dealers had come to a realization that had escaped most earlier reformers: that one can buy reforms with money, "just as an individual buys groceries, or a hat, or a ton of coal."[47]

Many of the liberals believed that, in addition to stepping up spending, the government should move vigorously to curb monopoly. Unlike the old trust busters, they were concerned neither with the moral intent of corporations nor with protecting small business against large.[48] Instead they focused on the way the inflexible "administered" prices of industrial monopolies impeded recovery.[49] The prophetic hero of the antimonopolists was Leon Henderson, who, it was said, had in the spring of 1937 predicted a recession within six months.[50] The burly Henderson, reputedly the only man in the administration who could outshout General Johnson, was, wrote *Fortune,* "a spittoon economist" who did not "care whether little or big businesses compete just so long as the competition is rough and tough. Not oftener than twice a day you can hear him say that no customer was ever hurt in a price war. . . ." As early as 1934, Henderson had urged an antitrust investigation; by 1936, he had won the support of Robert Jackson, who became convinced that an inquiry was an important first step in disciplining the trusts. Before you can win approval for an SEC, Jackson reasoned, you need a Pecora investigation.[51]

In the early months of the recession, the President's orthodox advisers had the upper hand. After indicating he was impressed by a memorandum that showed how reduced federal spending had helped

[47] Seymour Harris (ed.), *The New Economics* (New York, 1947), pp. 16–19; Eccles, *Beckoning Frontiers,* pp. 131–132; T.R.B., "Washington Notes," *New Republic,* LXXXIV (1935), 17.

[48] Wendell Berge later observed: ". . . Many of the antitrust offenses with which I have had contact are completely devoid of any moral element whatever." Berge to F. O. Berge, November 30, 1940, Wendell Berge MSS., Box 11.

[49] They were especially persuaded by Gardiner Means' study which pointed out that, as demand slackened, monopolists cut not prices but production and employment. *Industrial Prices and Their Relative Inflexibility,* Sen. Doc. 13, 74th Cong., 1st Sess. (Washington, 1935).

[50] In fact, Henderson's memo is more ambiguous than the legend suggests. "Boom and Bust," March 29, 1937, FDRL PSF 32.

[51] "SEC," *Fortune,* XXI (June, 1940), 139; John Chamberlain, *The American Stakes* (Philadelphia, 1940), p. 209.

precipitate the recession, Roosevelt turned around and approved a statement that the government would do everything possible to balance the budget by slashing spending. In Hooverian tones, he told his cabinet: "Everything will work out all right if we just sit tight and keep quiet." One of his cabinet officers even spoke of "corrective dips" in the market.[52]

Tired of waiting for Roosevelt to act, the liberals decided to go ahead on their own, leaving the President free to repudiate them. On December 26, 1937, Assistant Attorney General Robert Jackson, head of the Antitrust Division of the Department of Justice, delivered a radio address drafted by Cohen and Corcoran which pinned the blame for the recession on monopolists who "have simply priced themselves out of the market, and priced themselves into a slump." Three days later, he told the American Political Science Association in Philadelphia that business, in Sir Arthur Salter's words, was engaged in a "strike of capital" against reform. Although business, Jackson declared, had "let loose its bull dogs to bark at the public" that labor was to blame for high prices, labor had had "nowhere near the percentage advance that big business has given to its own darlings." Jackson ticked off the salaries of industrialists like Alfred P. Sloan, Jr., of General Motors; noted the sizable profits large corporations had made under the New Deal; and concluded: "The only just criticism that can be made of the economic operations of the New Deal is that it set out a breakfast for the canary and let the cat steal it. . . ." The very next day, Secretary Ickes charged that "America's Sixty Families" had resumed "the old struggle between the power of money and the power of the democratic instinct," an "irreconcilable conflict" between plutocracy and democracy that "must be fought through to a finish." Ickes, who had been reading Ferdinand Lundberg's *America's Sixty Families*, and had been impressed by Léon Blum's success in attacking the *deux cents familles* who owned voting stock in the Banque de France, suggested that Big Business purge itself of "its Fords, its Girdlers and its Rands." Monopoly, Ickes warned, threatened America with "big business fascism."[53]

[52] James M. Burns, *Roosevelt: The Lion and the Fox* (New York, 1956), p. 320; Ickes, *Diary*, II, 223; Blum, *Morgenthau Diaries*, pp. 349–350.

[53] *The New York Times*, December 27, 30, 31, 1937; Eugene Gerhart, *America's Advocate: Robert H. Jackson* (Indianapolis, 1958), pp. 125–127;

Jackson and Ickes were emboldened to go as far as they did without presidential authorization because they knew that Roosevelt seriously entertained the notion that the recession was the result of a capitalist conspiracy. In November, he told Morgenthau he thought some two thousand men had decided to thwart the New Deal by going on strike against the government. Worried by the gains the fascists were making all over the world, he feared that four or five men might get together in this country and decide they wanted their own man in the White House.[54] The trust busters were reasonably pleased with his response to the blasts by Jackson and Ickes. In his State of the Union Message on January 3, 1938, he included two paragraphs drafted by Tom Corcoran promising a special monopoly message. Businessmen, he told a Jackson Day dinner audience five days later, had no confidence in a people's government, yet "demanded that a people's government have confidence in them." He told businessmen he needed confidence as much as they did.[55]

Yet at the same time that he was using Brandeisian phrases such as "other people's money," he held a series of White House conferences with businessmen like Myron Taylor of U.S. Steel and lent an ear to the plea of Donald Richberg to resume planning along the lines of NRA. At a press conference on January 4, the President indicated he looked with favor on some kind of cartel approach resembling the NRA. Before long, Ickes was noting disgustedly: "The President, after letting Jackson and me stick our necks out with our antimonopoly speeches, is pulling petals off the daisy with representatives of big business."[56]

Marquis Childs, "Jackson versus Richberg," *Nation*, CXLVI (1938), 119–120; Ickes, *Diary*, II, 282–283.

[54] Blum, *Morgenthau Diaries*, pp. 293–294, 390; Ickes, *Diary*, II, 241, 253; F.D.R. to Walter Teagle, April 16, 1938, in Elliott Roosevelt, *F.D.R.: His Personal Letters, 1928–1945* (2 vols., New York, 1950), II, 775; F.D.R. to John Cudahy, April 16, 1938, *ibid.*, p. 776.

[55] Joseph Alsop and Robert Kintner, "We Shall Make America Over," *Saturday Evening Post*, CCXI (November 12, 1938), 112; Wendell Berge MS. Diary, January 3, 1938; *Public Papers*, VII, 42; Blum, *Morgenthau Diaries*, 402.

[56] *Public Papers*, VII, 11; Raymond Moley, *After Seven Years* (New York, 1939), p. 374; Donald Richberg, "A Suggestion for Revision of the Anti-Trust Laws," *University of Pennsylvania Law Review*, LXXXV (1936), 1–14; Christopher Lasch, "Donald Richberg and the Idea of a National Interest" (unpublished M.A. essay, Columbia University, 1955), pp. 91–92; Ickes, *Diary*, II, 326.

Roosevelt seemed nonplused by what course to take. He knew conditions were much better than in the crisis of 1932—the government was providing relief, farmers were supported by benefit payments, banks were not in trouble, national income was higher, and people felt "more perplexity than fear." Yet the economy continued to plunge downhill, and he did not know how to stop it. Until he did, he would not be bullied into adopting a position. He preferred to let the rival theorists war around him. It was almost as though he were watching himself, uncertain in his own mind and rather curious about which faction would win him over. At lunch with the President in mid-March, Morgenthau remarked: "As I see it, what you are doing now is just treading water . . . to wait to see what happens this spring."

"Absolutely," Roosevelt replied.[57]

In early 1938, many Americans once more neared starvation. In Chicago, children salvaged food from garbage cans; in Cleveland, families scrambled for spoiled produce dumped in the streets when the market closed. During the first six days of 1938, sixty-five thousand Clevelanders on the relief rolls went without food or clothing orders. One reliefer committed suicide. *Der Angriff* gloated that the misery of Cleveland offered yet another example of the impotence of democracy. Between October, 1937, and May, 1938, WPA rolls in the automobile towns of the Midwest increased at a staggering pace—194 per cent in Toledo, 434 per cent in Detroit. Yet, since the federal government assumed only part of the relief burden, many were in extreme want. In seventeen southern states, Hopkins reported in May, people were starving. When the Ohio legislature adjourned without providing for adequate relief, many people in Cleveland went hungry again that spring. In late May, the relief funds of St. Louis neared exhaustion; Omaha was making only token payments; and Chicago, whose relief rolls jumped from fifty thousand to one hundred and twenty thousand in the first five months of 1938, closed down all its nineteen relief stations. In New York City, Mrs. Sarah Goodman, the sixty-year-old mother of a WPA worker, died of malnutrition and anemia.[58]

[57] *Public Papers,* VII, 12; Blum, *Morgenthau Diaries,* p. 415; Leon Henderson MS. Diary, March 23, 1938; Moley, *After Seven Years,* pp. 364–365.

[58] Grace Abbott to Mary Dewson, February 3, 1938, Dewson MSS., Box 12; Miss Lonigan to Henry Morgenthau, Jr., May 23, 1938, FDRL PSF 26; Blum, *Morgenthau Diaries,* pp. 396–401, 434–435; *Time,* XXXI (January 17, 1938), 22; (May 16, 1938), 16; (May 20, 1938), 12; (June 13, 1938), 12; *New Republic,* XCV (June 1, 1938), 86.

Roosevelt found himself, like Hoover, the victim of his own *hubris*. In 1935, the President had remarked in Charleston: "Yes, we are on our way back—not just by pure chance, not by a mere turn of a wheel in a cycle; we are coming back soundly because we planned it that way. . . ."[59] Credited with the revival of 1936, he was now blamed for the downturn of 1937–38. Journalists began to write, as they once had of Hoover and hard times, of the "Roosevelt Depression." Bernard De Voto even claimed: "Tommy Corcoran's depression is a hell of a lot worse than Herbert Hoover's."[60] Roosevelt, like the business leaders of 1932, had lost his "magic." "The old Roosevelt magic has lost its kick," Hugh Johnson chortled. "The diverse elements in the Falstaffian army can no longer be kept together and led by a melodious whinny and a winning smile."[61]

The concatenation of the three events—the abortive attempt at Court reform, the sit-down epidemic, and the recession—undercut Roosevelt's authority with Congress. While the Supreme Court quarrel ground on, most other legislation had been becalmed like a schooner in a windless sea. When the contest finally ended, a resentful Congress was in no mood to give the President what he wanted. It passed a federal housing law, and approved a statute to aid sharecroppers and tenant farmers, but neither measure to aid the lower third of the nation even approached what needed to be done. Otherwise, Roosevelt got almost nothing he asked. Ickes noted of the President in August: "He is punch drunk from the punishment that he has suffered recently."[62]

[59] I have chosen the version reported in *The New York Times,* October 24, 1935, which is corroborated in essential details by the wire services, rather than the text in the *Public Papers.* Cf. *Public Papers,* IV, 425; San Francisco *Chronicle,* October 24, 1935.

[60] De Voto to Garrett Mattingly, June 15, 1938, De Voto MSS., Box 2.

[61] Burns, *Roosevelt,* p. 346. Many a congressman, pointed out one shrewd Washington observer, had never understood Roosevelt's economic policies. "What he did know was that, in dry spells, Mr. Roosevelt would go into a tent, make unintelligible noises and presently there would be rain.

"Last September saw the start of the severest dry spell in nine years. Mr. Roosevelt made no move to go into his tent, or reach for the bag of magic feathers. The rank-and-file Congressman jumped to the conclusion that his power was no longer intact." T.R.B., "Washington Notes," *New Republic,* XCIV (1938), 358–359.

[62] Ickes, *Diary,* II, 182; Francis P. Miller to Herbert Claiborne Pell, April 1, 1937, Pell MSS., Box 8.

When Congress adjourned, Roosevelt, who had been unusually ill-tempered in the closing weeks of the court strife, set out on a trip to the Northwest to revive himself and put his program before the people. In Boise, he told a crowd that he felt like Antaeus: "I regain strength by just meeting the American people."[63] Everywhere he was greeted with a remarkable display of affection. Congressmen who had opposed Court reform and other New Deal measures fought for a place on his train. Persuaded by his reception that he, rather than Congress, had the support of the country, he called Congress into special session in November to pass several items of "must" legislation, including the so-called "seven little TVA's."[64]

The session proved an almost total washout. Congress adjourned a few days before Christmas without passing a single bill the President requested, although some measures advanced toward final passage.[65] A year after his overwhelming triumph in the 1936 election, Roosevelt appeared to be a thoroughly repudiated leader.

[63] *Public Papers,* VI, 387.

[64] F.D.R. to John McCormack, September 24, 1937; William E. Leuchtenburg, "Roosevelt, Norris and the 'Seven Little TVAs,'" *Journal of Politics,* XIV (1952), 418–441; McAlister Coleman to George Norris, November 22, 1937, Norris MSS., Tray 80, Box 6; Benjamin Cohen to the writer, October 22, 1952.

[65] O. R. Altman, "Second and Third Sessions of the Seventy-fifth Congress, 1937–38," *American Political Science Review,* XXXII (1938), 1099–1123; *The New York Times,* December 22, 1937.

CHAPTER 11

Stalemate

I N THE summer of 1937, a group of conservative Democratic senators broke with President Roosevelt to forge a bipartisan bloc with
Republicans to stymie the New Dealers. "The Party is divided hope-
lessly," wrote North Carolina's Josiah Bailey; "the points of view and
the respective interests, the objectives, cannot be reconciled."[1] Un-
crowned head of the new coalition was Vice-President John Nance
Garner, who had been alienated by Roosevelt's failure to balance the
budget and by his refusal to chastise the sit-down strikers.[2] Led by
southern Democrats like Bailey, who was convinced he was "making
a battle for Constitutional Representative Government as opposed to
mass Democracy," the conservative alliance united fellow southerners
like Virginia's Harry Byrd, border-state Democrats like Maryland's
Millard Tydings, and northerners like New York's Royal Copeland.
The Democratic junto sought not only to combine with Republicans
to frustrate reform legislation but also to lay plans to wrest control of
the party from the New Dealers at the 1940 convention. They already
suspected that Roosevelt would run for a third term, and hoped that
by defeating the President's program on the floors of Congress they
would also end his tenure in the White House. "The decisive battle,"

[1] Bailey to Joseph O'Mahoney, August 22, 1937, O'Mahoney MSS. Some even
discussed a new conservative party. Frank Knox to Neil Carothers, July 16,
1937, Knox MSS.; Theodore Carlson to Stephen Chadwick, August 30, 1937,
Chadwick MSS.

[2] Garner to F.D.R., June 19, 1937, FDRL PSF 37; July 27, 1936, June 20,
1937, FDRL PPF 1416.

Bailey wrote in late August, 1937, "will be fought out in the next session."[3]

Veteran Democratic leaders believed that the party had been subverted by New Deal administrators whose ideas were alien to Democratic traditions. "They are a new type of person, not before encountered," wrote one critic of the Tugwells and Hopkinses, "and they do not inspire confidence."[4] Even ardent New Deal legislators found the bureaucrats trying. Maury Maverick, leader of the New Deal faction in the House, commented on TVA: "The atmosphere is—I cannot express myself exactly, since I am only a congressman, and therefore lack the cultural and scientific knowledge possessed by some of the T.V.A.'ers. But the air is somewhat rarefied, and I am sure I heard the swishing of long wings and saw Green Pastures and De Lawd (Morgan)."[5]

Part of the time, the old-time Democrats saw the New Dealers as bumbling professors, innocent of practical knowledge of the world. During the 1932 campaign, Senator Pittman wrote Raymond Moley: "Your understanding of political situations was indeed surprising to me in view of the fact that you are a professor." Congressmen scorned the White House circle as men who did not have to face an election— something like not having to meet a payroll—and who hence were irresponsible. Part of the time, they saw the New Dealers as worldly, power-hungry Machiavellians, bent on subverting the revered traditions of the Democratic party. "Democrats, Democrats," cried Carter Glass. "Why, Thomas Jefferson would not speak to these people." Congressmen protested that the New Dealers, while feigning political innocence, were, in fact, running a kind of "Phi Beta Kappa Tammany Hall" dominated by the eastern universities. "Unless an applicant can murder the broad 'a' and present a Harvard sheepskin he is definitely out," grumbled a Michigan congressman.[6] Conservative

[3] Josiah Bailey to W. W. Ball, September 24, 1937, Ball MSS.; Bailey to Joseph O'Mahoney, August 22, 1937, O'Mahoney MSS.

[4] Georgia Lindsey Peek MS. memoir, "Early New Deal."

[5] Maury Maverick, "T.V.A. Faces the Future," *New Republic,* LXXXIX (1936), 64. The liberal Illinois Representative T. V. Smith confided: "Between you and me, I don't like fanatics like Aubrey Williams, though I see their strident usefulness." T. V. Smith to S. Kerby-Miller, August 3, 1939, T. V. Smith MSS., Box VI.

[6] Pittman to Moley, October 4, 1932, Pittman MSS., Box 13; Arthur Schlesinger, Jr., *The Coming of the New Deal* (Boston, 1959), p. 483; John Dingell to F.D.R., June 26, 1937, James Roosevelt MSS., Box 36.

Democrats feared that the bureaucrats advanced humanitarian programs chiefly in order to expand their power; some even thought they detected a conspiracy to maintain a permanent army of reliefers who, in return for government largesse, would enable the bureaucrats to perpetuate themselves. Vice-President Garner told Hopkins pointedly that he did not like the implications of WPA officials calling reliefers "clients."[7]

The recession, which might have bred a radical movement of discontent with the workings of capitalism and liberal reform, served instead to strengthen the conservatives. Hard times halted the momentum of industrial unionism and dissipated its quasi-revolutionary potential. The S.W.O.C., thrown on the defensive by the cutback in steel, was not to score another victory until February, 1940.[8] Roosevelt's indecisiveness gave conservatives the opportunity to advance their own program for recovery. In December, 1937, they circulated secretly a manifesto, drafted chiefly by Senators Bailey and Vandenberg, calling for a balanced budget, states' rights, and the other staples of reaction.[9] Many in Congress were sympathetic to the contention that the downturn had been caused, in part, by government hostility to business; in particular, they sought to junk the capital gains and undistributed profits taxes. But only a few—Democrats like Byrd, Burke, Copeland, and Tydings, a Republican like Warren Austin—cared to sponsor the conservative manifesto. At the same time that they wanted to take steps to restore business "confidence," they had half a notion that renewed government spending might prove the right medicine. Most of all, they waited on Roosevelt for a cue.

With neither the President nor the Senate conservatives in command, the farm bloc seized the opportunity to advance its claims. Both New Dealers and Old Guard recognized that something had to be done to meet the farm crisis. The 1936 farm law had proven unworkable, for not enough farmers would co-operate voluntarily to limit production. The South staggered under an eighteen-million-bale crop which drove cotton prices down, and wheat growers faced a new glut.

[7] Bascom Timmons, *Garner of Texas* (New York, 1948), p. 220. See Robert Burke, *Olson's New Deal for California* (Berkeley and Los Angeles, 1953), p. 89.

[8] Walter Galenson, "The Unionization of the American Steel Industry," *International Review of Social History,* I, (1956), 36–37.

[9] *The New York Times,* December 16, 22, 1937; H. Styles Bridges to Bailey, November 29, 1937, Bailey MSS., Personal File.

As rural pressure groups cried for larger and larger bounties, the administration began to fear it had created a Frankenstein monster. The cotton South, strong in the councils of the Democratic party, pressed for unlimited production and higher subsidies, with the inevitable surpluses to be dumped overseas. Wallace continued to resist such proposals. Like Joseph of the Old Testament, he hoped to create an ever-normal granary by storing surpluses when yields were good and distributing them in lean years.[10]

After fighting off the more extravagant demands of farm lobbyists, Congress, in February, 1938, passed a new AAA statute which gave the farm bloc most of what it wanted. The act made a start toward an ever-normal granary, established soil conservation as a permanent program, authorized crop loans, offered crop insurance for wheat to give better protection against drought, and empowered the Secretary of Agriculture to assign national acreage allotments and subsidies to staple farmers. These new grants of authority, protested Senator McNary, made "the Secretary of Agriculture the autocrat of the breakfast table." A month later, farmers filed into grange halls and filling stations to ballot on the first AAA referendums under the 1938 act. Cotton farmers approved a compulsory marketing quota by the lopsided tally of 1,189,000 to 97,000; dark-tobacco growers, 38,000 to 9,000; flue-cured growers, 213,000 to 34,000.[11]

Still the surpluses mounted. Late in 1938 Wallace told the cabinet with unconscious irony that if the country had another good crop yield in 1939, "we would be sunk."[12] Even Wallace was forced to agree to export subsidies. From August, 1938, to December, 1939, the United States dumped 128,200,000 bushels of wheat abroad, during

[10] Joseph S. Davis, "The Economics of the Ever-Normal Granary," *Journal of Farm Economics,* XX (1938), 8–21; discussion by Mordecai Ezekiel, *ibid.,* 21–23; T.R.B., "Washington Notes," *New Republic,* XCII (1937), 186–187. The idea of the ever-normal granary came to Wallace in part from reading a Columbia dissertation on Confucian economics written by a Chinese student. Louis Bean, COHC, p. 150.

[11] *Congressional Record,* 75th Cong., 3d Sess., February 14, 1938, p. 1870; Dean Albertson, *Roosevelt's Farmer* (New York, 1961), pp. 109–117; Philip Glick, "The Soil and the Law," *Journal of Farm Economics,* XX (1938), 629–631; *Wallace's Farmer,* LXIII (1938), 152, 161; *Time,* XXXI (March 21, 1938), 17; Charles Hardin, "The Tobacco Program: Exception or Portent?" *Journal of Farm Economics,* XXVIII (1946), 920–937.

[12] Harold Ickes, *The Secret Diary of Harold Ickes* (3 vols., New York, 1954), II, 501.

the summer of 1939 at a loss of fifty cents on every bushel. It paid cotton growers an export subsidy of $7.50 a bale, and even concluded a barter deal with Britain to dispose of surplus cotton. All to no avail. In 1939, after six years of AAA subsidies to cut back production, the cotton carryover was three million bales greater than in 1932. Only the war rescued the New Deal farm program from disaster.[13]

Roosevelt went to Warm Springs late in March, 1938, an anxious man. The recession, now in its seventh month, had thrown four million people out of work; steel had lost two-thirds of its business; and the Federal Reserve Board index had fallen to 79, only ten points higher than it had been in 1932. When Roosevelt met with businessmen, they had little to offer; as late as mid-April, *Business Week* argued that, rather than resort to government spending, "it would be infinitely better for us . . . to rattle along as we are for a few months, and let the upward restoration come in the slow and hard way, by a restoration of confidence." The President, still unconvinced by his Keynesian advisers, would not pursue indefinitely a course of passivity. The hardships people were undergoing were too severe, the political consequences of hard times too dangerous. Moreover, at the very time of the Nazi conquest of Austria, the rest of the world was wondering whether the slump did not demonstrate the incapacity of the democracies to solve their internal problems. On March 25, the stock market cracked again. Harry Hopkins and Aubrey Williams, armed with memorandums from Leon Henderson, sped to Georgia to try once more to win the President to multibillion-dollar spending. This time they succeeded. On April 2, Roosevelt, at a lunch with Hopkins and Williams, announced he was willing to spend. "I'm awfully afraid that the cards are all stacked against us," Morgenthau told his staff a week later. "They have just stampeded him during the week I was away. He was completely stampeded. They stampeded him like cattle."[14]

---

[13] Gilbert Fite, *George N. Peek and the Fight for Farm Parity* (Norman, Okla., 1954), p. 300; Gunnar Myrdal, *An American Dilemma* (New York, 1944), pp. 255–256; Eliot Janeway, *The Struggle for Survival* (New York, 1951), pp. 151–152; Carl Schmidt, *American Farmers in the World Crisis* (New York, 1941), pp. 157–158. During this same period, the Federal Surplus Commodities Corporation "dumped" surpluses among America's poor. Russell Lord, *The Wallaces of Iowa* (Boston, 1947), pp. 470–472; John Blum, *From the Morgenthau Diaries* (Boston, 1959), pp. 435–438.

[14] *Business Week,* Apr. 16, 1938, p. 64; Joseph Alsop and Robert Kintner,

On April 14, the President asked Congress to embark on a large-scale lend-spend program. Roosevelt, who had liquidated Ickes' PWA some months before, now requested a huge fund for public works.[15] In June, Congress, with not much will to resist a President wanting to spend in an election year, voted a $3.75-billion omnibus measure. PWA got nearly a billion, and authority to lend millions more; Hopkins received over $1.4 billion for WPA; and assorted sums were granted for low-cost housing, parity payments, FSA, and NYA.

Once Roosevelt reversed himself on spending, the liberals expected that he would accept their antimonopoly recommendations too, even though the President sidled toward the trust question, as one columnist noted, "as gingerly as Hoover approached prohibition repeal." Cohen and Corcoran quickly threw together an antimonopoly prospectus, and Cohen and Jackson caught the President's northbound train at Atlanta. Before they reached Washington's Union Station, the President indicated he was ready to join the antitrust movement.[16] On April 29, Roosevelt asked Congress to vote funds for an investigation of the concentration of economic power. The President's message came at a time when several legislators had already filed similar proposals. While Roosevelt apparently envisioned a probe managed solely by the administration, Congress insisted the inquiry be conducted by a hybrid committee composed half of members of Congress and half of representatives of federal agencies. After perfunctory debate, Congress approved the first full-dress scrutiny of the economy since the Pujo study of 1912, and the most ambitious the government had ever undertaken.

The dominant figure in the Temporary National Economic Committee was its chairman, Senator Joseph O'Mahoney of Wyoming.[17]

"We Shall Make America Over," *Saturday Evening Post*, CCXI (November 19, 1938), 86; Blum, *Morgenthau Diaries*, p. 421.

[15] Samuel Rosenman (ed.), *The Public Papers and Addresses of Franklin D. Roosevelt* (13 vols., New York, 1938–50), VII, 221–233. The administration also reversed its policy of credit contraction; the Federal Reserve Board agreed to lower reserve requirements, and the Treasury desterilized gold.

[16] Raymond Clapper, *Watching the World* (New York, 1944), p. 129; Alsop and Kintner, "We Shall Make America Over," p. 86.

[17] David Lynch, *The Concentration of Economic Power* (New York, 1946), pp. 35–38; Edwin George, "The Personnel of Our National Economic Jury," *Dun's Review*, XLVI (September, 1938), 8–18. The dynamic executive secretary of the TNEC, Leon Henderson, soon found his energies diverted to other government assignments.

Although O'Mahoney had a Mountain State liberal's distrust of eastern monopoly, he disliked centralized bureaucracy too and sought "a formula which will set business free from regimentation by both monopoly and by government." He readily accepted Walton Hamilton's adjuration: "Impute no personal blame."[18] Indeed, O'Mahoney was so anxious to avoid any semblance of an inquisition that the committee never probed very deeply. O'Mahoney himself conceded: "We have not been making the headlines because we have been altogether too dull, sedate and conservative."[19] What conservatives feared would be a witch hunt served frequently as a public relations forum for business. Yet, although the TNEC insisted it was not a "grand jury" concerned with disclosing wrongdoing, inevitably it brought out a variety of malpractices, notably the operations of the vicious industrial life insurance business. It also led Congress to amend the patent statutes by pointing a finger at the Hartford-Empire Company which, through patent rights, controlled two-thirds of the country's glass-bottle industry, although it made neither bottles nor a single bottle-making machine.

The TNEC inquiry ran for almost three years, its public hearings for eighteen months. Five hundred and fifty-two witnesses produced thirty-one volumes of testimony, and 188 government economists prepared forty-three monographs under the committee's supervision. The TNEC's shelf of studies brought knowledge of business operations up to date, but the total yield of the investigation proved disappointing. The TNEC represented no commitment by Roosevelt to antibigness. His request for a study served to defer action and, as Moley has written sourly, constituted "the final expression of Roosevelt's personal indecision about what policy his administration ought to follow in its relation with business." The committee itself dodged the critical question of full employment and chose to concentrate on the single issue

[18] O'Mahoney to W. L. Batt, August 1, 1939, O'Mahoney MSS.; Walton Hamilton, "Memorandum to Assistant Attorney General Arnold," August 12, 1938, Arnold MSS., Box 1; Leon Henderson to F.D.R., March 7, 1939, "TNEC-Special-Important" folder, Henderson MSS., Box 25; Herbert Corey, "O'Mahoney Wants Facts—Not Scalps," *Nation's Business,* XXVI (September, 1938), 15–16, 72–74.

[19] O'Mahoney to Charles Collins, April 17, 1940, O'Mahoney MSS.; Dwight MacDonald, "The Monopoly Committee—A Study in Frustration," *American Scholar,* VIII (1939), 299. Cf. T. P. Gore to B. M. Anderson, Jr., July 21, 1938, Gore MSS.

of monopoly. Unwilling either to tackle the more difficult problems or to make recommendations which might disturb vested interests, the committee expressed the wistful hope that if it assembled enough facts, someone would be able to use them to solve the enigma of persistent unemployment.[20]

If the TNEC inquiry proved disappointing to the foes of monopoly, Thurman Arnold's tenure as Assistant Attorney General in charge of the Antitrust Division was a pleasant surprise. Roosevelt's appointment of the Yale Law School professor to replace Jackson in March, 1938, was regarded by old trust busters like Senator Borah as a cynical acceptance of the futility of antitrust prosecutions. Industrial combinations, Arnold seemed to imply in *The Folklore of Capitalism,* were a social necessity like houses of prostitution; the antitrust laws were society's way of expressing its disapproval. On pointless antitrust crusades, Arnold wrote, "men like Senator Borah founded political careers." Yet, as Fred Rodell observed, all that Arnold had stated was that antitrust laws were a farce; in view of their actual ineffectiveness, "Arnold's conclusion would seem more trite than profane or irreverent."[21]

Arnold turned out to be the most vigorous trust buster in history. In the next five years, he instituted 44 per cent of all the antimonopoly actions undertaken by the Justice Department since the passage of the Sherman Act. Under Arnold, the Division's personnel grew more than fivefold; two years after he took office, he had 190 antitrust lawyers, compared to the five in Teddy Roosevelt's reign. He declared war on excessively restrictive patents, on monopolistic practices in housing by unions as well as business, and on giants like General Electric and the Aluminum Company of America.[22] When the Supreme Court ruled that carpenters could use boycotts and other devices to win jurisdictional disputes, so long as they did not act in

---

[20] Raymond Moley, *After Seven Years* (New York, 1939), p. 376; U.S. Congress, Senate, Temporary National Economic Committee, *Investigation of Concentration of Economic Power,* Sen. Doc. 35, 77th Cong., 1st Sess. (Washington, 1941); Lynch, *Concentration of Power,* pp. 229–231, 262–279, 355–377; Robert Brady, *Business as a System of Power* (New York, 1943); Jerome Cohen, "The Forgotten T.N.E.C.," *Current History,* n.s. I (1941), 45–50.

[21] Thurman Arnold, *The Folklore of Capitalism* (New Haven, 1937), pp. 207–208, 217; Fred Rodell, "Arnold: Myth and Trust Buster," *New Republic,* XCV (1938), 178.

[22] Thurman Arnold, *The Bottlenecks of Business* (New York, 1940).

collusion with businessmen, Arnold felt handcuffed and abandoned his housing campaign.[23] When defense agencies protested that his prosecutions of firms like General Electric impeded mobilization, much of his antimonopoly program foundered. Although Arnold gave the antitrust drive a vigor it had not had at least since the Progressive era, he could claim few substantive gains.[24]

Congress was willing to sanction the Brandeisian antitrust program, for many legislators had been schooled in the evils of monopoly, but it had small liking for the Brandeisian undistributed-profits tax. Republicans insisted that the levy had precipitated the "Franklin D. Roosevelt depression," while Democrats like Representative Emanuel Celler of New York thought repeal of the tax would give hard-pressed businessmen "a word of cheer." Business Democrats like Bernard Baruch and Joseph Kennedy won unanticipated support from New Dealers like Marriner Eccles, who believed the tax had helped cause the downturn, and Harry Hopkins, who was accused of currying business support for the 1940 nomination. One Treasury official told Roosevelt he no longer knew whether he was being shot at by his friends or his enemies.[25] With the administration divided, the "Baruch men" in the Senate, Pat Harrison and Jimmie Byrnes, were able to push through a new measure which repealed the undistributed-profits tax and lowered the levy on capital gains. When Roosevelt threatened a veto, House conferees managed to restore the unpopular tax in principle but little more. The compromise version was still so unpalatable that the President seriously contemplated vetoing it. In the end, after scolding Congress for adopting a statute which discriminated in favor of large corporations and the wealthy, he permitted the Revenue Act of 1938 to become law without his signature.[26]

[23] United States *v.* Hutcheson, 312 U.S. 219 (1941); Maxwell Raddock, *Portrait of an American Labor Leader: William L. Hutcheson* (New York, 1955), pp. 239–250; Robert Christie, *Empire in Wood* (Ithaca, 1956), pp. 307–316.

[24] Corwin Edwards, "Thurman Arnold and the Antitrust Laws," *Political Science Quarterly,* LVIII (1943), 338–355; *Time,* XXXVI (September 9, 1940), 69–74; Thurman Arnold to William Castleman, January 4, 1940, Arnold MSS., Box 6; Wendell Berge MS. Diary, *passim;* Leon Henderson MS. Diary, Sept. 6, 1940.

[25] Randolph Paul, *Taxation in the United States* (Boston, 1954), pp. 210–211; Josiah Bailey to R. H. Gregory, Jr., February 19, 1938, Bailey MSS., Tax and Tariff, 1938; Blum, *Morgenthau Diaries,* p. 442; Roswell Magill to F.D.R., memoranda of March 19, April 25, 29, May 19, 1938, FDRL PPF 1820, Box 4.

[26] *Public Papers,* VII, 355–365; Robert Doughton to Ralph Hanes, April 23,

Congress granted the President one final reform, a federal wages and hours law, but it required almost every parliamentary weapon in the administration's arsenal to get it through. When the measure was introduced by Senator Hugo Black in May, 1937, it encountered bitter opposition from southerners who wanted to maintain the "differential" which enabled them to pay lower wages than other sections. South Carolina's Cotton Ed Smith, who was reported as saying that a worker could live comfortably in his state for fifty cents a day, fumed: "Any man on this floor who has sense enough to read the English language knows that the main object of this bill is, by human legislation, to overcome the splendid gifts of God to the South." Unmoved by such arguments, the Senate approved the bill at the end of July, in part because liberal senators shared Black's humanitarian sentiments, in part because northern businessmen hoped the act would protect them from southern cheap-labor competitors. The drive for wages and hours legislation, Walter Lippmann contended, was an attempt by northern industrialists to secure an internal protective tariff; only this, he wrote, explained the support of the proposal by Senator Henry Cabot Lodge of Massachusetts.[27]

By the time the bill reached the House, conservative organizations, having tasted blood in the rejection of the Supreme Court bill, had made the defeat of the wages and hours measure their next objective. The newspaper publisher Frank Gannett's National Committee to Uphold Constitutional Government decried the proposal as a stage in a conspiracy, of which Court packing had been the first step, to win dictatorial power for President Roosevelt. Southern Democrats like Howard Smith of Virginia, who worked with Gannett's organization, joined with Republicans to bottle up the bill in the House Rules Committee.[28] It took a determined campaign by its managers to win enough signatures to a discharge petition to get the bill to the House floor in December, 1937. Here, after having overcome some of the

1938, Doughton MSS., Drawer 10; "Statement of Senators Walsh and George," April 22, 1938, David I. Walsh MSS.; Paul, *Taxation,* pp. 208–220; Roy and Gladys Blakey, *The Federal Income Tax* (New York, 1940), pp. 436–463.

[27] *Congressional Record,* 75th Cong., 1st Sess., p. 7882; New York *Herald-Tribune,* August 3, 5, 1937.

[28] Frank Gannett to James Truslow Adams, August 4, 1937, J. T. Adams MSS.; Amos Pinchot to Max Eastman, June 15, 1937, Amos Pinchot MSS., Box 60; Howard Smith to Amos Pinchot, August 13, 1937, Amos Pinchot MSS., Box 62.

insularity of the southerners, it collided with the parochialism of organized labor. More concerned with union prerogatives than with the lot of millions of unprotected workers, Federation leaders like Big Bill Hutcheson and John Frey forced William Green to oppose the bill, while John L. Lewis gave only grudging support. On December 17, the House, by a narrow margin, voted to recommit the measure. In April, 1938, the House Labor Committee reported out a new version, more acceptable to the A.F. of L., but since it now lacked a wage differential, southerners once more buried the bill in the Rules Committee. A year had passed since the measure had first been introduced, and it now seemed unlikely that it would ever be passed.

The tussle over the wages and hours bill came to symbolize a joust not only of conservatives against liberals, of unionized northern manufacturers against southern industrialists, of "labor statesmen" like Sidney Hillman who supported the measure against Federation spokesmen like Green, but, perhaps most important, of southern conservatives like Cotton Ed Smith against southerners of advanced views like Hugo Black. This clash came sharply into focus in Florida's senatorial primary, where Claude Pepper, who favored the wages and hours legislation, dueled with the anti-New Deal Congressman James Wilcox. On May 3, 1938, Pepper, who claimed to represent the thinking of the "new" South, scored an overwhelming victory. Three days after Pepper's stirring triumph, the discharge petition on the wages and hours bill was opened for signatures. It took only two hours and twenty-two minutes to win enough names; the proceedings of the House were disrupted by congressmen clamoring to sign the "honor roll." On May 24, the House passed the wages and hours measure. By mid-June, managers of the bill had worked out a compromise with the Senate draft, steered it past threats from both the South and the A.F. of L., and won approval from both houses for the conference report. The final version allowed two years to reach the standards of a forty-cent minimum and forty-hour maximum; permitted numerous exemptions; and, in effect, left the question of differentials to the administrator of the law. The most memorable provision forbade the employment of child labor in interstate commerce. On June 25, the President signed the bill. "That's that," he said with a sigh.[29]

[29] James Burns, *Congress on Trial* (New York, 1949), pp. 68–82; Paul Douglas and Joseph Hackman, "The Fair Labor Standards Act of 1938 I," *Political Science Quarterly*, LIII (1938), 491–515; J. S. Forsythe, "Legislative History

The Fair Labor Standards Act was a highly unsatisfactory law. So many exemptions had been written into the measure that at one point Representative Martin Dies filed a satirical amendment: "Within 90 days after appointment of the Administrator, she shall report to Congress whether anyone is subject to this bill. . . ." Each congressman, even if he approved the purpose of the legislation, first sought to secure immunity for the industries of his particular district. A Florida congressman asked to exempt "persons employed in forestry or in the taking of fish, sea food, and sponges, or in the turpentine industry." Yet, weak as it was, the law established a foundation on which later Congresses could build. Even at the time, it forced immediate pay raises for an impressive number of workers. In 1937, twelve million employees in industries affecting interstate commerce earned less than forty cents an hour; no southern textile hand got that much, and New England shoe workers averaged but $853 a year.[30]

In the summer of 1938, the economy began to show signs of improvement. At Bowling Green, Kentucky, in July, Roosevelt crowed: "We are again on our way."[31] Yet the gains came painfully slowly. Even when productivity and national income rose, the number of unemployed remained exasperatingly high. In 1937, Harry Hopkins had declared it "reasonable to expect a probable minimum of 4,000,000 to 5,000,000 unemployed even in future 'prosperity' periods," and by 1939 many had come to conclude that Hopkins was right—that millions of jobless would never find jobs again, not even if "recovery" was finally achieved. Fiorello La Guardia confided his anxiety about the unemployment situation to Senator Byrnes: "I believe we can start on common ground, and that is, that instead of considering the situation as an emergency, we accept the inevitable, that we are now in a new normal."[32] Ten years after the Wall Street crash, there were still nine million without work.

of the Fair Labor Standards Act," *Law and Contemporary Problems*, VI (1939), 464–490.

[30] Burns, *Congress on Trial*, p. 77; O. R. Altman, "Second and Third Sessions of the Seventy-fifth Congress, 1937–38," *American Political Science Review*, XXXII (1938), 1105; Jerry Voorhis, *Confessions of a Congressman* (Garden City, N.Y., 1948), p. 88.

[31] Louisville *Courier-Journal*, July 9, 1938.

[32] Harry Hopkins, "The Future of Relief," *New Republic*, XC (1937), 8; Robert Burke, *Olson's New Deal for California* (Berkeley and Los Angeles, 1953), pp. 12, 89; Fiorello La Guardia to James Byrnes, April 5, 1939, La Guardia MSS.

While Roosevelt had broken with the budget-balancers and resumed spending, he still had not embarked on the kind of massive spending which the Keynesians called for. He seems to have been even less impressed with Keynes than the British economist was with the President's grasp of economics, and at no point did he, in fact, embrace Keynesianism.[33] "It is not clear," reflected Keynes's official biographer, "that he acted on the principle that it was the deficit, rather than the public works themselves, that was the potent agency in reducing unemployment." Roosevelt trusted in common sense and in finding a middle way between extremes. These attitudes were often sources of strength, but in this instance they proved disadvantageous. Total commitment was required, and total commitment he would not give.[34] The Keynesian formula for gaining prosperity by deliberately creating huge deficits year after year seemed to defy common sense. Roosevelt was willing to countenance limited, emergency spending, but halfway measures of this sort antagonized business and added to the public debt without giving a real fillip to the economy.[35]

[33] *New Republic*, CXIV (1938), 345–346; Leon Henderson to Harry Hopkins, June 9, 1938, James Roosevelt MSS., Box 36. Although accounts vary, most confirm Frances Perkins' report that Roosevelt complained that Keynes, who had "left a whole rigamarole of figures," seemed more a mathematician than a political economist, and that Keynes had confided that he had "supposed the President was more literate, economically speaking." Frances Perkins, *The Roosevelt I Knew* (New York, 1946), pp. 225–226; Alvin Johnson, *Pioneer's Progress* (New York, 1952), p. 244. But see F.D.R. to Felix Frankfurter, June 11, 1934, FDRL PPF 140; Rexford Tugwell, *The Democratic Roosevelt* (Garden City, N.Y., 1957), p. 375.

[34] R. F. Harrod, *The Life of John Maynard Keynes* (New York, 1951), p. 449; James Burns, *Roosevelt: The Lion and the Fox* (New York, 1956), pp. 330–336. Burns speculates: "If Roosevelt had urged spending programs on Congress rather than the court plan and certain reform measures in 1937, he probably could have both met his commitments to the one-third ill-housed, ill-fed, and ill-clothed and achieved substantial re-employment." *Ibid.*, p. 331.

[35] The timidity of New Deal fiscal policy is examined in E. Cary Brown, "Fiscal Policy in the 'Thirties: A Reappraisal," *American Economic Review*, XLVI (1956), 857–879. Cf. Eccles, *Beckoning Frontiers*, pp. 182–183, 311, 317; Tugwell, *Roosevelt*, p. 374; Seymour Harris (ed.), *The New Economics* (New York, 1947), pp. 17–19. It has been argued that since Keynes's major work did not appear until 1936, it is unreasonable to expect Roosevelt to have pursued Keynesian policies as early as 1937. In fact, many in Washington were familiar with Keynesian doctrine in a general way, if not in specific detail, before 1937, and they pressed their views vigorously on the President. But the demand for a balanced budget had far wider support; one poll revealed that even a majority of reliefers favored a cutback in federal spending. Under the circumstances, it is understandable, albeit regrettable, that Roosevelt did not take up the new

The sluggishness of the economy stripped the New Dealers of much of their self-confidence. The recession demonstrated that they had done nothing far-reaching enough to change the cyclical character of the economy, yet they recoiled from the thought of more drastic action. The resort to the old panacea of antimonopoly crusades seemed to men like Adolf Berle to be entering a plea of intellectual bankruptcy. When Berle resigned as Assistant Secretary of State in the summer of 1938, he charged that Robert Jackson was "working on the progressive thesis of the nineties, which was old in the time of McKinley," and underscored the need for original economic thinking. "Unless the government nationalizes great blocks of industry," Berle had pointed out that spring, "reliance must be had on at least seven times as much private activity as government activity."[36] No one contemplated nationalization, yet it was not clear how government could get private business to take over. Business confidence had been badly shaken by the depression and by the challenge to its prerogatives from government and labor; no embryonic industry like radios or automobiles had emerged to spark a new wave of business expansion. Business publicists disputed the view that the country had reached a state of "stagnation," but their protests were so vocal because this was precisely what many businessmen themselves deeply believed. Roosevelt neither embraced a bold program of national planning which would assume some of the traditional business prerogatives to decide how funds would be invested nor could he devise a way to revive private initiative. Alvin Hansen queried: "Has the government intervention, made inevitable by the distress incident to vast unemployment, created a hybrid society, half-free and half-regimented, which cannot operate at full employment?"[37] By the end of 1938, even its supporters sensed that much of the impetus of the New Deal had died.[38]

---

economics. The President's reluctance to abandon the economic beliefs of a lifetime, ideas which most of the country shared, is less remarkable than the resiliency he did show in adapting himself to new ways of thinking.

[36] Berle to F.D.R., August 16, 1938; Berle, Memorandum to the President, April 18, 1938, FDRL PPF 1306.

[37] Hansen, *Full Recovery or Stagnation?* (New York, 1938), p. 8. Hansen's answer was that public investment was not only compatible with private development but indispensable to the economy in an era of stationary population. "Fortune Round Table," *Fortune*, XX (October, 1939), 113.

[38] "I am hanging on to the principles of the New Deal," said one Akron

For the New Dealers, 1938 was a discouraging year. They had scored some successes in Congress, a surprising number given the strength of the bipartisan conservative coalition, but they had suffered some mean setbacks too. The rebuffs would have been hard to take under any circumstance, but what exasperated the New Dealers most was the sense that Congress did not reflect the sentiments of the country, and that many Democrats had won office on Roosevelt's popularity only to knife him as soon as the returns were counted. By the end of 1937, Harry Hopkins had pulled together a council of liberals, including Corcoran, Ickes, and the President's son and secretary James Roosevelt, to explore the possibility of purging the party of its conservatives. Dubbed the "elimination committee," they wished not only to discipline party irregularity but to assure liberal control of the 1940 Democratic convention.

When in January, 1938, Representative Lister Hill, an ardent New Dealer, won an impressive victory in the Democratic senatorial primary in Alabama over the conservative demagogue, Tom Heflin, the New Dealers were encouraged to intervene in the Florida primary. On February 6, in Palm Beach, Jimmy Roosevelt announced the administration's support of Claude Pepper.[39] When Pepper scored his smashing triumph, the "White House Janizaries," as Hugh Johnson called them, felt emboldened to undertake a full-scale campaign against the party's conservatives. Roosevelt neither committed himself to the "purge" nor restrained his lieutenants from carrying it out. He gave Senator James Pope of Idaho a note of commendation, wrote Senator McAdoo a "Dear Mac" letter, told his aides to work for the re-election of the Senate Majority Leader, Alben Barkley, but refused

---

worker, "but I really know that it won't solve our problems." Alfred Winslow Jones, *Life, Liberty, and Property* (Philadelphia, 1941), pp. 271–272. Bernard De Voto wrote: "The New Deal is just obscene and there is nothing else." De Voto to Garrett Mattingly, June 15, 1938, De Voto MSS., Box 2. Some writers have suggested that Roosevelt sacrificed domestic reform because he needed to win conservative southern Democratic votes for his foreign policy. Little evidence has been cited for this view. The New Deal seems to have sputtered out for quite other reasons.

[39] J. B. Shannon, "Presidential Politics in the South: 1938, I," *Journal of Politics,* I (1939), 150; Joseph Keenan to James Roosevelt, December 11, 1937, James Roosevelt MSS., Box 12; Bibb Graves to F.D.R., January 4, 1938, FDRL OF 300, Box 10; *The New York Times,* February 7, 1938. Cf. J. Mark Wilcox to F.D.R., February 7, 1938, FDRL OF 300 (Florida).

to join the vendetta against foes of the New Deal.[40]

On June 24, the President, out of patience with party conservatives and perhaps nettled by the victory of the anti-New Deal Senator Guy Gillette in Iowa, broke his silence. He denounced the "Copperheads" in the party, and claimed the right to intervene in elections wherever there was a clear issue of principle.[41] Roosevelt followed up this fireside chat by barnstorming the country. Under a burning sky at Kentucky's Latonia Race Track, he backed Senator Barkley over his opponent Happy Chandler. In Oklahoma he gave a pat to Senator Elmer Thomas. The victories of both Barkley and Thomas were counted as triumphs for the Old Campaigner, although the President's influence, at least in Kentucky, was negligible.[42]

The rest of Roosevelt's trip was a disaster. Chary of opposing incumbents he felt he could not lick, he ducked fights with men like Pat McCarran in Nevada and Alva Adams in Colorado.[43] Not until he reached Georgia, on his way home, did he feel confident enough to tackle a major opponent. On August 11, in Barnesville, he declared war on Senator Walter George; on most public questions, the President stated, he and George did "not speak the same language."[44] In South Carolina, he fired a shot at Senator Cotton Ed Smith. In Maryland, he and his aides attempted to unseat Senator Millard Tydings. Even Jim Farley, who harbored doubts about the purge, approved the sortie against Tydings, who had opposed the NRA, AAA,

[40] F.D.R. to Pope, February 22, 1938, FDRL PPF 5209; F.D.R. to McAdoo, March 16, 1938, FDRL PPF 308; Ickes, *Diary,* II, 342.

[41] *Public Papers,* VII, 391–400.

[42] *Ibid.,* p. 438; Louisville *Courier-Journal,* July 9, 1938; August 7–9, 1938; "Memorandum for Senator Barkley from Mr. John L. Lewis," June 6, 1938, Barkley MSS., Politics-Primary, Senate, Barkley-Chandler; J. B. Shannon, "Presidential Politics," pp. 168–169; "40 Years a Legislator," p. 423, Elmer Thomas MSS. Mrs. Chandler announced she hoped the Governor would quit politics. "You know, you can't make money in politics, especially when you're a psychopathic case of honesty such as Happy is." Louisville *Courier-Journal,* August 9, 1938.

[43] Roosevelt did give a nod of approval to the liberal San Antonio Congressman Maury Maverick, but Maverick went down to defeat. Ickes, *Diary,* II, 421; Maverick to O. G. Villard, August 8, 1938, Villard MSS.

[44] *Public Papers,* VII, 469. In late March at Gainesville, Roosevelt had ignored George, and denounced the feudal system of the South. "When you come down to it," remarked the President, "there is little difference between the feudal system and the Fascist system. If you believe in the one, you lean to the other." *Ibid.,* p. 168.

TVA, the Wagner Act, and the Court bill. Roosevelt was beaten in Georgia. He was beaten in South Carolina. He was beaten in Maryland. Only in New York City, where the administration purged the conservative chairman of the House Rules Committee, John O'Connor, was an anti-New Dealer ousted. O'Connor's removal, Roosevelt insisted, made the whole effort worthwhile. "Harvard," he said, "lost the schedule but won the Yale game." But no one doubted that Roosevelt had, in fact, suffered a humiliating drubbing. Some even believed that only Roosevelt's intervention had saved Cotton Ed Smith from defeat in South Carolina.[45]

Roosevelt had hoped that, by distinguishing between liberal and conservative representatives, he could win popular support for the creation of a liberal Democratic party. Unhappily, ideological issues that seemed clear in Washington blurred in South Carolina. When Cotton Ed Smith waged a white-supremacy campaign, the New Deal's candidate, Governor Olin Johnston, replied in kind. "Why, Ed Smith voted for a bill that would permit a big buck Nigger to sit by your wife or sister on a railroad train," the Governor cried.[46] While the C.I.O. backed Johnston, both the A.F. of L. and the Railroad Brotherhoods supported Smith. Only South Carolina's conservatives saw the same implications in the race that the New Dealers did. A violently anti-New Deal editor wrote: "As for old Smith, monstrous wind-bag, he has been invaluable—we have used him for the undoing of Blease and now of Roosevelt who is an edition de luxe of Blease."[47] But liberalism and conservatism became confused in an encounter which pitched the might of the federal machine against the State House crowd, a contest in which each candidate sought to outdo the other in whipping up race hatred. In the end, Smith won by splitting the vote of the millhands. "It takes a long, long time," the President commented wearily, "to bring the past up to the present."[48]

The New Dealers had attempted to nationalize the party institution, to transform a decentralized party, responsible only to a local elec-

[45] Breckinridge Long MS. Diary, August 22, 1938; Report of "The Group," September 12, 1938, FDRL OF 300 (Maryland), Box 36; *Time*, XXXII (October 3, 1938), 9; F.D.R. to James H. Fay, September 23, 1938, FDRL PPF 5538.

[46] *The New York Times*, August 23, 1938.

[47] W. W. Ball to "Sara," September 2, 1938, Ball MSS.

[48] J. B. Shannon, "Presidential Politics in the South: 1938, II," *Journal of Politics*, I (1939), 289.

torate, into an organization responsive to the will of the national party leader and the interests of a national electorate.[49] But in the primaries, Roosevelt faced not "the people" but the enrolled Democratic voters, and the President could not shake the local loyalties party voters had to the national figures from their states. Newspapers played on states' rights sensitivities with editorials bearing titles like "Marching Through Georgia!" or "Upon What Meat Doth This, Our Caesar, Feed?" In Maryland, Tydings shrewdly refused to be drawn into a debate on his conservative record, and instead denounced federal carpetbaggers for invading the Free State of Maryland. "Maryland will not permit her star in the flag to be 'purged' from the constellation of the states," Tydings proclaimed.[50]

In deciding to intervene in "local" elections, the President placed himself on what seemed the wrong side of a "moral" question. The word "purge," which quickly became a generic term for Roosevelt's role in the 1938 primaries, summoned up images of the bloody extermination of Roehm and other Nazi leaders by Hitler in 1934. That summer, as the Czech crisis built toward the showdown at Munich, it was easy to represent Roosevelt's move as the act of a man with the same unquenchable appetite for power that possessed the European dictators.

The purge heightened fears that Hopkins was using the WPA to build a national political machine. During the campaign, his lieutenant, Aubrey Williams allegedly urged WPA workers to vote to "keep our friends in power."[51] Hopkins himself asked Iowans to oust Senator Gillette. In states like Pennsylvania, Democrats shamelessly mulcted defenseless relief workers and maced federal employees.[52] Yet it was not true that Hopkins had a national machine. Frequently, New

[49] It is even possible that Roosevelt thought of the purge as the first step toward creating the nucleus for a future liberal party. Thomas Stokes, *Chip Off My Shoulder* (Princeton, 1940), p. 504.

[50] Raymond Clapper, "Roosevelt Tries the Primaries," *Current History,* XLIX (1938), 16; *The New York Times,* August 22, 1938. It is also true that the purge was maladroitly handled, and executed in an amateurish and only half-committed fashion. Neither George's nor Tydings' opponents entered the race until the incumbents had already pre-empted many of the Roosevelt backers.

[51] *The New York Times,* June 29, 1938.

[52] Richard Keller, "Pennsylvania's Little New Deal" (unpublished Ph.D. dissertation, Columbia University, 1960), pp. 226–227, 307, 314–315, 348–350; David I. Walsh to Rev. Francis X. Talbot, December 2, 1938, Walsh MSS.

Deal opponents controlled the local WPA organization, and, even when they did not, reliefers generally voted as they pleased. In Sioux City, where there were four thousand WPA clients, Gillette's opponent polled only three hundred votes. Although Roosevelt was not so pristine a politician as to fire relief recipients on the eve of an election, he often moved vigorously when he learned of corruption in welfare machinery, as in 1935, when he ordered Hopkins to remove the Democratic governor of Ohio, Martin Davey, from all connection with the administration of federal relief. Moreover, the cry of "Keep the WPA out of politics" frequently meant that Republicans wanted to be free to use county courthouse machines for campaign funds and votes, but deny the Democrats the right to counter.

The most damaging shot fired against Hopkins came not from a Republican but from a New Deal sympathizer, the Scripps-Howard writer, Tom Stokes. In a series of eight articles, Stokes charged that Barkley was manipulating the WPA in Kentucky to defeat Chandler. Although Hopkins denied all but two of the twenty-two counts, the accusation was well founded.[53] WPA corruption proved insignificant in the race, and there was no correlation between WPA rolls and the Barkley vote, but it provided the occasion for a Senate investigation into the Kentucky contest.[54] Out of this probe came the so-called Hatch Act of 1939, which began as a measure to ban the exploitation of reliefers in campaigns but ended as an omnibus law to prohibit any political activity by federal employees. The Hatch Act weakened the hold of factions of federal office holders on the Democratic party, and permitted other groups—local bosses, rural conservatives entrenched in state legislatures, or labor unions—to deal themselves into a share of control.[55]

[53] Stokes, *Chip Off My Shoulder,* pp. 534–536; Louisville *Courier-Journal,* July 1, 1938; Ernest Rowe, District Director, WPA, to George Goodman, State Administrator, June 16, 1938. An accompanying report reads: "We have a splendid organization in this county and the WPA is giving 100% cooperation. . . ." See also Goodman to Alben Barkley, May 26, 1938, Barkley MSS., Politics—Primary, Senate, Barkley-Chandler.

[54] J. B. Shannon, "Presidential Politics in the South: 1938, I," pp. 166–170.

[55] Bascom Timmons, *Garner of Texas* (New York, 1948), p. 259; Burke, *Olson,* p. 140; Elmer Schattschneider, *Party Government* (New York, 1942), pp. 166–168. Despite the enormous sums that passed through the hands of the New Dealers, no corruption was ever traced back to any national New Deal administrator. The New Deal, as Roosevelt proudly claimed, had "no Teapot Dome." *Public Papers,* V, 484.

The Roosevelt coalition received another heavy blow in the November, 1938, elections. Republicans picked up eighty-one seats in the House, won eight in the Senate, and captured a net of thirteen governorships. Although the G.O.P. polled only 47 per cent of the House vote compared to 48.6 for the Democrats, it made a decided gain over its 39.6 per cent showing in 1936. The "New Faces" movement in the Republican party introduced Robert Taft as the new senator from Ohio, John Bricker as governor of that state, and the thirty-one-year-old Harold Stassen as governor of Minnesota. In New York, the young racket-buster Thomas Dewey narrowly missed ousting Governor Herbert Lehman. Some of the most prominent liberal governors—Frank Murphy in Michigan, Elmer Benson in Minnesota, and Philip La Follette in Wisconsin—went down to defeat. It was the recession, more than any other issue that hurt the Democrats.[56] On Election Day, millions of jobless men and women still walked the streets. The low price of wheat and corn turned Midwestern farm votes against the Democrats. Irritated at relief spending and alarmed by the burgeoning power of labor, many middle-class voters lost confidence too in the administration's ability to restore prosperity.[57]

It is by no means clear that the 1938 elections marked a rejection of the New Deal. Despite their gains, the Republicans lost twenty-four of the thirty-two Senate contests. Moreover, they proved wary of de-

[56] A defeated Pennsylvania Democratic congressman, asked to explain the party's ill fortunes, responded: "The main reason is the Democrats thus far have failed in their major objective. The New Deal was set up not only to arrest the depression but to lay the foundation for permanent recovery. The prosperity for which the American people have been yearning for more than a decade has failed to make its appearance. It is still around the corner. Truly has it been said, 'The Republicans wrecked the Country and the Democrats are at a stalemate with reference to the problem of recovery and reconstruction.' " Charles R. Eckert to James A. Farley, February 2, 1939, Joseph O'Mahoney MSS. Cf. Frank Murphy to James A. Farley, December 7, 1938, FDRL OF 300, Election Forecasts (Michigan); "Confidential Memorandum," David I. Walsh to Clifton Carberry, November 10, 1938, Walsh MSS.; Wesley Clark, *Economic Aspects of a President's Popularity* (Philadelphia, 1943); Philip La Follette, Elmer Benson, and Frank Murphy, "Why We Lost," *The Nation,* CXLVII (1938), 586–590.

[57] The issue of local corruption cost the Democrats in Pennsylvania, New Jersey, and Massachusetts. The party was also badly damaged by the division in labor, particularly by the conservative orientation of the A.F. of L. William Green backed the reactionary Republican Senator James Davis in Pennsylvania and the conservative Governor Frank Merriam in California, and boasted of the Federation's role in unseating the liberal Maury Maverick in Texas.

manding outright reversal of Roosevelt policies. The liberals scored some striking triumphs, notably in California, where the progressive Culbert Olson, pledged to free Tom Mooney, ousted Republican Governor Frank Merriam by a convincing 221,000-vote margin. The elections also revealed new sources of Democratic strength—among lower-income groups, especially union labor, and Negroes, Italians, and Poles.[58]

Yet the consequences of the elections were disastrous for the New Dealers. Without enough leverage to carry the President's program through Congress before November, they now faced the prospect of a greatly strengthened Republican–conservative Democratic coalition. The G.O.P. had been resuscitated as a national power. Roosevelt's political career seemed all but ended. The 1938 outcome, wrote the usually astute newspaperman Raymond Clapper, showed "clearly, I think, that President Roosevelt could not run for a third term even if he so desired."[59]

As Roosevelt's power waned, Congress began to handle him more roughly. Since he was not expected to run again in 1940, congressmen no longer had anything to fear, nor, since they could not ride on his coattails next time, and since he had less patronage left to dispense, did they have much to hope. As ardent a New Deal senator as Montana's James Murray could exclaim in a moment of pique over factional politics: "There was a time when I would have bled and died for him, but in view of the way he has been acting I don't want to have any more dealings with him and I just intend to stay away from him and he can do as he pleases."[60]

Through 1938, conservatives had fought chiefly a defensive war against the expansion of the New Deal. "For God sakes, don't send us any more controversial legislation," a group of congressmen asked in the spring of 1938.[61] By 1939, Congress was moving aggressively to dismantle the New Deal. It slashed relief appropriations, killed Roo-

[58] Burke, *Olson*, pp. 23–35; Samuel McSeveney, "The Michigan Gubernatorial Campaign of 1938," *Michigan History*, XLV (1961), 122–127; Henry Morrow Hyde MS. Diary, January 29, 1939.

[59] Raymond Clapper, "Return of the Two-Party System," *Current History*, XLIX (December, 1938), 14.

[60] John Chamberlain, *The American Stakes* (Philadelphia, 1940), pp. 76–79; James Murray to Hugh Daly, March 23, 1939, Murray MSS.

[61] Memo to Stephen Early, April 6, 1938, FDRL PPF 3458.

sevelt appointments, and, in a deliberate slap at the Keynesians, eliminated what was left of the undistributed-profits tax. Faced by the need to make concessions, Roosevelt threw his opponents some of his least popular experiments. On June 30, 1939, the Federal Theatre Project died without a fight. While departments with powerful pressure groups massed behind them held their own—"the silly Congress gave me three hundred million more than I wanted for farm subsidies," Roosevelt wrote in 1939—agencies with weak constituencies were snuffed out.[62]

The more successful the New Deal was, the more it undid itself. The more prosperous the country became, the more people returned to the only values they knew, those associated with an individualistic, success-oriented society. When the Lynds returned to Muncie to study the impact of the depression on "Middletown," they concluded that essentially it had changed nothing at all: "Middletown is overwhelmingly living by the values by which it lived in 1925.[63] The New Dealers were never able to develop an adequate reform ideology to challenge the business rhetoricians. During the upturn of 1935–37, conservatives argued that, since the crisis had passed, reforms were no longer appropriate. When the recession struck, this plea had even greater force; as the nerve of business opposition revived, the old conviction that business could run the economy with greater efficiency than bureaucrats reappeared.

In the early years of the depression, the nation was united by a common experience. People felt genuine compassion for the victims of hard times. By Roosevelt's second term, as it seemed that the country might never wholly recover, the burden of the unemployed

[62] F.D.R. to Joseph Kennedy, August 5, 1939, in Elliott Roosevelt (ed.), *F.D.R.: His Personal Letters, 1928–1945* (2 vols., New York, 1950), II, 911; Emmett Lavery, "Who Killed Federal Theatre?" *Commonweal*, XXX (1939), 351–352; Floyd Riddick, "First Session of the Seventy-sixth Congress," *American Political Science Review*, XXXIII (1939), 1022–1043.

[63] Robert and Helen Lynd, *Middletown in Transition* (New York, 1937), p. 489. Farmers, in particular, felt the pull of the twenties. Laetitia Conard, "Differential Depression Effects on Families of Laborers, Farmers, and the Business Class: A Survey of an Iowa Town," *American Journal of Sociology*, XLIV (1939), 526–533. The more the New Deal helped the farmer, the more he tended to return to the Republican party, where he found a home for his sentiments as a property owner.

had become too exhausting a moral and economic weight to carry. Those who held jobs or drew income from other sources could hardly help but feel that the depression had been a Judgment which divided the saved from the unsaved. Increasingly, the jobless seemed not merely worthless mendicants but a menacing *Lumpenproletariat.* America, Harry Hopkins wrote in 1937, had become "bored with the poor, the unemployed and the insecure."[64]

Yet if the country no longer quickened to the promise of the New Deal, and had wearied of assaults on business, it still placed more faith in the Democrats than in the Republican party, and it wanted none of the Roosevelt reforms undone. Even after the 1938 elections, the Democrats held more than two-thirds of the Senate and a comfortable House majority. Roosevelt pointed out that he was the first two-term President since Monroe who had not lost control of Congress before the end of his second term. Moreover, the resurgence of the G.O.P. cut two ways. If it magnified the President's dependence on old-line Democratic support, it also meant that conservative Democrats needed to unite with the New Dealers to stave off defeat in 1940. Ambitious party leaders knew that no Democrat could be elected President that year without Roosevelt's backing.

Roosevelt was willing to mollify congressmen by promising to keep his bright young men off the hill, but he would not surrender his party to his adversaries. He gave a fighting Jackson Day speech in 1939; made a series of liberal appointments—Harry Hopkins as Secretary of Commerce, Frank Murphy as Attorney General, Felix Frankfurter and William Douglas to the Supreme Court—and won from Congress a modified reorganization act and a liberalization of the social security law. When irreconcilables attempted to amend the Wagner Act, he beat them back. If Roosevelt could get little vital legislation through Congress, and seemed uncertain of his direction, neither could the conservatives scuttle what had already been achieved. The result was stalemate.[65]

[64] Hopkins, "The Future of Relief," *New Republic,* XC (1937), 8–9.
[65] Ernest Lindley, "The New Congress," *Current History,* XLIX (February, 1939), 15–17; Burns, *Roosevelt,* 367–368; Ickes, *Diary,* II, 531; John Coffee to F.D.R., January 25, 1939, FDRL PPF 2581.

CHAPTER 12

The Fascist Challenge

IN THE late 1930's, liberal democracy experienced one of its sorriest hours. Palsied by fear of war and by internal divisions, the western powers watched passively as the Nazis annihilated the republics of Central Europe. Unable to solve the riddle of unemployment, the United States writhed under the taunts of dictators who claimed they had provided both guns and butter. Fascism had little appeal for most Americans, but as the shadow of depression lengthened, as the streets of Vienna and Prague echoed to the marching boots of fascism triumphant, it was not always possible to overcome the cold fear that democracy might fail, that fascism might hold the key to the future.

Americans scanned the horizon for portents of fascism in this country, and many were not comforted by what they saw. The success of Huey Long seemed evidence that fascism could come from within, not through a *coup d'état,* but with the acquiescence of the people.[1] Others saw fascism in the social order of the South, in the company towns, in the violence industrialists employed against unions, in the suppression of civil liberties by local tyrants like Jersey City's Mayor Frank Hague. "As soon as they begin to shout about 'free speech' and 'free press' and 'civil rights,' I know they are Communists," Hague fumed. "They are

---

[1] One writer noted: "Kingfish may be the Mississippi valley rendering of Il Duce." Forrest Davis, *Huey Long* (New York, 1935), p. 286. "In Huey Long was a potential American Hitler," wrote another. "I say that advisedly. I knew the man." Thomas Stokes, *Chip Off My Shoulder* (Princeton, 1940), p. 399.

not really fighting for rights, but raising this cry as a subterfuge."[2]

In retrospect, we can see that Americans fretted too much about the danger of fascism in the thirties. Long no doubt constituted a menace, but he was hardly a fascist.[3] Yet fascism did muster a disturbing amount of support. Anti-Semites flocked into organizations like William Dudley Pelley's Silver Shirts—the initials aped those of Hitler's S.S.—which claimed to represent "the cream, the head and flower of our Protestant Christian manhood." Pelley, obsessed with the phantasm of an international Jewish conspiracy, even warned against the "half-Jew" Alfred E. Schmidt. In 1938, Gerald Winrod, a fundamentalist evangelist who blamed "the international Jew" for "the scourge of international communism," polled 53,000 votes in the Republican senatorial primary in Kansas.[4] In February, 1939, 22,000 members and sympathizers of the German-American Bund packed Madison Square Garden.[5] The country found Sinclair Lewis's ironically titled *It Can't Happen Here* an unsettling anticipation that it could; Buzz Windrip, an amalgam of Huey Long and Gerald Winrod, suggested that America could manufacture its own native-born Hitlers.

Spawned by the anxieties of a depression that it seemed might never end, many of the hate groups took a particularly ugly character. When Detroit police investigated a ritual slaying in 1936, they uncovered a

[2] Frederic Howe to F.D.R., June 30, 1936, FDRL PPF 3702; Raymond Gram Swing, *Forerunners of American Fascism* (New York, 1935); *Jersey Journal,* January 13, 1938. Cf. Dayton McKean, *The Boss* (Boston, 1940), pp. 227–228.

[3] Long himself dismissed the question of what he represented by remarking: "Oh, hell, say that I'm *sui generis* and let it go at that." Forrest Davis, *Huey Long,* p. 22. Cf. Louis Hartz, *The Liberal Tradition in America* (New York, 1955), pp. 275–276; V. O. Key, Jr., *Southern Politics* (New York, 1949), p. 164.

[4] Arthur Schlesinger, Jr., *The Politics of Upheaval* (Boston, 1960), pp. 80–81; "Civil Liberties Committee. 'Silver Shirt' Charges," Elbert Thomas MSS., Box 19; Donald Strong, *Organized Anti-Semitism in America* (Washington, 1941), pp. 40–56; F.D.R. to William Allen White, June 8, 1938, White MSS., Box 324.

[5] *The New York Times,* February 21, 1939; Alson J. Smith, "I Went to a Nazi Rally," *Christian Century,* LVI (1939), 320–322. The Bund appeared to be more powerful than it was. It attracted so few members and aroused so much antagonism that the Third Reich severed its ties with this "stupid and noisy" organization, which it recognized to be a liability. Joachim Remak, " 'Friends of the New Germany': The Bund and German-American Relations," *Journal of Modern History,* XXIX (1937), 38–41; John Hawgood, *The Tragedy of German-America* (New York, 1940), pp. 301–308.

secret terror group called the Black Legion which intimidated a wide area of the industrial Midwest; by 1939, thirteen of its members had received life sentences for murder.[6] In New York, young hoodlums roved subway platforms assaulting Jews; often they would insult a Jewish girl, provoking her escort to attack; he would be beaten by the gang. Many of the anti-Semites found a home in the Christian Front, an organization which flourished in Irish-Catholic neighborhoods in New York and Boston under the paternal eye of Father Coughlin, who had abandoned "social justice" for Jew-baiting.[7]

The domestic fascists execrated Franklin Roosevelt and the "Jew Deal." At a camp at Narrowsburg, New York, Christian Fronters engaged in rifle practice used a likeness of President Roosevelt's head as a target.[8] Such attacks only succeeded in rallying much of the nation to the President. Yet many Americans, especially conservatives, while in no way accepting the fascist creed, concurred on this one point: that the real peril to the country lay not without but within, not in the augmented power of the Axis but in Roosevelt's consular ambitions.

The unreasoning suspicion of President Roosevelt reached a point of near hysteria in the debate over the reorganization bill in 1938. When the proposal to empower the President to reshuffle agencies in the interest of efficiency had first been presented early in 1937, it had engendered relatively little controversy, for conservatives from William Howard Taft to Herbert Hoover had pressed for government renovation.[9] But by early 1938 the same elements which had fought Court packing had stamped reorganization as yet another attempt by Roosevelt to subvert democratic institutions.

[6] Morris Janowitz, "Black Legions on the March," in Daniel Aaron (ed.), *America in Crisis* (New York, 1952), pp. 305–325; Elmer Akers, "A Social-Psychological Interpretation of the Black Legion," Wagner MSS.

[7] Dale Kramer, "The American Fascists," *Harper's*, CLXXXI (1940), 384; A. J. Smith, "Father Coughlin's Platoons," *New Republic*, C (1939), 96–97; James Shenton, "Fascism and Father Coughlin," *Wisconsin Magazine of History*, XLIV (Autumn, 1960), 6–11; John Roy Carlson, *Under Cover* (New York, 1943), pp. 54–69; "Christian Front etc." folder, Fiorello La Guardia MSS.; Leo Lowenthal and Norbert Guterman, *Prophets of Deceit* (New York, 1949); Alfred McClung Lee and Elizabeth Briant Lee, *The Fine Art of Propaganda* (New York, 1939).

[8] *The New York Times,* April 9, 1940.

[9] From the first, however, some congressmen suspected a challenge to their prerogatives. On the origins of the bill, see Louis Brownlow, *A Passion for Anonymity* (Chicago, 1958), pp. 313–398.

A hazy knowledge of the history of the Weimar Republic led to the conclusion that dictatorships arise from a steady accretion of power to the Executive, although as a Rhode Island congressman protested: "I have yet to learn of one case where dictatorship was established because of a strong Government." "Germany once had a good government, but, little by little, they gave all power and authority to their President," admonished Representative Bert Lord of New York. "If we pass this bill [Roosevelt] will have powers to correspond to the powers given to President Hindenburg. Hindenburg did not become a dictator, but Hitler did." Senator Vandenberg conceded that the reorganization bill was "more sinister as a symbol than as a reality. But we are dreadfully sensitive these days to symbols—whether they be fasces or swastikas or hammers and cycles [sic] or new blue eagles over the White House."[10]

Some members of Congress genuinely feared a Roosevelt despotism; others sounded the alarm to cloak resentment at the loss of still more congressional prerogatives to the President, for the bill would mark a shift of power from a Congress subject to pressure groups and sectional interests to a President claiming to speak for the national interest. Hamilton Fish asked: "Why should Congress continue to surrender and abdicate its legislative functions to the Chief Executive and leave itself with no more legislative authority than Gandhi has clothing?"[11] They particularly objected to the loss of power to the eastern intellectuals Roosevelt had brought into government. The actual reshuffling, Senator Wheeler anticipated, would be carried out by "some little fellow," "some professor from Dartmouth, or Yale, or Harvard, or Columbia." Yet it was the President's own lieutenants who came to Wheeler's aid. Henry Wallace, for one, surmised that "reorganization" would encourage incursions from Harold Ickes, who ran his department like a Chinese warlord who coveted the contiguous lands of other border ruffians. Wallace, Joseph Eastman of the ICC, and other New Deal administrators worked with the President's enemies in Congress to mutilate the bill.[12]

[10] Aime J. Forand to Herbert Claiborne Pell, May 26, 1938, Pell MSS., Box 4; *Congressional Record*, 75th Cong., 3d Sess., pp. 2813, 4599.

[11] *Congressional Record*, 75th Cong., 3d Sess., p. A1556. See "Statement of David I. Walsh on the Reorganization Bill," March 23, 1938, Walsh MSS.

[12] *The New York Times*, March 9, 1938; Joseph Harris, "The Progress of Administrative Reorganization in the 75th Congress," *American Political*

Frank Gannett's National Committee to Uphold Constitutional Government mobilized a powerful lobby against the measure, and Father Coughlin, who had broken his pledge to retire from the air, stirred up Catholic anxieties by the baseless charge that the President planned to seize all the church schools in the country, although, as Roosevelt plaintively remarked, "I don't know what the President of the United States was going to do with them when he did grab them."[13] Coming at the very time Hitler was devouring Austria, such imputations intensified worry that Roosevelt was importing European totalitarianism into the United States. "You talk about dictatorship," declared Congressman Charles Eaton. "Why, Mr. Chairman, it is here now. The advance guard of totalitarianism has enthroned itself in the Government in Washington. . . ."[14] Ironically, Coughlin had just come out in favor of authoritarian government, while Edward Rumely, secretary of Gannett's committee, worked the fringes of fascist sentiment. Antidemocratic forces succeeded in convincing many Americans that Roosevelt, in carrying through a routine operation to make democratic government more effective, aspired to be a dictator. As in the Weimar Republic, foes of democracy aroused suspicion of its leaders and cast democratic institutions in disrepute.

The tide of fear swept all before it. The Senate approved the reorganization bill by the narrowest of margins; even Senator Wagner, buffeted by thousands of telegrams from conservatives, from Catholic groups, and from the A.F. of L., voted against the measure.[15] On April 8, the House rejected the bill, 204–196; 108 Democrats bolted the President. One of the greatest defections in history, it marked the worst rebuff Roosevelt was ever to suffer in the House. "Jim," the President said to Farley, "I'll tell you that I didn't expect the vote. I can't under-

Science Review, XXXI (1937), 867; Joseph O'Mahoney to Harry Illsley, March 30, 1938, O'Mahoney MSS.

[13] Samuel Rosenman (ed.), The Public Papers and Addresses of Franklin D. Roosevelt (13 vols., New York, 1938–50), VII, 251; Samuel T. Williamson, Frank Gannett (New York, 1940), pp. 198–205; Gannett to James Truslow Adams, July 22, 1937, Amos Pinchot to F.D.R., July 26, 1937, Adams to Edward Burke, July 30, 1937, J. T. Adams MSS.

[14] Congressional Record, 75th Cong., 3d Sess., p. 4641.

[15] Harold Ickes, The Secret Diary of Harold Ickes (3 vols., New York, 1954), II, 349; William Green to Robert Wagner, March 21, 1938, Wagner MSS.

stand it. There wasn't a chance for anyone to become a dictator under that bill."[16]

Nothing divulged the sourish spirit of 1938 better than the creation of the House Committee on Un-American Activities. The chairman of the Committee, Texas Democrat Martin Dies, had earlier agitated to investigate Ogden Mills and the international bankers; now he drummed the communist menace. Common to both conceptions was the conviction that the crisis of the thirties had been caused by conspiratorial elements whose suppression would quickly restore the nation to "a normal condition." In the ninth year of the depression, the country was half-persuaded that Dies might be right.[17]

When the Dies Committee began hearings in the summer of 1938, it largely ignored the Nazis, although much of the pressure for the inquiry had come from antifascist congressmen, and made itself a forum for allegations of communist infiltration.[18] Dies permitted witnesses to make unsupported charges of the most fantastic character, and rarely accorded the accused the right to reply. In the first few days, witnesses branded as communistic no less than 640 organizations, 483 newspapers, and 280 labor unions. One witness implicated in subversion a series of Roman Catholic organizations, the Boy Scouts, and the Camp Fire Girls. When a witness made the reasonable point that movie stars, even Shirley Temple, had been careless in lending their names to communist causes, he opened the committee to ridicule. At a luncheon in Tacoma, Secretary Ickes sniggered at "a burly Congressman leading a *posse comitatus* in a raid upon Shirley Temple's nursery to collect her dolls as evidence of her implication in a Red Plot."[19] The antics of

[16] James A. Farley, *Jim Farley's Story* (New York, 1948), p. 130.

[17] August Raymond Ogden, *The Dies Committee* (Washington, 1945), pp. 38–46; William Gellermann, *Martin Dies* (New York, 1944), pp. 54–64.

[18] Congressmen like Samuel Dickstein and Adolph Sabath had pressed for an investigation of Nazis. The most formidable opposition to the proposal had come from Midwestern congressmen, who charged that the probe was aimed chiefly at Germans. Ogden, *Dies Committee*, pp. 32–34, 41–45; D. A. Saunders, "The Dies Committee: First Phase," *Public Opinion Quarterly*, III (1939), 223–238; Samuel Dickstein to Fiorello La Guardia, May 14, 1938, La Guardia MSS.

[19] Ickes, *Diary*, II, 455. Some months later, when the director of the Federal Theatre Project referred in passing to the Elizabethan dramatist Christopher Marlowe, Representative Joseph Starnes, a committee member, demanded to know whether Kit Marlowe was a Communist. Hallie Flanagan, *Arena* (New York, 1940), p. 342.

witnesses before the Dies Committee, and the yahooism of some of its members, influenced a whole generation of Americans to dismiss any alarum about a communist conspiracy as ludicrous.[20]

Ostensibly nonpartisan, the Dies Committee served the purposes of those who claimed that the New Deal was a Red stratagem. Republican committee member Noah Mason charged that federal money had been "generously used to advance the cause of communism in the United States." Shortly before the 1938 elections, testimony stigmatized liberal candidates in several states as pawns of the Communists. Especially damaging were calumnies noised against Governor Frank Murphy of Michigan. Murphy was taxed with working with communist elements in the C.I.O. and even of "treasonable action" in not using force to dislodge sit-down strikers.[21]

President Roosevelt had little liking for the Dies Committee because he had heard the cry of "Red" raised falsely so often, because he sometimes thought that radicals were moving in the same direction he was, and because he had an old-fashioned distaste for public probing into private affairs. Nevertheless, he hoped to avoid an open fight with Dies. The assault on Murphy stung him to speak. In a carefully prepared statement, he arraigned the committee for allowing itself to be used "in a flagrantly unfair and un-American attempt to influence an election."[22] Dies retorted that the President had been "wholly misinformed." A week later, he implied that Ickes had canceled two PWA projects in his district as a reprisal. Queried about this, the President responded: "Ho-hum."[23]

So long as the Communists reviled Roosevelt as a "social fascist," they had not won much of a following. Not until they changed their line in

---

[20] Ogden, *Dies Committee,* pp. 52–66; Saunders, "Dies Committee," pp. 228–229; William Bonsor to John Frey, August 17, 1938, Frey MSS.; Raymond Brandt, "The Dies Committee: An Appraisal," *Atlantic Monthly,* CLXV (1940), 232–237.

[21] Mason, "It Is Happening Here," *Proceedings of the National Education Association,* LXXVII (1939), 64; Ogden, *Dies Committee,* p. 77.

[22] "On the threshold of a vitally important gubernatorial election, they permitted a disgruntled Republican judge, a discharged Republican City Manager and a couple of officious police officers to make lurid charges against Governor Frank Murphy," the President declared. *Public Papers,* VII, 559. See Thomas Greer, *What Roosevelt Thought* (East Lansing, Mich., 1958), pp. 41–44.

[23] Ogden, *Dies Committee,* p. 81; *The New York Times,* November 2, 1938.

1935, in order to advance the interests of Soviet foreign policy, did they have any considerable appeal. To entice liberals into a common front against fascism, the Communists de-emphasized revolutionary dogma and claimed that they represented twentieth-century Americanism. The Communist party, said Earl Browder soberly, "is really the only party entitled by its program and work to designate itself as 'sons and daughters of the American Revolution.' "[24]

In the "Popular Front" era, the Communists made their greatest gains. They captured peace groups and youth organizations, infiltrated church associations, and played a fateful role in third parties in Minnesota and New York. Impressed by the zeal of the Communists, convinced that the Soviet Union was a bulwark against Hitler, finding in the Cause a kind of personal salvation, some intellectuals joined the party or parroted the party line.[25] When John L. Lewis needed a tough cadre of C.I.O. organizers, he turned to the Communists. By the end of the decade, the Communists sat in control of more than one-third of the C.I.O. and had men in key places in national headquarters.[26]

Yet what is really impressive is how little allegiance the Communists won at a time when conditions could hardly have been more propitious. Many intellectuals—men like John Dewey and Sinclair Lewis—spoke out forcefully against them. Few intellectuals of distinction could tolerate for long the rigid behavior of party members; many were annoyed, as Mary McCarthy recalls, by the Communists' "lack of humor, their fanaticism, and the slow drip of cant that thickened their utterance, like a nasal catarrh."[27] Most New Dealers, while they often had a covert admiration for men who could "go all the way," found specific questions of tobacco quotas and utility rates much more absorbing than

[24] Eugene Lyons, *The Red Decade* (Indianapolis, 1941), p. 173.

[25] Daniel Aaron, *Writers on the Left* (New York, 1961), pp. 149–396; Robert Iversen, *The Communists and the Schools* (New York, 1959), pp. 32–193.

[26] Max Kampelman, *The Communist Party vs. the C.I.O.* (New York, 1957), pp. 18–19; Irving Howe and Lewis Coser, *The American Communist Party* (Boston, 1957), pp. 368–386. Cf. Julius Emspak, COHC.

[27] Mary McCarthy, "My Confession," *The Reporter*, IX (December 22, 1953), 32; Grace Adams, "Comrades, Lay Off!" *Harper's*, CLXXVI (1938), 218–220; Stuart Browne, "A Professor Quits the Communist Party," *Harper's*, CLXXV (1937), 133–142; Iversen, *The Communists and the Schools*, pp. 194–223.

subterranean meetings.[28] Of the small number of Communists in the federal government, few held strategic places and fewer still had any knowledge of Soviet espionage. Although they wielded too much power in agencies like the National Labor Relations Board, the Communists did not shape a single piece of legislation, nor did they play a decisive role in any significant administrative policy.

The brackish discontents of 1938 made enemies of progressives who had once been allies. Wisconsin's Governor Philip La Follette grumbled that the recession demonstrated the failure of the New Deal's domestic program, and Senator Bob, his brother, argued that Roosevelt should be concentrating not on building battleships but on massive low-cost housing. At a conference at the University of Wisconsin Live Stock Pavilion on April 28, 1938, Phil La Follette launched a new national party, the National Progressives of America. "Make no mistake, this is NOT a *third* party," the Governor told delegates. "As certain as the sun rises, we are launching THE party of our time." Typically Midwestern in its view of reform, the party attacked the money power, declared its abhorrence of restricting production, implied that distress was the result of a flaw in character, and combined a call for public ownership and control of money and credit with a denunciation of "every form of coddling or spoon-feeding the American people." It reflected too the isolationism of the La Follette dynasty. The Western Hemisphere, its foggy foreign policy statement concluded, had been "set aside for the ultimate destiny of man. . . . Here it was ordained that man should work out the final act in the great drama of life. From the Arctic to Cape Horn, let no foreign power trespass."[29]

At a time when neither left nor right seemed certain of its direction, La Follette hoped to create a Center party which would be, as Elmer Davis explained, "a Party of National Concentration, THE party of America." Yet if liberals were vexed with an indecisive Roosevelt, they did not intend to abandon him for a cranky La Follette. Moreover, they were disturbed by the steel-helmeted National Guardsmen who "pro-

[28] Schlesinger, *Politics of Upheaval,* pp. 191–197; Joseph L. Rauh, Jr., to the editor, *Commentary,* XIII (1952), 493–494.

[29] Donald McCoy, *Angry Voices* (Lawrence, Kan., 1958), pp. 163–172; *The New York Times,* April 29, 1938; Robert Morss Lovett, "April Hopes in Madison," *New Republic,* XCV (1938), 13; Heywood Broun, "Shoot the Works," *ibid.,* p. 16; Adolf Berle to F.D.R., April 26, 1938, FDRL PSF 24.

tected" the Madison conference; by the presentation of La Follette standing alone on a stark platform like a fascist dictator; and by the party's symbol of a red, white, and blue cross and circle, which one writer called a "circumcized swastika." The National Progressives made a poor showing in Iowa and California in 1938, and when Philip La Follette was trounced in the gubernatorial race in Wisconsin the party folded.[30]

The same suspicion of Roosevelt's designs that poisoned domestic politics in 1938 hampered his direction of foreign policy. When, in the year that Franco's forces neared final victory in Spain and Nazi tanks rolled into Austria, Roosevelt sought to rebuild American defenses, critics hooted that he was trying to cover up domestic failures by embroiling the country in foreign adventures. "We Democrats have got to admit that we are floundering," declared Maury Maverick. "The reason for all this battleship and war frenzy is coming out: we have pulled all the rabbits out of the hat and there are no more rabbits."[31] The historian Charles Beard asked: "If we cannot solve even the problem of putting 10,000,000 of our own citizens to work on the lavish resources right at hand, or have collective security at home, how can we have the effrontery to assume that we can solve the problems of Asia and Europe, encrusted in the blood-rust of fifty centuries? Really, little boys and girls, how can we?"[32]

The President cautiously explored the possibility of taking the initiative in calling an international conference, but, in part because he was manacled by the isolationists, he never made firm commitments to Britain and France in the event of war. "It is always best and safest to count on nothing from the Americans but words," Neville Chamberlain commented. Roosevelt, who could hardly be expected to assume responsibilities Chamberlain shirked, stressed not collective security but

[30] Elmer Davis, "The Wisconsin Brothers: A Study in Partial Eclipse," *Harper's*, CLXXXVIII (1939), 277; McCoy, *Angry Voices*, pp. 172–181; Russel Nye, *Midwestern Progressive Politics* (East Lansing, Mich., 1951), p. 373.

[31] *Time*, XXXI (March 7, 1938), 13. Cf. John T. Flynn, "Other People's Money," *New Republic*, XCVII (1938), 17; Charles Beard, "Giddy Minds and Foreign Quarrels," *Harper's* CLXXIX (1939), 338–351.

[32] Beard, "A Reply to Mr. Browder," *New Republic*, XCIII (1938), 358–359.

hemispheric defense. "All signs," observed one interviewer, "point 'Home'."[33]

In September, 1938, Europe approached the very edge of war. That month, as tension built in London, in Paris, in Rome, in Prague, the universe, as Ellen Glasgow wrote, came to seem like a "vast lunatic asylum."[34] For two decades, men had said that another world war was inconceivable; now it appeared it could come within hours. When word arrived that a settlement had been reached at Munich, most Americans felt a sense of relief. President Roosevelt, who played an inconsequential role in the negotiations, confided to Ambassador Phillips: "I want you to know that I am not a bit upset over the final result." Sumner Welles declared after the dismemberment of Czechoslovakia that the world had the best opportunity in two decades to achieve "a new world order based upon justice and upon law."[35]

The sense of relief was short-lived. In November, 1938, a young Polish Jewish refugee, Herschel Grynszpan, murdered the third secretary of the German Embassy in Paris, Ernst vom Rath. In retaliation, the Nazis assaulted Jews in Germany and burned their synagogues; the Reich fined them a billion marks, barred them from high schools and universities, and forbade them to attend theaters or concerts or even to drive cars. Americans were sickened by the new pogrom. President Roosevelt recalled our ambassador from Berlin for consultation, and told a news conference: "I myself could scarcely believe that such things could occur in a twentieth century civilization."[36] In a nation-wide broadcast, Herbert Hoover, Harold Ickes, and leaders of various religious faiths expressed their horror.

[33] Keith Feiling, *The Life of Neville Chamberlain* (London, 1946), p. 325; Anne O'Hare McCormick, "As He Sees Himself," *The New York Times Magazine,* October 16, 1938, p. 19.

[34] Ellen Glasgow to Mrs. Howard Mumford Jones, September 27, 1938, Glasgow MSS.

[35] FDR to William Phillips, October 17, 1938, in Elliott Roosevelt (ed.), *F.D.R.: His Personal Letters, 1928–1945* (2 vols., New York, 1950), II, 818; *Department of State Press Releases,* XIX (1938), 240. "Let no man say that too high a price has been paid for peace in Europe," reflected *The New York Times,* "until he has searched his soul and found himself willing to risk in war the lives of those who are nearest and dearest to him." *The New York Times,* September 30, 1938.

[36] *Public Papers,* VII, 597; "Le conflit entre le Reich et les États-Unis," *L'Europe Nouvelle,* XXI (1938), 1436–1437.

To flee the Hitler terror, a new influx of Jews and other opponents and victims of fascism entered the United States. During the decade, such distinguished German and Austrian refugees as Albert Einstein, the composers Paul Hindemith and Kurt Weill, the noted Bauhaus architect Walter Gropius, the expressionist painter George Grosz, and the novelists Thomas Mann and Lion Feuchtwanger found refuge in the United States.[37] Yet even a senator like Elbert Thomas, regarded as a spokesman for the Jewish cause in Congress, opposed changing the quota system to give asylum to more refugees. Robert Wood, the president of Sears, Roebuck, wrote Senator Borah: "I hope that our natural sympathy with the plights of the Jews in Germany and Austria will not allow us to relax our immigration laws for one moment." Borah, for his part, was against lowering immigration barriers when there were millions of jobless in the United States.[38] For every refugee who came to this country, many more who could have been saved died in Hitler's extermination chambers.[39]

A few days after the Munich accord, Hitler brandished a new German arms program on the ground that Chamberlain could not be relied upon, since he might be replaced by critics of Munich like Churchill and Duff Cooper. On October 11, 1938, in response to the Führer's speech, Roosevelt announced a buildup of $300 million in American armaments. In January, 1939, he spoke to Congress of measures "short of war, but stronger and more effective than mere words." That same month, the Atlantic Squadron of the Navy was formed. By 1939, the President had all but renounced the posture of isolationism. For many Europeans, he had come, more than their own heads of state, to represent democratic resistance to fascism. "For Roosevelt is a leader in a world where leadership is badly needed," wrote a Tory M.P. If Britain could not herself produce such a statesman, she should "profit by the example of an alien Pitt."[40]

[37] Donald Peterson Kent, *The Refugee Intellectual* (New York, 1953). A small number of Italian intellectuals, notably Enrico Fermi and Giuseppe Borgese, came to America from Mussolini's Italy.

[38] Elbert Thomas to Rabbi Louis Gross, April 25, 1938, Thomas MSS., Box 21; Isaiah Bowman to F.D.R., November 25, 1938, FDRL PSF 34; Wood to Borah, November 22, 1938, Borah to Sidney Fox, November 28, 1938, Borah MSS., Box 416.

[39] Dwight MacDonald, "Old Judge Hull & the Refugees," *Memoirs of a Revolutionist* (New York, 1957), pp. 154–158; Dorothy Detzer, *Appointment on the Hill* (New York, 1948), pp. 226–227.

[40] William Langer and Everett Gleason, *Challenge to Isolation* (New York,

Nationalist as well as internationalist senators were willing to vote money for defense, so long as rearmament did not form part of a co-operative undertaking with European powers. When, on January 23, 1939, one of the newest American bombers crashed in California on a test flight and a French officer was pulled from the burning wreckage, it gave away what Roosevelt had hoped to conceal: that he had been permitting foreigners to familiarize themselves with secret American equipment and to fly our military craft. When the President called in members of the Senate Military Affairs Committee to quiet fears that he had forged a secret military alliance with France, the meeting back-fired when one of the senators told newsmen that Roosevelt had said the American frontier was on the Rhine. When the President derided this report as a "deliberate lie" concocted by some "boob," senators became still more indignant.[41]

In September, 1938, Hitler had promised that once the Sudeten issue was settled, there would be "no further territorial problem in Europe." He added: "We want no Czechs." Less than six months later, at 6 A.M. on March 15, 1939, German troops invaded Czechoslovakia, after Hitler had offered the elderly Czech President the choice of surrender or annihilation. He called it his "last good turn" for the Czech people. That night the swastika was raised over the palace of the kings of Bohemia in Prague.[42] In one stroke Hitler had destroyed the last re-maining illusion that his ambitions were limited by a desire to create a state embracing only Germans. War now seemed inevitable. Within a month Nazi troops entered Memel, Franco captured Madrid, Japan claimed the Spratly Islands, and Mussolini grabbed Albania. Senator Borah expressed his openmouthed admiration: "Gad, what a chance Hitler has! If he only moderates his religious and racial intolerance, he would take his place beside Charlemagne. He has taken Europe without firing a shot."[43]

1952), pp. 37–39; *Public Papers,* VIII, 3; Richard Law, "Notes on the Way," *Time and Tide,* XX (1939), 39.

[41] Langer and Gleason, *Challenge to Isolation,* pp. 48–49; *The New York Times,* February 4, 1939.

[42] Alan Bullock, *Hitler: A Study in Tyranny* (London, 1952), p. 444; *The New York Times,* March 16, 1939; Memorandum, March, 1939, Wilbur Carr MS. Diary Notes; report of a conference of March 8, 1939, of Nazi leaders, FDRL Tully Safe File, Bullitt.

[43] William K. Hutchinson, "News Articles on the Life and Works of Honor-able William E. Borah," U.S. Senate, 76th Cong., 3d Sess., *Sen. Doc. 150* (Washington, 1940), p. 39.

The fascist successes deepened the disgust the United States felt toward the leaders of England and France; people came to speak of the "so-called democracies." It was hard to persuade Americans that the conflicts in Europe constituted a crucial struggle between totalitarianism and democracy when Neville Chamberlain had done nothing to halt the Nazi conquest of Austria; had surrendered his Foreign Minister, Anthony Eden, to placate Hitler and Mussolini; and had displayed such callous indifference to the betrayal of the Czechs at Munich.[44] "I know it to be a fact as much as I ever will know anything . . . that Britain is behind Hitler," Senator Borah reasoned.[45] Chamberlain, who, six days before the Nazis marched into Czechoslovakia, had assured newsmen, "Europe was settling down to a period of tranquility," seemed a Pangloss. In this period Margaret Halsey's *With Malice Toward Some,* an anti-British tirade so devastating that it made up "for the American travel books of at least six traveling British lecturers," sold several hundred thousand copies.[46]

Many influential Americans hoped for some accommodation with the fascist states. Both Ambassadors Joseph Kennedy and William Bullitt believed the great danger to be not Nazi Germany but Soviet Russia. Kennedy, who appears to have had some sympathy toward the Nazis, wanted them to have a free hand in eastern Europe, while Bullitt exhorted the United States to press for a Franco-German *rapprochement.*[47] Breckinridge Long, the former American ambassador to Italy, thought the only threat to the peace of Europe lay in Britain's unfair treatment of Mussolini.[48] Such convictions were by no means limited to the diplomatic corps or to men of conservative persuasion. The

[44] When the news of Munich was broadcast in Prague, some of the chiefs of mission, with a sense of undefined shame, avoided meeting their Czech friends. Memorandum, n.d., Wilbur Carr MS. Diary Notes.

[45] Hutchinson, "Borah," p. 38.

[46] Raoul de Roussy de Sales, "L'Amérique et le rapt de la Tchécoslovaquie," *L'Europe Nouvelle,* XXII (1939), 351–352; Marquis Childs, "Home Truths from America," *Spectator,* CLXII (1939), 123–124; Carrie Chapman Catt to Josephine Schain, March 21, 1939, Catt MSS.; Earl Johnson to D. A. Ellsworth, December 5, 1938, William Allen White MSS., Columbia University.

[47] William Kaufman, "Two American Ambassadors: Bullitt & Kennedy," in Gordon Craig and Felix Gilbert (eds.), *The Diplomats, 1919–1939* (Princeton, 1953), pp. 651–657; Germany, Auswärtiges Amt, *Documents on German Foreign Policy 1918–1945, Series D* (11 vols., Washington, 1949–60), I, 713–718.

[48] Breckinridge Long MS. Diary, February 21, 23, 1938.

radical editor of *Common Sense,* Alfred Bingham, observed: "I am inclined to feel however that Fascism is not a world menace by way of conquest and that it will ultimately evolve along lines similar to the Soviet Union, and both will probably become much more liberalized forms of collectivism. . . . It is far preferable to give Fascism room at present, than to fight it."[49]

In mid-April, 1939, as the Führer was making menacing gestures toward Danzig and the Polish Corridor, President Roosevelt made a last plea to avert war. He asked Hitler and Mussolini if they would promise not to attack thirty-one nations he listed by name. Hitler responded with a mocking harangue. As he called the roll of states: "Finland, Latvia, Lithuania . . . Palestine . . ." the Reichstag burst into derisive laughter.[50] Göring, visiting Rome, ventured that Roosevelt's message suggested a brain malady; Mussolini retorted that it might be rather a case of creeping paralysis. At home, the Protestant *Christian Century* complained that Roosevelt had "taken his stand before the axis dictators like some frontier sheriff at the head of a posse," and the Catholic *Commonweal* objected that his appeal ignored "the wrongs committed by post-war England and France, what they have contributed to the impoverishment of the axis powers. . . ." Isolationists gloated that Hitler had gotten the better of the President. "He asked for it," Nye commented.[51]

Many of the same men who upbraided Roosevelt for meddling in European affairs chastised him for not intervening more forcefully in Asia. In 1938, the Japanese, taking advantage of the preoccupation of the western powers with the crisis in Europe, opened a drive to subdue China and establish a "new order" in the Orient. The tone of Japanese statements changed markedly. Instead of claiming sententiously that it wished to return to the status quo as soon as the Chinese unpleasant-ness ended, Tokyo now bluntly announced its intention to create a Greater East Asia Co-Prosperity Sphere that would unite Japan, China,

---

[49] Alfred Bingham to Herbert Claiborne Pell, June 14, 1938, Pell MSS., Box 2.
[50] *Public Papers,* VIII, 209–216; *The New York Times,* April 29, 1939; Norman Baynes, *The Speeches of Adolf Hitler* (2 vols., London, 1942), II, 1605–1656; Bullock, *Hitler,* pp. 461–464.
[51] James Burns, *Roosevelt: The Lion and the Fox* (New York, 1956), pp. 391–392; *Völkischer Beobachter,* April 15, 1939; *Giornale d'Italia,* April 17, 1939; "Mr. Roosevelt Intervenes," *Christian Century,* LVI (1939), 536; "R.S.V.P.," *Commonweal,* XXX (1939), 3.

and Manchuria under Japanese suzerainty. When Grew protested the new departure, the Japanese Foreign Minister made no pretense of upholding the Open Door; he "wondered" indeed whether it applied anywhere any more save possibly in the Congo Basin. The United States responded by extending a loan of $25 million to the Chinese. Early in 1939, the Japanese demonstrated the futility of such limited economic gestures when they seized the Chinese island of Hainan, and the following month occupied the Spratly Islands, only four hundred miles west of the Philippines. Unless effective action was taken to curb the Japanese, they threatened to turn the entire western Pacific into a Nipponese lake.[52]

Both left-wing organizations and collective-security advocates like Henry Stimson urged a boycott of Japanese silk and an embargo on war materials to Japan. At an American Student Union convention, a coed took off her shoes, stripped her silk stockings from her legs, and, barefoot in the snow, flung her hose into a bonfire. Other girls not only followed her example but snatched silk ties from male delegates to add to the blaze.[53] To leftist critics, Roosevelt's refusal to embargo goods to Japan, when these materials were being used in the ruthless war against China, could only be explained by cowardice in the face of business interests.

In fact, Roosevelt was playing a cautious game. He sought to exert just enough pressure to indicate displeasure with the new Japanese truculence, but not so much as to undermine the moderates in Tokyo, nor to push the militarists into new adventures to the south. American military leaders cautioned against provoking a conflict with Japan, since the United States could not fight an effective two-front war, and Hitler was the greater menace. Although State Department officers like Stanley Hornbeck and Herbert Feis argued that economic sanctions would throttle the Japanese economy and dynamite the war chest the militarists needed, Grew warned that, after all Japan had already sunk into China, it would not now admit defeat, no matter how great the financial stringency. "I should add, parenthetically," Grew remarked, "that these colleagues had been confidently predicting for the past two years that the economic collapse of Japan was about to occur."[54]

[52] Joseph Grew, *Ten Years in Japan* (New York, 1944), pp. 270–272; Langer and Gleason, *Challenge to Isolation,* pp. 42–45.

[53] *Time,* XXXI (January 10, 1938), 42.

[54] Langer and Gleason, *Challenge to Isolation,* pp. 52–53; Charles Tansill, *Back Door to War* (Chicago, 1952), p. 501. Just a few months before Japan

Yet if moral persuasion was ineffective, and economic coercion was not even to be tried, Roosevelt was left with the alternatives of acquiescence or war. In the spring of 1939, the Japanese Army gave signs of getting out of hand; it blockaded the British and French concessions at Tientsin and maltreated British nationals. Senators Pittman and Schwellenbach demanded firm American action, but the administration continued to mark time. When, on July 18, 1939, Republican Senator Arthur Vandenberg filed a resolution asking the President to give Japan notice of the termination of the commercial treaty of 1911, Roosevelt's hand was forced. On July 26, Secretary Hull notified Tokyo that in six months the pact of 1911 would expire. This constituted a big stride in the direction of economic warfare, yet Roosevelt heeded Hull's and Grew's advice not to impose an embargo or raise duties sharply when the treaty expired. The President insisted, however, that Japan understand that this policy might be changed at any moment. As the United States continued negotiations with Tokyo, Roosevelt kept the threat of sanctions hanging over the Japanese.[55]

In the spring of 1939, as the world edged toward war, Roosevelt once again sought to free himself from the shackles of the Neutrality Act. The President objected that cash and carry, while it worked all right in the Atlantic, where the British controlled the seas, worked all wrong in the Pacific, where the Japanese were dominant. He wanted a flexible law that would permit him to discriminate between aggressor and victim; but since he knew he could not get this, he preferred no embargo at all. When Senator Pittman refused to sponsor outright repeal, the President was forced to compromise on a bill which would renew the cash and carry principle, but apply it to arms and munitions as well as other exports. Roosevelt rehearsed once more the objections to cash and carry, but Pittman would not be budged. Pittman's wariness proved justified when Senate isolationists in the Foreign Relations Committee refused to report out even his compromise measure.[56]

---

opened new hostilities in the Far East in 1937, the U.S. ambassador to China observed: "It begins to look very much as though old man 'Economic Law' were catching up with the Japanese." Nelson Trusler Johnson to William Castle, January 26, 1937, Johnson MSS.

[55] Tansill, *Back Door to War,* pp. 505–508; Langer and Gleason, *Challenge to Isolation,* pp. 147–159, 302–311; "Correspondence, 1939. Incoming" folder, Lewis Schwellenbach MSS.; Frederick Moore, *With Japan's Leaders* (New York, 1942), pp. 111–114.

[56] Donald Drummond, *The Passing of American Neutrality 1937–1941*

Roosevelt now turned to the House, where Representative Sol Bloom, chairman of the House Foreign Affairs Committee, sponsored a bill for outright repeal of the arms embargo. The proposal ran into stout antipathy from isolationists in both parties. Representative Martin Kennedy, New York Democrat, opposed the Bloom bill because his constituents were "scared to death" it would lead to war. "God knows they have enough to worry about without this problem," he added. Representative J. M. Vorys, Ohio Republican, declared: "The President has no more intention of taking us into war than he had 6 years ago of taking us into debt; but we have learned that despite good intentions, if you spend enough you get into debt, and if you threaten enough you get into war." When on June 30 the House passed the Bloom bill by the slim margin of twelve votes, 200–188, it gutted the measure by adding the confusing Vorys amendment which continued an embargo on "arms and ammunition" but not on "implements of war."[57]

Once more Roosevelt paid court to the Senate, and this time Pittman agreed to sponsor outright repeal of the arms embargo. On July 11, in a showdown vote, the Senate Foreign Relations Committee decided, 12–11, to defer consideration of neutrality legislation until the next session. A week later, Roosevelt met with Senate leaders at the White House. He explained that war might be imminent in Europe, and that he needed all the power he could muster to avert it. "I've fired my last shot," he told them. "I think I ought to have another round in my belt." The President's appeal met a cold response. After polling the senators in the room on the prospects of rescinding the embargo, Garner turned to the President and said: "Well, Captain, we may as well face the facts. You haven't got the votes, and that's all there is to it."[58]

(Ann Arbor, 1955), pp. 87–89; Fred Israel, "Nevada's Key Pittman" (unpublished Ph.D. dissertation, Columbia University, 1959), pp. 311–314; F.D.R. to Cordell Hull and Sumner Welles, March 28, 1939, PSF 33; Press Conference 540A, April 20, 1939, FDRL; Wayne Cole, "Senator Key Pittman and American Neutrality Policies, 1933–1940," *Mississippi Valley Historical Review*, XLVI (1960), 657–658.

[57] *Congressional Record*, 76th Cong., 1st Sess., pp. 8151, 8173; Francis Wilcox, "The Neutrality Fight in Congress: 1939," *American Political Science Review*, XXXIII (1939), 818–823.

[58] Robert Divine, *The Illusion of Neutrality* (Chicago, 1962), pp. 229–285; Joseph Alsop and Robert Kintner, *American White Paper* (New York, 1940), pp. 44–46; Cordell Hull, *The Memoirs of Cordell Hull* (2 vols., New York, 1948), I, 648–653; Memorandum, Wallace White MSS., Box 58; "Interview with Senator Walsh, Clinton, Mass., July 16, 1939," David I. Walsh MSS.

In the summer of 1939, people tuned in the newly popular portable radios to hear the reports of H. V. Kaltenborn and Raymond Gram Swing on the impending war in Europe. When, on August 24, the Nazis signed a pact with the Soviet Union which separated Russia from the West, war could no longer be very far off, yet people still could not believe that the nightmare of World War I would be repeated. One American diplomat noted in his diary: "These last two days have given me the feeling of sitting in a house where somebody is dying upstairs. There is relatively little to do and yet the suspense continues unabated."[59] On the hot Saturday of September 2, crowds hemmed in parked taxicabs to catch the latest radio bulletins; almost every broadcast was interrupted by a fresh dispatch on Danzig and the Corridor. Through the day, fears that war would come alternated with suspicions that Britain and France would abandon Poland and there would be no war. Sunday morning at 6:10 A.M. Eastern Standard Time, the waiting ended. Neville Chamberlain announced to the British people over the radio: "I have to tell you now that this country is at war with Germany."[60]

No one doubted where American sympathies lay; a Gallup poll revealed that 84 per cent of respondents wanted an Allied victory, 2 per cent a Nazi triumph, while 14 per cent had no opinion.[61] So strong was the identification with the Allied cause, so faithfully did the alignments of 1939 appear to repeat those of 1917, that many were overcome by the fatalistic feeling that, whether they willed it or not, the United States would inevitably be drawn into the war on the Allied side. One observer noted, "But everyone—everyone—little people like taxi drivers and drugstore clerks and farmers down in the Shenandoah and the like are saying we'll sure get into it."[62] Indeed, the country

---

Montana Senator Murray's political confidant had urged him to dodge a vote on the proposal. "The Irish, Germans, Austrians and Italians are strong against it becoming a law," he explained. Hugh Daly to James Murray, June 15, 1939, Murray MSS.

[59] Jay Pierrepont Moffat MS. Diary, August 26–27, 1939. Moffat noted: "The issues involved are so terrible, the outlook so cloudy, the probability of ultimate Bolshevism so great, and the chances of a better peace next time are so remote that if one stopped to think one would give way to gloom." *Ibid.*, September 1, 1939.

[60] "A.D. 1939," *Fortune,* XXI (January, 1940), 93.

[61] *Public Opinion Quarterly,* IV (1940), 102.

[62] Maxine Davis to Florence Jaffray Harriman, September 7, 1939, Harriman MSS., Box 4.

seemed to fear nothing so much as its own sentiments. Raoul de Roussy de Sales wrote: "The average American apparently visualized himself as the Greeks did: his will was set one way, but he felt that the gods would be stronger than his will. . . . Isolationists and their opponents acted as if the American people might be tempted to go to war the way a man who was once addicted to drink is suddenly seized with terror at the sight of a bottle of whiskey."[63]

Three weeks after the invasion of Poland, the President asked Congress, summoned into special session, to repeal the arms embargo and apply the cash-and-carry principle to munitions as well as raw materials. The President's request touched off an explosion of isolationist propaganda, much of it sparked by Coughlinites, which was remarkable for the violence of its imagery and the bitterness of its Anglophobia. "Hitler is an Angel compared to John Bull," a Massachusetts woman wrote Senator Taft. Hundreds of thousands of letters implored senators: "Keep America Out of the Blood Business."[64] Roosevelt, impressed by the intensity of isolationist passions, bent his argument to emphasize not aid to the Allies but keeping America out of war. He dismissed the contention that repeal of the embargo would lead to armed intervention as "one of the worst fakes in current history." The President stated: "The simple truth is that no person in any responsible place . . . has ever suggested . . . the remotest possibility of sending the boys of American mothers to fight on the battlefields of Europe."[65]

So persistent was the conviction that the real menace confronting the country lay not abroad but in Roosevelt's desire for dictatorial power that the President felt compelled to heed the advice of administration

---

[63] Raoul de Roussy de Sales, "America Looks at the War," *Atlantic Monthly*, CLXV (1940), 153. Alistair Cooke has pointed out that "Americans were more afraid of their sympathies than of a military threat to the continental United States if the Nazis won. They had to keep up a separate annoyance at the British and French as potential seducers who have the run of the house." Cooke, *A Generation on Trial* (New York, 1950), p. 29.

[64] Mrs. Ann Gath to Robert Taft, September 24, 1939, Taft MSS.; Mrs. Andrew Dillman to Taft, September 25, 1939, Taft MSS. Cf. Maxine Davis to Florence Jaffray Harriman, September 27, 1939, Harriman MSS., Box 4. "The sole purpose of amending the Neutrality Act," Henry Ford insisted, "is to enable munitions makers to profit financially through what is nothing less than mass murder." *The New York Times*, September 21, 1939. Many of the letters in the Taft manuscripts at Yale University are striking for their curious unrelatedness; they speak, for example, of saving "the flower of American manhood."

[65] *Public Papers*, VIII, 557.

senators that he stay in the background during the debate and let them
lead the fight. He kept his New Deal advisers off the Hill and deliberately cultivated conservative Democratic senators who would support his international position. Senator Bailey confided: "I have rather
important assurances that he is not going to run for a third term." He
asked Lord Tweedsmuir, the Governor General of Canada, to postpone
a visit to the United States until the debate ended. "I am almost
literally walking on eggs," he explained.[66] By pursuing such cautious
tactics, the President carried the day. After six weeks of acrimonious
controversy, Congress, over substantial Republican opposition, repealed
the arms embargo.

Roosevelt got what he wanted; the Allies could buy the munitions
they needed. But the isolationists won significant concessions too. The
Allies would have to pay cash and haul the arms away themselves;
United States citizens could not sail on belligerent vessels; American
ships could not sail to belligerent ports and might be forbidden to enter
combat zones. When on November 4 the President defined the entire
Baltic and the Atlantic from southern Norway to the northern coast of
Spain as a combat zone, Hitler was relieved of the threat of conflict with
the most important neutral carrier, could brush aside the rights of the
smaller neutrals, and was free to launch unrestricted submarine warfare with little risk. "By taking our ships off the seas," noted one
writer, "the bill aided the German blockade of Britain as effectively as
if all our ships had been torpedoed." In the midst of war, the Neutrality
Act of 1939 marked one final effort at neutrality by statute to apply the
lessons of World War I.[67]

The United States hoped—by neutrality legislation and by inter-
American agreement—to insulate the Western Hemisphere from the
European war. At the First Meeting of Foreign Ministers of the Ameri-
can Republics, the United States scored a notable triumph. On Oc-

[66] Burns, *Roosevelt,* 396; Langer and Gleason, *Challenge to Isolation,* pp.
227–231; Theodore Carlson to Stephen Chadwick, November 2, 1939, Chadwick MSS.; Bailey to Dr. Julian Ruffin, September 18, 1939, Bailey MSS.,
Personal File; F.D.R. to Tweedsmuir, October 5, 1939, in *Personal Letters,* II,
934.

[67] Denna Frank Fleming, "Arms Embargo Debate," *Events,* VI (1939), 342;
Langer and Gleason, *Challenge to Isolation,* pp. 231–235; Divine, *Illusion of
Neutrality,* pp. 286–335; Philip Jessup, "The Neutrality Act of 1939," *American
Journal of International Law,* XXIV (1940), 95–99; L. E. Gleeck, "96 Con-
gressmen Make up their Minds," *Public Opinion Quarterly,* IV (1940), 3–24.

tober 3, 1939, delegates adopted unanimously the Declaration of Panama, which created a zone around the Americas, south of the Canadian border, from 300 to 1,000 miles wide, in which belligerents were warned not to undertake naval action. Since neither side would recognize this "chastity belt," and since the American republics were not prepared to command recognition by force, the Declaration of Panama had little practical effect, although Hitler did give his commanders secret orders not to initiate action within this zone.[68] The Declaration marked an important step in collective action by the American nations and in the "multilateralization" of the Monroe Doctrine; it signified too the hold of the ancient illusion that the United State could, simply by fiat, isolate itself from Europe's quarrels.[69]

Even the Soviet invasion of Finland on November 30, 1939, failed to break the spell of isolationism, despite the nearly unanimous sympathy of the country for the Finns. Finland had won affection by being the one nation which had paid its war debts to the United States, thereby tapping the vast well of sentimentality of the American people. "By last week," observed *Time* in March, 1940, "a U.S. Citizen who had neither danced, knitted, orated, played bridge, bingo, banqueted or just shelled out for Finland was simply nowhere socially." President Roosevelt, who referred privately to "this dreadful rape of Finland," denounced the Russian incursion as a "wanton disregard for law." He placed Finland's annual debt installment in a separate account, extended the Finns $10 million in credits to buy farm surpluses and supplies in the United States, and applied a "moral embargo" against the U.S.S.R.[70]

But what the Finns really needed was not agricultural credits but,

---

[68] H. L. Trefousse, *Germany and American Neutrality, 1939–1941* (New York, 1951), p. 41; U.S. Navy Department, *Fuehrer Conferences on Matters Dealing with the German Navy 1940* (5 vols., Washington, 1947), II, 46.

[69] Langer and Gleason, *Challenge to Isolation*, pp. 210–212; Drummond, *Passing of Neutrality*, pp. 119–121. The foreign ministers met under authority of the Declaration of Lima of 1938, which gave substance to the principle of consultation adopted at Buenos Aires in 1936 by providing that, in the event of war or other threats to security, any American foreign minister could call an emergency meeting.

[70] *Time*, XXXV (March 11, 1940), 16; Max Jakobson, *The Diplomacy of the Winter War* (Cambridge, 1961), pp. 191–192; F.D.R. to Lincoln Mac-Veagh, December 1, 1939, FDRL PPF 1192; *The New York Times,* December 2, 1939; Langer and Gleason, *Challenge to Isolation,* pp. 329–331.

as the Finnish minister explained, cold cash to buy arms in Sweden. Roosevelt refused to antagonize isolationists by asking for an unrestricted loan, and Congress quickly proved that the President's caution was well founded. When Senator Prentiss Brown of Michigan filed a bill to lend the Finns $60 million to buy arms, Congress so mutilated the measure that in the end the "Finnish aid bill" did not mention Finland at all, and limited the Finns to another $20 million credit for nonmilitary supplies. The Senate approved the bill on February 13, as the Red Army was breaking through the Mannerheim Line; the House passed it on February 28, as the Russians started mopping-up operations. In the end the Finns, by capturing Russian arms on the battlefield, received more military aid from the Soviet Union than from their friends. In March, 1940, a Finnish leader commented that the sympathy of the United States and its other well-wishers proved so great that "it nearly suffocated us."[71]

America's commitment to "stormcellar" neutrality rested on a series of assumptions which seemed wholly reasonable—that the war would quickly settle down to a protracted state of siege, with the Germans unable to crack the Maginot Line, and the Allies unable to pierce German defenses; that the British would maintain control of the seas; and that Hitler would be defeated in a war of attrition. In the first months of the war, this appeared to be precisely what was happening; some Americans even protested the lack of excitement in the "bore" war. The war itself, and the threat of Hitler, seemed far distant. When the British cruiser *Orion* chased the German merchantman *Arauca* along the Florida coast in late December, 1939, a newspaper ran the headline: "SEA INCIDENT AT MIAMI'S FRONT YARD DRAWS GAY HOLIDAY CROWD TO DOCK." [72]

The unreal quality of the war left Roosevelt in a quandary. How, he asked William Allen White, could he persuade the nation of the danger it faced without "scaring the American people into thinking that they are going to be dragged into this war?"[73] As time passed, not only the

[71] *The New York Times,* March 17, 1940; Langer and Gleason, *Challenge to Isolation,* pp. 334–340; Mrs. Florence Jaffray Harriman to F.D.R., February 27, 1940, Harriman MSS.; W. R. Castle to David I. Walsh, March 1, 1940, Walsh MSS. For a different view, see Robert Sobel, *The Origins of Interventionism* (New York, 1960).

[72] *Time,* XXXV (January 1, 1940), 8.

[73] F.D.R. to White, December 14, 1939, White MSS., Box 324.

country, but Roosevelt too, came to lose the fatalistic feeling of September that the country would inevitably be drawn into the conflict. The President came to think of the United States as the Great Neutral, giving its aid to one side but determined to stay out of the war. In the early months of 1940, an eerie stillness fell over the White House. "Things here are amazingly quiet," Roosevelt wrote in March, 1940.[74] That same month, as American diplomats traveled by train through Germany, they could see both French and German trenches within a short distance of one another. "Not a gun was fired, not a disturbing sound broke the stillness," one envoy noted in his diary. "Men, women, and children were in the fields, or tying up the vines in the vineyards." In Paris, fowl was plentiful, and one drank champagne as an apéritif; in London, Englishmen enjoyed a warm spring afternoon in the park.[75] The character of the European war served to reinforce the illusion of isolation. Only a revolutionary change in the nature of the European struggle could explode that illusion.

[74] F.D.R. to John Boettiger, March 7, 1940, in *Personal Letters*, II, 1006.
[75] Nancy Harvison Hooker (ed.), *The Moffat Papers* (Cambridge, 1956), pp. 295–298. Cf. Sumner Welles, *The Time for Decision* (New York, 1944), pp. 73–147.

CHAPTER 13

An End to Isolation

AT DAWN on April 9, 1940, Nazi troops crossed the Danish border and emptied out of barges and troopships all along the Norwegian coast. They overran Denmark in a matter of hours, Norway in a few weeks. On May 10, the Germans struck swiftly at the Low Countries and, through them, into France. In five days Holland surrendered, in eighteen days Belgium; in less than three weeks, the Nazis had driven the British armies out of France. As the Panzer divisions, supported by siren-blowing dive bombers, careened toward Paris, Premier Reynaud pleaded vainly with the United States for "clouds of planes." A week later, on June 22, 1940, in the very same railway car in the Compiègne forest where Germany had agreed to the 1918 armistice, France capitulated to Hitler.

In the United States, the national mood changed overnight. In Times Square, great crowds, grim-faced, awesomely silent, blocked traffic as they spilled into the middle of Broadway to stare up at the bulletins on the Times Building. Suddenly the country felt naked and vulnerable. Only the British stood between Hitler and the United States. If Germany struck across the Channel with the same ferocity and cunning that had annihilated opposition in western Europe in a few weeks, Britain might well be flying the swastika by the end of the year, and Admiral Raeder might be commanding the British fleet.

The Nazi blitzkrieg revolutionized American thinking about defense. On May 16, 1940, President Roosevelt, after explaining how enemy bombers could leave West Africa and emerge in the skies over Omaha, requested vast sums to mechanize and motorize the Army (it had only 350 serviceable infantry tanks) and funds for 50,000 planes a year (it then had a total of 2,806 aircraft, most of them outmoded).[1] By the end of May, Congress, acting with uncommon speed, had granted the President even more money than he wanted. Yet this did not begin to be enough, and by May 31 Roosevelt was back asking for more. In less than a month Congress responded by appropriating another billion dollars for the Army and some $700 million for the Navy.[2] On June 15, the President set up a National Defense Research Committee under Dr. Vannevar Bush of the Carnegie Institution to co-ordinate research of military significance, a decision which would lead five years later to the explosion of the atomic bomb.[3] When Admiral Harold Stark recommended a two-ocean Navy, Congress promptly authorized 1,325,-000 tons of new fighting ships. By October, 1940, Congress, which in early spring had planned to slash War Department appropriations, had voted over $17 billion for defense.[4]

The shock of realizing that the country might experience a foreign invasion led to panicky efforts to throw together homemade defenses. The California legislature approved a proposal to organize a corps patterned after the Swiss Home Guards, Chicago rifle clubs began to mobilize "a civilian army of modern minute men," and in New York the National Legion of Mothers of America formed the Molly Pitcher Rifle Legion to pick off descending parachute troops. Unable to agree on what to do about the enemy without, the nation hunted fifth-columnists within. The same month that France fell to the Nazis, Federal Judge Michael Roche canceled the citizenship of Communist William Schneiderman in California; the House voted to deport Harry Bridges; Connecticut residents set up the North Stamford Breastwork and

---

[1] *Public Papers and Addresses of Franklin D. Roosevelt* (13 vols., 1938–50), IX, 200–201; Louis Johnson to F.D.R., May 15, 1940, FDRL PSF 27; Harry Woodring to Representative Joseph Bryson, June 5, 1940, Bryson MSS.

[2] William Langer and Everett Gleason, *The Challenge to Isolation, 1937–1940* (New York, 1952), pp. 473–476; Mark Watson, *Chief of Staff: Prewar Plans and Preparations* (Washington, 1950), pp. 169–171.

[3] James Phinney Baxter, 3rd, *Scientists Against Time* (Boston, 1946), pp. 11–18; Geoffrey Hellman, "The Contemporaneous Memoranda of Dr. Sachs," *New Yorker*, XXI (December 1, 1945), 69–76; Albert Einstein to F.D.R., August 2, 1939, FDRL Tully Safe File.

[4] Samuel Eliot Morison, *The Battle of the Atlantic* (Boston, 1947), pp. 27–28; Langer and Gleason, *Challenge to Isolation*, p. 676; Whitney Shepardson and William Scroggs, *The United States in World Affairs, 1940* (New York, 1941), Ch. 5.

Overhead Menace Committee; and the Supreme Court, speaking through Mr. Justice Frankfurter, sustained Pennsylvania school authorities who expelled two children who refused to salute the flag, because this violated their religious scruples.[5] In July, President Roosevelt signed the Smith Act, which required some 3,500,000 aliens to register at their local post offices and be fingerprinted, and stipulated fines and imprisonment for written, spoken, or printed words which "in any manner cause insubordination, disloyalty, mutiny or refusal of duty by any member of the military or naval forces. . . ." "If the clause were strictly construed," warned *The New York Times,* "several of the leading speakers at last week's Republican National Convention might be in danger."[6]

In the spring of 1940, Roosevelt moved toward a crisis government of national unity. On June 19, he startled the nation by naming two prominent Republicans to the key defense posts in his cabinet: Henry Stimson as Secretary of War and Frank Knox as Secretary of the Navy. In making these appointments, Roosevelt not only stole a march on the Republicans on the eve of their national convention but also shored up the foreign-policy militants within his administration. Knox and Stimson had both taken positions well in advance of any yet assumed by Roosevelt; Stimson, in fact, had just come out in favor of using the American fleet to convoy munitions to Britain.[7]

Although Roosevelt did not go as far as Stimson, he was just as determined to speed all possible aid to Britain, a resolve he proposed to expound to the nation in his address to the graduating class of the University of Virginia. Just before departing for Charlottesville, the

[5] *Time,* XXXV (June 3, 1940), 12; Minersville School District *v.* Gobitis, 310 U.S. 586; Robert Morss Lovett to Mildred Brady, August 9, 1940, Lovett MSS., Box 1, Folder 3. Cf. West Virginia State Board of Education *v.* Barnette, 319 U.S. 624 (1943). The Supreme Court later reversed Judge Roche's verdict. Schneiderman *v.* United States, 320 U.S. 118 (1943). After litigation that ran through all the war years, Bridges survived the attempt to deport him. *The New York Times,* June 19, 1945.

[6] Zechariah Chafee, Jr., *Free Speech in the United States* (Cambridge, 1941), pp. 440–490; *The New York Times,* June 30, 1940.

[7] Elting Morison, *Turmoil and Tradition* (Boston, 1960), pp. 477–486; Louis Brownlow, *A Passion for Anonymity* (Chicago, 1958), pp. 433–454; Henry Stimson MS. Diary, June 25, 1940; Langer and Gleason, *Challenge to Isolation,* pp. 509–511. The G.O.P. national chairman promptly read Knox out of the party, although no one could recall when he had been vested with the power of excommunication.

President received word that Italy was about to declare war on France. Roosevelt was infuriated, not least because he had labored so assiduously to detach the Duce from Hitler. Only a few days earlier, he had sent a personal message to Mussolini with an offer to mediate between Italy and the Allies and to hold himself personally responsible for the execution of any agreements. Count Ciano noted that he had told Ambassador Phillips: "It takes more than that to dissuade Mussolini. In fact, it is not that he wants to obtain this or that; what he wants is war, and, even if he were to obtain by peaceful means double what he claims, he would refuse." On May 29, Bullitt cabled Roosevelt: "Al Capone will enter war about the fourth of June, unless you can throw the fear of the U.S.A. into him."[8] Three days before, Mussolini told his generals that he had already promised Hitler he would intervene. "I need only a few thousand dead to enable me to take my seat, as a belligerent, at the peace table," the Duce explained.[9]

On the train to Charlottesville, the President's anger at the fascists mounted. As he afterward said, he became "so hot" he was "grumbling about it." In his formal oration, he pledged that America would not only step up its own defense preparations but would "extend to the opponents of force the material resources of this nation." The United States, he premonished, could not hope to remain an island of democracy in a totalitarian sea, "a people lodged in prison, handcuffed, hungry, and fed through the bars from day to day by the contemptuous, unpitying masters of other continents." At the conclusion of his speech, Roosevelt ad-libbed one stinging sentence: "On this tenth day of June, 1940, the hand that held the dagger has struck it into the back of its neighbor."[10]

In making the Charlottesville commitment to aid Britain, the President risked a good deal. Polls revealed that only 30 per cent of the nation still believed an Allied victory possible. Senator Pittman, chairman of the Senate Foreign Relations Committee, announced it was futile for Britain to continue to fight on in England, and even a pro-

---

[8] Hugh Gibson (ed.), *The Ciano Diaries* (New York, 1946), p. 255; Langer and Gleason, *Challenge to Isolation*, p. 462. Cf. Bullitt to R. Walton Moore, April 18, 1940, Moore MSS., Box 3.

[9] Pietro Badoglio, *L'Italie dans la Guerre Mondiale* (Paris, 1946), pp. 38–39.

[10] Henry Stimson MS. Diary, December 29, 1940; *Public Papers*, IX, 261–264; William Bullitt to Cordell Hull, May 31, June 10, 1940, FDRL, Tully Safe File; William Phillips MS. Diary, June 11, 1940, XV, 4010.

British senator like Josiah Bailey wrote that under no circumstances could Britain hope to dislodge Hitler from his conquests on the Continent.[11] The Joint Planners of the War and Navy Departments pointed out the slim prospects for Britain's survival, and General Marshall warned that if the United States stripped its defenses to help Britain, and the British capitulated, leaving this country to face unarmed an Axis invasion, Roosevelt's policy could never be justified to the American people. The United States had only 160 P-40 planes for the 260 pilots waiting for them; it had no antiaircraft ammunition, and would not for six months.[12]

All that summer the Nazis rained bombs on British cities as a prelude to the invasion scheduled for September. Alone, beleaguered, the British desperately needed to maintain control of the seas to keep military supplies flowing into the country. By the end of July, they had lost nearly half of the hundred destroyers available for home waters. Prime Minister Winston Churchill beseeched Roosevelt to let him have fifty or sixty "overage" World War I destroyers to use against the Nazi submarines.[13]

Roosevelt hesitated. He had already turned down similar entreaties from Norway and France—and for compelling reasons.[14] Not only might the ships be needed for American defense, but the lease of ships

[11] *The New York Times*, June 26, 1940; Bailey to Dr. Julian Ruffin, July 24, 1940, Bailey MSS., Personal File. The former ambassador to Italy, Breckinridge Long, commented on the Charlottesville address: "If we are not very careful we are going to find ourselves the champions of a defeated cause. . . . France has utterly collapsed. It may not be but a few weeks before England collapses and Hitler stands rampant across the continent of Europe." Breckinridge Long MS. Diary, June 13, 1940. Long noted that Ambassador Phillips believed Germany would win, and hence there should be no more provocative statements; that Ambassador Kennedy was "quite a realist and he sees England gone"; and that Ambassador Cudahy was "sure Germany will win. . . ." Kennedy, Long set down, thought the United States should approach Germany and Japan for "an economic collaboration," while Ambassador Cudahy "sees no disadvantage to us if Germany wins. . . ." Breckinridge Long MS. Diary, August 16, November 6, 1940.

[12] Langer and Gleason, *Challenge to Isolation,* pp. 487–488, 566–569, 654.

[13] Winston Churchill, *Their Finest Hour* (Boston, 1949), p. 402; Langer and Gleason, *Challenge to Isolation,* pp. 744–749. Cf. William Allen White to Fiorello La Guardia, July 29, 1940, La Guardia MSS.

[14] F.D.R. to Mrs. Florence Jaffray Harriman, January 9, 1940, Harriman MSS., Box 4; William Bullitt to Cordell Hull, May 18, 1940, FDRL, Tully Safe File; F.D.R. to Sumner Welles, June 1, 1940, FDRL PSF 24.

of war to a belligerent would almost certainly violate international law. Even if he overcame these objections, he faced nearly insurmountable difficulties. The Walsh Act of June 28, 1940, stipulated that equipment could be released for sale only if the Navy certified it was useless for defense; and naval officials had just testified to the value of the destroyers in order to prevent Congress from junking them. The President was convinced it would require an act of Congress to permit even their indirect sale; yet if he went to Congress, Senator David I. Walsh and the isolationists might very well block the proposal. Such a rebuff, taken as an index to the views of the American Congress, would deal a mean blow to Allied morale.[15]

Roosevelt found a way out of one part of his dilemma by taking an old isolationist scheme and giving it a new twist. He suggested a deal with Great Britain in which the United States would exchange the destroyers for bases in the British possessions in the Western Hemisphere.[16] Unhappily, he could not obtain a firm warranty of Republican support for a destroyer-base bill, and without such assurance he risked a costly defeat in Congress. As Nazi bombers blackened the skies over Britain, Roosevelt made a bold decision: to bypass Congress entirely and negotiate a compact with Churchill on his own. On the strength of an opinion from Attorney General Robert Jackson, military leaders were persuaded they could certify the destroyers were not essential to national security, since the bases would be still more valuable. "This is the time," Admiral Stark penciled at the end of his memorandum approving the transfer, "when a 'feller' needs a friend."[17]

Yet this did not end Roosevelt's troubles. Winston Churchill now insisted that the United States make a free gift of the destroyers, and Britain make a present of the bases. He feared that if the transaction was presented as a *quid pro quo*, Britons would think he had made a bad bargain in bartering historic possessions for overage vessels. More-

[15] Benjamin Cohen to F.D.R., July 19, 1940, F.D.R. to Frank Knox, July 22, 1940, FDRL PSF 19; Langer and Gleason, *Challenge to Isolation,* pp. 744–746.

[16] Memorandum, August 2, 1940, FDRL PSF 19; Henry Stimson MS. Diary, August 2, 1940; *The Secret Diary of Harold Ickes* (3 vols., New York, 1953–54), III, 292–293.

[17] Harold Stark to F.D.R., August 21, 1940, FDRL PSF 19. Jackson's opinion was roughly handled by international lawyers. See especially Herbert Briggs, "Neglected Aspects of the Destroyer Deal," *American Journal of International Law,* XXXIV (1940), 569–587.

over, he was unwilling to have the agreement pivot on a public pledge that the British fleet would not be surrendered in the event of a Nazi conquest, since even to discuss such an eventuality would be demoralizing. Roosevelt believed he could avoid congressional wrath only by presenting the exchange as a sharp bargain, and asked Hull to explain to "former Naval person" that he had no authority to give away any ships. The United States proposed a compromise, which Churchill reluctantly accepted. The bases would be divided into two lots; the British would lease Bermuda and Newfoundland as outright gifts; leases on the remaining bases in the Caribbean would be traded for the fifty destroyers. At a later time, Churchill would reaffirm his vow to Parliament of June 4 that the British fleet would not be surrendered or scuttled under any circumstances. On September 3, the President electrified the country by announcing the details of the deal. "This is the most important action in the reinforcement of our national defense that has been taken since the Louisiana Purchase," Roosevelt declared.[18]

The parallel to Jefferson and Louisiana was less than perfect. While the response to the disclosure of the transfer was generally enthusiastic —even the Chicago *Tribune* approved it—many were disturbed by the method Roosevelt had chosen. Critics charged that he had flouted the authority of Congress, whose rights in the matter he had previously conceded; that he had acted without taking the people into his confidence; and that he had transgressed international law.[19] At a time when Britain faced annihilation, Roosevelt and his supporters had little patience with animadversions based on Edwardian conventions of international law. To his admirers, the transaction was a

[18] F.D.R. to Hull, August 28, 1940, in Elliott Roosevelt (ed.), *F.D.R., His Personal Letters, 1928–1945* (2 vols., New York, 1950), II, 1061; *Public Papers,* IX, 391. See F.D.R. to David I. Walsh, August 22, 1940, Walsh MSS.; Langer and Gleason, *Challenge to Isolation,* pp. 764–770; Eugene Gerhart, *America's Advocate: Robert H. Jackson* (Indianapolis, 1958), pp. 216–217; Sumner Welles to F.D.R., August 8, 1940, FDRL PSF 19. The deal also stipulated that the United States would give twenty torpedo boats, five heavy bombers, thirty million pounds of ammunition and other supplies. By a fantastic inadvertent error these were left out of the agreement.

[19] "Mr. Roosevelt today committed an act of war," the St. Louis *Post-Dispatch* declared in a full-page advertisement in newspapers across the country. "He also became America's first dictator." St. Louis *Post-Dispatch,* September 3, 1940.

magnificent stroke of daring statesmanship. They rejoiced in the abandonment of traditional neutrality and the fusion of British and American interests. Henceforth, as Churchill told the Commons, the United States and Great Britain would be "somewhat mixed up together."[20]

Fresh from his triumphs in Europe, Hitler warned Latin American governments that by autumn Britain would be vanquished, and Latin America would have to depend on the Reich for markets. To appease Berlin, Brazil, Argentina, Chile, and other countries refused to send their foreign ministers to the Havana Conference in July. Roosevelt was compelled to recognize that repentance did not constitute a Latin American policy. By the end of 1940, the United States had worked out defense agreements with every South American nation save Argentina, and Congress had expanded the lending authority of the Export-Import Bank to finance exports to countries south of the border. In August, 1940, an Office for Co-ordination of Commercial and Cultural Relations between the American Republics was established under the direction of Nelson Rockefeller, although it was not until after Pearl Harbor that we began to send professors, tennis players, and ballet companies to Latin America to prove that we were a cultured people. The Havana Conference, too, scored another success for the Good Neighbor policy. The delegates accepted the historic "no transfer" principle of the Monroe Doctrine; agreed to Hull's appeal to take steps to establish a "collective trusteeship" over European possessions in the Caribbean; and, during the emergency, gave the United States a blank check to take what action it deemed necessary.[21]

The American Army of 1940 could hardly cope with an invasion, let alone protect Latin America. In August, 1940, when the Army conducted its greatest peacetime maneuvers in history in St. Lawrence County, New York, the troops were virtually weaponless. Pieces of

[20] "I do not view the process with any misgivings. I could not stop it if I wished; no one can stop it. Like the Mississippi, it just keeps rolling along. Let it roll. Let it roll on—full flood, inexorable, irresistible, benignant, to broader lands and better days." James Burns, *Roosevelt: The Lion and the Fox* (New York, 1956), p. 441. Cf. 365 H. C. DEB 5S 39.

[21] Langer and Gleason, *Challenge to Isolation,* pp. 629–637, 689–700, and *The Undeclared War* (New York, 1953), pp. 151–158, 162–169; John A. Logan, Jr., *No Transfer* (New Haven, 1961), pp. 330–345; Percy Bidwell, *Economic Defense of Latin America* (Boston, 1941); Arthur Whitaker, *The Western Hemisphere Idea* (Ithaca, 1954), pp. 147–149. The Export-Import Bank had been created in 1934.

stovepipe served as antitank guns, beer cans as ammunition, rainpipes as mortars, and broomsticks as machine guns. The attempt to simulate conditions of blitzkrieg warfare failed miserably. The Army had only one fully armored unit, and it was tied down at training centers in the South; no combat planes were available; and, for Panzer divisions, men maneuvered trucks bearing placards with the word "TANK."[22]

The United States Army numbered little more than a half-million men, including the National Guard. Chief of Staff General George Catlett Marshall wanted a force of at least two million, yet enlistments had been falling off sharply. In May, 1940, a group of eastern gentry, who had voluntarily undertaken military training at a camp in Plattsburg, New York, in 1915, turned the occasion of their twenty-fifth reunion into the start of a campaign for peacetime conscription. Neither Roosevelt, in an election year, nor Marshall, who did not want to jeopardize War Department appropriations or break up seasoned units in order to train draftees, would support such an unpopular proposal. The Plattsburg group, headed by Grenville Clark, a prominent New York attorney, and Colonel Julius Ochs Adler, publisher of *The New York Times,* framed their own bill and had it introduced late in June by Senator Edward Burke, an anti-New Deal Democrat from Nebraska, and Representative James Wadsworth, a New York Republican.[23]

Opposition to the selective service bill crossed a wide spectrum from conservatives like Robert Taft to progressives like George Norris. Senator Wheeler admonished: "If this Bill passes—it will slit the throat of the last Democracy still living—it will accord to Hitler his greatest and cheapest victory. On the headstone of American Democracy he will inscribe—'Here Lies The Foremost Victim Of The War Of Nerves.' "[24] A peacetime draft, opponents charged, could only signify a step toward involving the United States in the European war, since this country was not imperiled.[25] A conscript force, military experts

[22] *The New York Times,* August 1–27, 1940.

[23] Langdon Marvin, COHC; James Wadsworth, COHC, pp. 434 ff.; Henry Hooker to Missy LeHand, May 23, 1940, FDRL PPF 482; Langer and Gleason, *Challenge to Isolation,* p. 680; Mark Watson, *Chief of Staff,* pp. 189–192.

[24] Burton K. Wheeler, "Marching Down the Road to War," *Vital Speeches,* VI (1940), 692.

[25] "Whom do we fear?" Hamilton Fish asked. "Do we fear Hitler, who seems afraid to attack England over 20 miles of sea, when he would have 3,000 miles to cross over here?" *Congressional Record,* 76th Cong., 3d Sess., p. 11361.

like Hanson Baldwin argued, would be useless for modern warfare, which required not mass armies but skilled, mobile striking forces. Senator James Murray of Montana protested: "A conscript army made up of youths trained for a year or two, compared to Hitler's army, is like a high school football team going up against the professional teams they have in Chicago and New York."[26]

Over the summer, as the Nazis poised to invade Britain, support for the bill mounted. Although mail to congressmen ran 90 per cent against the measure, polls indicated a rapid shift of opinion in favor of selective service.[27] Secretary Stimson, another Plattsburg alumnus, converted President Roosevelt and General Marshall to the conscription cause.[28] In mid-September, Congress passed the Burke-Wadsworth bill. On October 16, men between the ages of twenty-one and thirty-five registered for America's first peacetime draft. Ten days later, Secretary Stimson, blindfolded, reached into a fishbowl to pluck out the first of the numbers which would determine the order men would be called into the armed services.

The Japanese viewed the Allied reversals in Europe with unconcealed delight. With France and Holland overrun, and Britain threatened with extinction, Japan might now free itself of dependence on America for vital resources by reaching into the rice paddies of French Indochina, the oil fields of the Netherlands East Indies, and the rubber plantations of Malaya. On June 25, 1940, the War Minister, General Shunroku Hata, one of the more moderate of the Japanese military men, declared: "We should not miss the present opportunity or we shall be blamed by posterity." In July, Tokyo compelled the British to close the Burma Road for three months, cutting off the trickle of munitions and gasoline to China. That same month, the military precipitated the fall of the Yonai cabinet and persuaded Emperor Hirohito, against his better judgment, to name the Army's

[26] Hanson Baldwin, "Wanted: A Plan for Defense," *Harper's*, CLXXXI (1940), 225–238; James Murray to Hugh Daly, August 31, 1940, Murray MSS. Senator Gillette of Iowa declared: "This idea of letting the boys sit around for a year playing stud poker and blackjack is poppycock." *Time*, XXXVI (August 12, 1940), 13. John L. Lewis denounced the proposal as "a fantastic suggestion from a mind in full intellectual retreat." *The New York Times*, June 20, 1940. See National Council for the Prevention of War files, Drawer 84.

[27] Rowena Wyatt, "Voting Via the Senate Mailbag," *Public Opinion Quarterly*, V (1941), 373–374.

[28] Stimson MS. Diary, July 9, 1940, *et seq.*

choice, Prince Fumimaro Konoye, as Premier. Konoye picked one of the Army hotspurs, General Hideki Tojo, as his new War Minister, and another of the Army's favorites, Yosuke Matsuoka, as his Foreign Minister. The Japanese policy of moderation had been unable to survive the Nazi triumphs in Europe, which, as Grew later noted, had "gone to their heads like strong wine."[29]

Japan's new militance sharpened a division within the administration over Far Eastern policy. One faction, led by Morgenthau and Stimson, pressed for the imposition of economic sanctions on Japan, while Hull and Sumner Welles in the State Department and Ambassador Grew in Tokyo warned that an embargo would not only destroy what little strength the moderates in Japan still held but would drive the Nipponese to attack Singapore and the Indies. Through the first eight months of 1940, Hull held the upper hand. When in July the President signed a proclamation, drafted by Morgenthau, banning the export of oil and scrap metals, Welles persuaded Roosevelt to modify the edict sharply to restrict it to aviation gasoline and only high-grade iron and steel scrap.[30]

Yet as the Japanese in the late summer of 1940 moved toward an alliance with the fascist powers, and tightened the vise on French Indochina, even their warmest American friends lost faith in their peaceful intentions. On September 12, Ambassador Grew sent his "Green Light" message. Long an advocate of patient negotiations with Tokyo, and an eloquent opponent of sanctions, Grew had come to feel that Japan could be deterred only by "a show of force, together with a determination to employ it if need be." Less than two weeks later, Nipponese soldiers invaded Indochina. Roosevelt responded on September 26 by slapping a total embargo on the shipment of iron and steel scrap to Japan. The following day, at a great ceremony in Berlin, Japan signed a treaty of alliance with the Rome-Berlin Axis. The Tripartite Pact, one section of which appeared to be aimed directly at the United States, explicitly linked events in Europe and the Far East. There was now only one war, Roosevelt concluded, and the de-

[29] Langer and Gleason, *Challenge to Isolation*, pp. 599, 603, 719–720; F. C. Jones, *Japan's New Order in East Asia* (London, 1954), pp. 165–170; *Japan Chronicle*, July 4–19, 1940; Joseph Grew, *Turbulent Era*, ed. Walter Johnson (2 vols., Boston, 1952), II, 1125.

[30] Langer and Gleason, *Challenge to Isolation*, I, 721–722; Herbert Feis, *The Road to Pearl Harbor* (Princeton, 1950), pp. 89–93.

mocracies had to unite against a common peril, worldwide in its scope.[31]

Concern over the Axis conquests stepped up the tempo of the great debate over foreign policy in the United States. The "internationalists," who favored aid to Britain, found their stanchest support on the two coasts and in the South; in university faculties and among cosmopolitan businessmen; with old-stock Americans, who had ties to Britain; and with anti-Hitler ethnic groups, particularly Jews but including Poles and others whose countrymen had been overrun by the Nazis. Although the pro-Allied bloc boasted Republican leaders like Henry Stimson, its strength centered in the Democratic party.

In May, 1940, internationalists organized the Committee to Defend America by Aiding the Allies. Its cautious title reflected the views not only of its chairman, William Allen White, but of most of its members, who sought to help the European democracies but to stay out of war. Yet the rhetoric of the White Committee, especially of its New York branch, often suggested greater militancy. In a slashing advertisement, the playwright Robert Sherwood charged that anyone who claimed the Nazis would wait until America was ready to fight them was "either an imbecile or a traitor."[32] That same month, June, 1940, thirty Americans, known as the Century Group, after the New York club at which they met, published "A Summons to Speak Out," which urged an immediate declaration of war on Germany.[33]

The "isolationists," or "noninterventionists," had their largest following among Irish and German ethnic groups, among college students, and in the Republican party. Although the Midwest was not as isolationist as it was commonly thought to be, it provided a dispro-

---

[31] Grew, *Turbulent Era,* II, 1229; Robert Aura Smith, "The Triple-Axis Pact and American Reactions," *Annals of the American Academy of Political and Social Science,* CCXV (1941), 127–132; "Axis Alliance, 1940," Hugh Byas MSS. For the thesis that the United States misinterpreted the Pact, see Paul Schroeder, *The Axis Alliance and Japanese-American Relations 1941* (Ithaca, 1958).

[32] Walter Johnson, *The Battle Against Isolation* (Chicago, 1944), p. 85. Sherwood, who in 1936 had mocked war as an *Idiot's Delight,* was now the most effective adversary of the peace groups.

[33] *The New York Times,* June 10, 1940; St. Louis *Post-Dispatch,* September 22, 1940. One of the thirty was Walter Millis. Cf. *Magazine of Wall Street,* LXVI (1940), 333–334.

portionate number of the isolationist leaders.[34] Leadership of the isolationists had shifted from radical agrarians to conservatives like Senator Robert Taft of Ohio, who distrusted the President's policies both foreign and domestic. The progressives and radicals in their camp were almost all men who had already broken with Roosevelt on domestic policy—John L. Lewis, Hugh Johnson, Burton Wheeler, Philip La Follette, and Norman Thomas—in addition to the Communists, who had turned an ideological somersault when the Nazi-Soviet pact was signed.[35]

In July, 1940, a group of Midwestern businessmen, including Robert Wood of Sears, Roebuck and Jay Hormel, the meat packer, established the America First Committee; its chief organizer, R. Douglas Stuart, Jr., was a Yale Law School student and the son of the first vice-president of Quaker Oats Company in Chicago. Nearly two-thirds of its members lived in the "Chicago *Tribune* belt" within three hundred miles of the Windy City. While the committee embraced responsible noninterventionists like Chester Bowles and Philip Jessup, many of its supporters believed that a Jewish-British-capitalist-Roosevelt conspiracy aimed to plunge the country into war, and General Wood and Mrs. Burton Wheeler publicly welcomed the Coughlinites into their ranks. A Kansas leader wrote: "Roosevelt and his wife are Jewish and this goes for 90% of his Administration."[36]

The isolationists argued that the war in Europe involved, not an apocalyptic struggle between democracy and fascism, but, in Senator

[34] On the "isolationism" of the Middle West, see George Grassmuck, *Sectional Biases in Congress on Foreign Policy* ("The Johns Hopkins University Studies in Historical and Political Science," LXVIII [Baltimore, 1950]); Jeannette Nichols, "The Middle West and the Coming of World War II," *Ohio State Archaeological and Historical Quarterly*, LXII (1953), 134–135.

[35] Eric Goldman, *Rendezvous with Destiny* (New York, 1952), p. 382; Theodore Carlson to Stephen Chadwick, October 28, 1939, Chadwick MSS.; Norman Holmes Pearson, "The Nazi-Soviet Pact and the End of a Dream," in Daniel Aaron (ed.), *America in Crisis* (New York, 1952), pp. 327–348. By the summer of 1941, business had abandoned its earlier isolationism, in part because of alarm at fascist successes in Europe, in part because Roosevelt's conservative handling of defense mobilization had quieted business fears of wartime controls. Roland Stromberg, "American Business and the Approach of War, 1935–1941," *Journal of Economic History*, XIII (1953), 72–75; *Textile World*, XC (May, 1940), 51; *Business Week* (September 7, 1940), p. 64.

[36] Wayne Cole, *America First* (Madison, Wis., 1953), pp. 10–30, 137–138.

Borah's words, "nothing more than another chapter in the bloody volume of European power politics" that had cursed the Old World "from the Spanish succession to the present hour." "There is little more democracy in a British election today than in one of Hitler's fake plebiscites," wrote Harry Elmer Barnes. They thought it the better part of wisdom to learn to do business with Hitler; an accommodation with Germany, reasoned Colonel Charles A. Lindbergh, "could maintain peace and civilization throughout the world as far into the future as we can see." Since the United States had no stake in the conflict, the President's actions could be explained only as those of a man "whose lust for power," asserted Congressman Carl Curtis of Nebraska, made him "a rival of Stalin, Hitler, and Mussolini."[37]

In the summer of 1940, the quarrel over foreign policy was exacerbated by the quadrennial presidential campaign. Neither of the front runners for the Republican nomination seemed powerful enough to defeat Franklin Roosevelt, who many G.O.P. leaders anticipated would make an unprecedented bid for a third successive term in the White House. New York's young district attorney, "Racket Buster" Thomas Dewey, made an impressive showing in the primaries, but his youth and inexperience led observers to brush him off as "Buster" Dewey. Dewey, Secretary Ickes commented, had thrown his diaper in the ring.[38] Dewey's chief rival, Robert Taft, had made his mark quickly in the Senate, but his conservatism and his glacial personality made him an easy target for a captivating campaigner like Roosevelt.

As early as February, 1939, a new element in the Republican party —the country would one day know them as "Eisenhower Republicans" —began to beat the drums for a dark-horse candidate: the utility executive, Wendell Willkie. As president of the Commonwealth and Southern Corporation, Willkie had fought the Tennessee Valley Authority, in the courts and on the public platform. He presented himself as the most prominent businessman victim of the New Deal, and in debates with New Deal officials he had acquitted himself well. An Indiana lawyer who, it was noted, had two farms he actually farmed,

[37] Paul French, *Common Sense Neutrality* (New York, 1939), pp. 19, 70–72; *The New York Times*, August 5, 1940; *Congressional Record*, 76th Cong., 3d Sess., p. A6253. Cf. Oswald Garrison Villard to Charles Beard, July 29, 1940, Villard MSS.; David I. Walsh to Herbert Hoover, October 18, 1939, Walsh MSS.

[38] Ickes, *Diary*, III, 92.

and who still had his hair cut country style, Willkie might appeal to the folks back home. In the next year, Willkie's friends—Wall Street investors, anti-New Deal publicists, renegade Democrats—worked quietly to win him a following. Willkie himself, especially after a sparkling performance on the popular radio quiz program "Information Please," attracted younger Republicans like Oren Root, Jr., by his eloquent candor and his disheveled charm. Root, a twenty-nine-year-old New York lawyer, the grandnephew of Elihu Root, set up the Associated Willkie Clubs of America, a grassroots fraternity of Ivy League graduates.[39]

Yet it hardly seemed credible that the Republicans would choose as their presidential nominee a man most of the country had never heard of at the beginning of 1940, a lifelong Democrat who had voted for Franklin Roosevelt in 1932 and had but recently changed his party affiliation, a magnate of one of the hated utility companies whose office, at 20 Pine, was, as Willkie himself pointed out, just "one block off Wall."[40] As late as April, 1940, two months before the G.O.P. convention, Willkie still did not have a single delegate.

Those who discounted Willkie's chances did not know that the Republican nomination would be decided not in Philadelphia but on the battlefields of France. On May 7, on the eve of the German invasion of the Netherlands, Willkie held the support of only 3 per cent of respondents to a national poll. By the end of May, as the Nazi battalions drove the British to the sea, he jumped to 10 per cent; on June 12, as Hitler's legions crossed the Marne, he climbed to 17 per cent; and eight days later, as the French explored terms of surrender, Willkie leaped to 29 per cent.[41] The deeper the Panzer divisions drove toward Paris, the less did young Dewey seem the man for the job. He had held no office more important than a county post and his views on foreign policy had jibed with the prevailing winds. One commentary observed acidly: "He is a hard man to imagine in a toga."[42] Nor, as

[39] Mary Earhart Dillon, *Wendell Willkie* (Philadelphia, 1952), pp. 122–132.

[40] *Fortune*, XXI (April, 1940), 47.

[41] *Public Opinion Quarterly*, IV, 537.

[42] Wolcott Gibbs and John Bainbridge, "St. George and the Dragnet," *New Yorker*, XVI (May 25, 1940), 24. "Dewey's emotional immaturity, his remarkably limited experience, the whole unfinished effect of his personality, are the most striking things about him," noted one writer. "He is psychically, mentally, and even physically the typical Young Man of Forty." Benjamin Stolberg, "Thomas E. Dewey, Self-Made Myth," *American Mercury*, L (1940), 136–137.

the mood of the country shifted in response to the menace of Nazi power, did it seem wise to risk nominating a man as relentlessly isolationist as Senator Taft.

The G.O.P. convention blew up in the faces of the Old Guard. Two weeks before the Philadelphia gathering, Willkie still did not have the vote of a single state. Yet by then, Samuel Pryor, Jr., a businessman who was a power in Connecticut politics, had used Willkie's Wall Street connections to line up delegates, especially in New England, and some adroit professional politicians had moved into the Willkie camp. They were buttressed by members of the amateur Willkie clubs who came to Philadelphia inspired with a zealotry that dumfounded veterans of past Republican parleys. Not only was it impossible for delegates to read their mail without going through thousands of telegrams and letters for Willkie, not only were they unable to find a quiet moment in the convention hall without hearing the ceaseless roar of "WE WANT WILLKIE" from the young club members in the galleries, but even when they sent their suits to the hotel valet to be pressed, they returned with Willkie literature in every pocket. The combined professional-amateur operations overwhelmed the convention. Dewey jumped to his expected big lead on the first ballot with 360 votes, followed by Taft's 189 and Willkie's 105, but Dewey soon cracked, and when he did, Willkie men like Minnesota's Harold Stassen moved around the floor picking off Dewey delegates with remarkable ease. On the sixth ballot, the convention named Wendell Willkie the Republican presidential nominee. To balance their internationalist utility tycoon candidate, the delegates chose Senator Charles McNary, an isolationist from public power-conscious Oregon, as their vice-presidential candidate.[43]

Franklin Roosevelt's political expectations present an enigma. Well into 1940, discerning observers believed he would not be a candidate for a third term.[44] Harold Ickes has speculated that, even at the last moment, Roosevelt may have been trying to escape renomination.[45] He appears to have been deeply divided. He sincerely hoped to return

[43] Dillon, *Willkie,* pp. 133–171; Donald Bruce Johnson, *The Republican Party and Wendell Willkie* ("Illinois Studies in the Social Sciences," XLVI [Urbana, Ill., 1920]), pp. 67–70, 74–108.

[44] Harry Truman to Wilbert McCune, February 14, 1940, Truman MSS., Third Term Correspondence; Washington *Post,* March 15, 1940. Cf. Leon Henderson MS. Diary, December 11, 1939.

[45] Ickes, *Diary,* III, 264–265.

to Hyde Park and its hilltop "dream house" which was nearing completion, yet he was reluctant to surrender the prerogatives of office. No candidate he encouraged to run—a Harry Hopkins, a Robert Jackson, a William O. Douglas—ever developed a following. Critics have charged that the President, ravenous for power, planned all along to seek another term and deliberately killed off possible rivals. Perhaps Frances Perkins is closer to the mark in reflecting simply: "No head rose high enough and big enough . . . to offer any alternative."[46] Roosevelt insisted that the Democratic nominee in 1940 be a New Dealer, and he rejected each of the formidable candidates around whom the conservatives grouped: Cordell Hull, John Garner, and Jim Farley.[47] Even more crucial was the international situation. As early as the spring of 1939, a veteran Democrat observed after a talk with Secretary of Commerce Daniel Roper: "We also agreed that if there were to be another European war the President would be renominated and reelected."[48] Meanwhile, Roosevelt kept alternative lines of action open, and calculatedly picked Chicago as the convention site. "I am not overlooking the fact that Kelly could pack the galleries for us," the President remarked.[49]

Sometime toward the end of May, 1940, Roosevelt appears to have decided, because of the crisis in Europe, to run for re-election. For some months, liberal senators like George Norris and James Murray, intimates like Ickes and Corcoran, who wished to preserve both the New Deal and their own power, and northern bosses like Frank Hague and Ed Kelly, who wanted a proven vote-getter, had jumped on the third-term bandwagon. Still the President confided in no one.[50] He

[46] Frances Perkins, COHC, VII, 377. Rexford Tugwell has written: "Yet I suggest that Franklin, like others in a similar position, even if almost unconsciously, had managed, as 1940 had to be planned for, not to have any acceptable successor." Tugwell, *The Democratic Roosevelt* (Garden City, N.Y., 1957), p. 490.

[47] Bascom Timmons, *Garner of Texas* (New York, 1948), pp. 246–271; Cordell Hull, *The Memoirs of Cordell Hull* (2 vols., New York, 1948), I, 855–860; T. V. Smith to Chas. I. Francis, February 21, 1940, T. V. Smith MSS., Box V; Edward Seay to Stephen Chadwick, January 8, 1940, Chadwick MSS.; F.D.R. to Henry Horner, March 27, 1940, FDRL PPF 2422.

[48] Memorandum, May 19, 1939, Robert Woolley MSS., Box 18.

[49] Ickes, *Diary*, III, 122. There is an excellent discussion of the whole question of Roosevelt's intentions in Burns, *Roosevelt*, pp. 408–428, 532–533.

[50] George Norris to Claude Pepper, August 28, 1939, Lowell Mellett MSS.; Hugh Daly to James Murray, June 15, 1939, Murray MSS.; Ickes, *Diary*, III, 158.

hoped to meet the third-term issue, and perhaps his own lingering indecision, by being drafted, ostensibly against his will, by a spontaneous uprising of the delegates. To the very last, Roosevelt refused to lift a finger to win renomination. When F.D.R. men, anxious about the listless, leaderless convention, sought to throw together an organization to press for the President's nomination, Roosevelt killed the plan. When Ickes wired from Chicago, "THIS CONVENTION IS BLEEDING TO DEATH," Roosevelt ignored him.[51]

Roosevelt preferred to carry on the charade of the draft. Alben Barkley began the proceedings when, at an appropriate moment, he read to the bewildered delegates a message from the President stating he did not desire to run. As the convention sat, stunned, a voice roared through loudspeakers around the hall: "WE WANT ROOSEVELT!" As a few delegates started to march down the aisles, the loudspeakers bellowed: "EVERYBODY WANTS ROOSEVELT!" As the parade line lengthened and gallery spectators began to surge on to the floor, Kelly's superintendent of sewers, stationed in the convention basement, chanted: "THE WORLD WANTS ROOSEVELT!" In a few moments, it was pandemonium: hundreds of wild, cheering, screeching Democrats snake-danced up and down the aisles, while from the loudspeakers came the hoarse cry of "the voice from the sewers": "ROOSEVELT! ROOSEVELT! ROOSEVELT!" All to no avail. An angry, stubborn Jim Farley insisted that Roosevelt be denied designation by acclamation, and that his own name be presented to the delegates. The next day, the convention nominated Roosevelt on the first ballot with 946 votes to 72 for Farley, 61 for Garner, 9 for Tydings, and 5 for Hull. Unquestionably, the vast majority wanted Roosevelt, but the stratagem of the "draft" and the chicanery of the Chicago machine embittered many of the delegates.[52]

[51] Harold Ickes to F.D.R., July 16, 1940, FDRL PSF 17. Only Hopkins had a direct line to the White House from his bathroom in the Blackstone Hotel, but not even he had any instructions. Ickes hid in his hotel room, because "after all, it was embarrassing for me, in view of my supposed prominent connection with the third-term movement, to have to confess that I did not know what was going on." Ickes, *Diary*, III, 246. Cf. Robert Sherwood, *Roosevelt and Hopkins* (New York, 1948), pp. 176–177; Paul Appleby, "Roosevelt's Third Term Decision," *American Political Science Review*, XLVI (1952), 754–765.

[52] James A. Farley, *Jim Farley's Story* (New York, 1948), pp. 271–272; Burns, *Roosevelt*, pp. 427–430; *Time*, XXXVI (July 29, 1940), 14. Thomas Gore, former Democratic senator from Oklahoma, wrote: "You know I have

Roosevelt and Vice-President John Garner had long since parted ways.[53] For his new running mate, the President selected a committed New Dealer, his Secretary of Agriculture, Henry Wallace. Wallace, Roosevelt believed, would appeal to the Corn Belt and, as a trenchant antifascist, would clarify issues for the country. Party regulars were aghast at the choice of an ex-Republican with little political savvy and a reputation for mysticism. Even more to the point, a small army of men had been working to secure the vice-presidential nomination for themselves. For a time it appeared that the querulous delegates, testy over the displacement of old-time Democrats by New Dealers like Hopkins, would kick over the traces and nominate Speaker Bankhead of Alabama instead. As Roosevelt listened to the proceedings from Chicago on the radio, he scribbled out his declination of the nomination. On the convention floor, Jimmie Byrnes moved from delegation to delegation, saying: "For God's sake, do you want a President or a Vice-President?" Bullied by the threat Roosevelt might not run, the convention designated Wallace with 628 votes to 329 for Bankhead, but Wallace did not dare make an acceptance speech to the irate delegates. Apparently checkmated by the failure of the purge in 1938, Roosevelt and the New Dealers had demonstrated convincingly that they still controlled the Democratic party.[54]

Willkie entered the campaign like a heavyweight challenger preparing for the big bout. He was eager, he announced, to take on "the Champ." A little-known figure but a few months before, Willkie quickly elicited an enthusiastic response from millions of middle-class Americans tired of alphabet agencies, wearied by years of strife, and vexed by the failure of the New Dealers to end the depression. "For the first time since Teddy Roosevelt," observed *Time*, "the Republicans had a man they could yell for and mean it."[55] With compelling direct-

never doubted for one moment since he was nominated in 1932 that he would seek a third term. . . . Caesar thrice refused the kingly crown—but this Caesar, never!" Gore to Mary Wyatt, May 16, 1940, Gore MSS. See too John W. Davis to David Reay, July 2, 1940, Reay MSS.

[53] It was widely noted that Garner did not vote at all in 1940. John Garner to Jesse Jones, November 19, 1940, Jones MSS., Box 8.

[54] Samuel Rosenman, *Working with Roosevelt* (New York, 1952), pp. 215–218; Burns, *Roosevelt*, p. 429; *The New York Times*, July 19, 1940; Henry Wallace to Harry Hopkins, August 1, 1940, "Confidential Political File," Hopkins MSS.

[55] *Time*, XXXVI (July 8, 1940), 14.

ness, Willkie charged that the New Deal had lost faith in the potential of American society. Born of the panic of the depression, it was oriented to security rather than new enterprise, committed to regulating business rather than liberating it, worried about overproduction when the nation was at the threshold of an era of abundance. Willkie's message scored a hit with many independents: in August, he won the backing of the Cleveland *Plain Dealer,* which had never before endorsed a Republican presidential candidate; in September, of *The New York Times,* which had supported Roosevelt in 1932 and 1936.[56]

Willkie cleverly related his assault on the President's failure to restore prosperity to the current uneasiness over the state of our defenses. The New Deal, critics charged, had left the nation vulnerable by promoting internal disunity, squandering billions on unproductive projects, and shamefully neglecting defenses. Roosevelt, wrote one disgruntled Democrat, "is our Léon Blum."[57] In harping on the defense issue, the G.O.P. hit the President on an exposed flank, even if he could retort that he had been more foresighted than they about the need to rearm. Burned by his experience in World War I, the President was determined that this new crisis would not spawn a swarm of profiteers; and fearful that big business might get too much power, he refused to heed demands that he appoint a defense czar. Even when he brought men like William Knudsen of General Motors and Donald Nelson of Sears, Roebuck to Washington to run the mobilization, he insisted he would do as little as possible to upset "the normal processes of life."[58] In September, 1940, Leon Henderson noted in his diary: "Nelson troubled. . . . Soon Willkie is going to ask—'how are we fixed for war? What have we got[?]' And we haven't even a plan."[59]

But the tide of events was running against Willkie. Even though the economy responded to war orders with remarkable sluggishness, the

[56] Cleveland *Plain Dealer,* August 20, 1940; *The New York Times,* September 19, 1940. He also won a goodly number of Willkie Democrats who were "somewhat a reminder of the McKinley Gold Democrats in 1896." Theodore Carlson to Stephen Chadwick, June 10, 1940, Chadwick MSS.

[57] Theodore Carlson to Stephen Chadwick, May 29, 1940, Chadwick MSS. Cf. *Food Industries,* XI (1939), 667.

[58] "I am not looking at anybody," he joshed reporters in May, 1940, "I am looking at the ceiling—the answer is that this delightful young lady will not have to forego cosmetics, lipsticks, ice cream sodas and—[Laughter]." *Public Papers,* IX, 242.

[59] Leon Henderson MS. Diary, September 10, 1940.

country sensed in 1940 that the war boom which would terminate the depression was now well under way. As Willkie belabored Roosevelt for not eradicating mass unemployment, thousands of workers streamed into California aircraft plants. As Willkie chided the President for inattention to defense, a flood of dollar-a-year men—notably Hopkins' "tame businessmen" like W. Averell Harriman of Union Pacific—moved into Washington. When Roosevelt, his long cape flapping in the breeze, received the salutes of naval officers to the nation's Commander in Chief, he stole the defense issue away from Willkie. In campaign jaunts disguised as "inspection" tours of factories and military sites, Roosevelt punched home the theme of defense gains while appealing to local pride: "You citizens of Seattle who are listening tonight —you have watched the Boeing plant out there grow. . . . You citizens of Southern California can see the great Douglas factories. . . . You citizens of Hartford, who hear my words: look across the Connecticut River at the whirring wheels and the beehive of activity which is the Pratt and Whitney plant which I saw today."[60]

Willkie had a hard time finding a viable political position. Old Guard Republicans instinctively distrusted his mercurial temperament and resented the fact that he had been a Democrat such a short time before. "I don't mind the church converting a whore," former Indiana Senator Jim Watson had told him bluntly at the G.O.P. convention, "but I don't like her to lead the choir the first night!"[61] On the other hand, voters with harsh memories of Hoover and the Republican party questioned the claims to liberalism of a utility tycoon. Democrats sneered that he was "Hopson's Choice"; that he had "an electric background, an electric personality and an electric campaign chest"; that he was, in Ickes' telling phrase, a "simple, barefoot Wall Street lawyer."[62]

When Willkie was simply a spokesman for business, he had been free to flay the New Deal with abandon, but, as he stumped the country as a candidate, he found that much of the New Deal pattern was

[60] *Public Papers,* IX, 518.

[61] Dillon, *Willkie,* p. 143. Cf. Henry Evjen, "The Willkie Campaign: An Unfortunate Chapter in Republican Leadership," *Journal of Politics,* XIV (1952), 241–256.

[62] *Time,* XXXVI (July 8, 1940), 10; (August 26, 1940), 16. The head of a large utility empire, Howard Hopson was sentenced to prison for fraud in 1941.

too well accepted to reward criticism. The 1940 Republican platform had embraced many of the Roosevelt achievements. As Frances Perkins read it, she exclaimed: "Well, God's holy name be praised. No matter who gets elected we've won."[63] Willkie went well beyond the G.O.P. platform. In Oregon, the leading foe of public power upheld the people's right "to take over the private utilities"; elsewhere he promised to maintain the Wagner Labor Relations Act, to give government aid to agriculture, to extend social security, and "to provide jobs for every man and woman in the United States willing to work and to continue public relief to those who could not work." Such statements helped win him the backing of disaffected union leaders like John L. Lewis and Harry Bridges, but they did not cut very deeply into Roosevelt's labor support. In industrial areas like Chicago's Union Stock Yards, workers greeted Willkie with stony silence. In Michigan, they booed him from factory windows; he was egged in Pontiac, and, at Grand Rapids, a rock was hurled through a train window. In Toledo, workers cursed him and shook their fists at him; they held up signs reading "Win What with Willkie?" or "To Hell with Willkie."[64]

By early autumn Willkie still had not found a winning issue. Even the third-term question won few voters who did not oppose Roosevelt on other grounds. In late September, he struck out in a new direction; he claimed for himself the role of guarantor of peace and excoriated Roosevelt as a warmonger. Willkie, who, with typically disarming indiscretion, had made clear earlier in the year that he endorsed Roosevelt's internationalist policies, now shouted: "If his promise to keep our boys out of foreign wars is no better than his promise to balance the budget they're already almost on the transports." In a savage speech in Baltimore at the close of the campaign, Willkie flatly declared: "On the basis of his past performance with pledges to the people, if you re-elect him you may expect war in April, 1941."[65]

Willkie's charges stung Roosevelt to extravagant replies. "While I am talking to you mothers and fathers, I give you one more assurance," the President told a Boston crowd. "I have said this before, but I shall say it again and again and again: Your boys are not going to be

[63] Frances Perkins, COHC, VII, 15.

[64] Time, XXXVI (September 30, 1940), 13; Johnson, Willkie, pp. 138–140; Dillon, Willkie, pp. 201–202, 214–216.

[65] The New York Times, October 23, 31, 1940.

sent into any foreign wars."[66] On November 2, in Buffalo, Roosevelt declared: "Your President says this country is not going to war."[67]

Roosevelt won re-election with 27 million votes to Willkie's 22 million. His margin was much wider in the Electoral College, where he captured 449 to the Republican's 82. Willkie had cut Roosevelt's plurality to the smallest of any winner since 1916. He ran best in the Midwest, where he carried the farms and the country towns, and where he picked up the antiwar ballots of German-Americans. He cut into Roosevelt's Irish following; Irish-Americans, who had backed him more strongly than any other ethnic group in 1932, had grown more prosperous and tended to take on some of the conservatism of anti-New Dealers. They also thought Roosevelt's foreign policy too Anglophile. Some Italian-Americans, who had been Roosevelt's most ardent supporters in 1936, turned against him in 1940 because they resented the contemptuous implications of the Charlottesville speech. Democratic workers could enter Italian sections of the Bronx only under police protection.[68]

Roosevelt's victory was won in the cities. His plurality in New York City swung the Empire State into the Democratic column, just as Chicago gave him the measure of victory in Illinois, Cleveland in Ohio, and Milwaukee in Wisconsin. Balloting divided sharply on class lines. While the President's vote fell off in almost every group, he ran almost as well as he had in 1936 in working-class and lower-middle income districts in the cities. Pittsburgh wards with rentals below $40 a month gave Roosevelt nearly 75 per cent of the vote; wards with rentals of $65 and up went only 40 per cent for the President. Roosevelt also picked up some internationalist ballots: Jews, Yankees on the eastern

---

[66] *Public Papers*, IX, 517. Robert Sherwood, who helped write the Boston speech, later confessed: "For my own part, I think it was a mistake for him to go so far in yielding to the hysterical demands for sweeping reassurance; but, unfortunately for my own conscience, I happened at the time to be one of those who urged him to go the limit on this. . . . I burn inwardly whenever I think of those words 'again—and again—and again.' " Sherwood, *Roosevelt and Hopkins*, p. 201. Yet it should be noted that in the very same address, Roosevelt, speaking in a city known for the Anglophobia of its Irish-American citizens, outlined his plans for stepped-up aid to Britain.

[67] *Public Papers*, IX, 543.

[68] Burns, *Roosevelt*, pp. 454–455; William Allen White, "Thoughts After the Election," *Yale Review*, XXX (1941), 225; Eugene Keogh, COHC, pp. 64–65; Victor Sholis, Memorandum, n.d. (*c.* October, 1940), FDRL PSF 17.

seaboard—the President actually registered slight gains in northern New England—and voters with ties to occupied countries like Norway and Poland. Polish precincts in Buffalo ran as high as 25–1 for Roosevelt, Polish precincts in Hamtramck 30–1.[69] Although Willkie was a strong foe of racial discrimination, he had little appeal to Negroes. Of fifteen Negro wards in nine northern cities studied by Gunnar Myrdal, Roosevelt captured four in 1932, nine in 1936, and fourteen in 1940.[70] Finally, the President attracted many who were impressed by the humanitarianism of the New Deal. On a nationwide broadcast on the eve of the election, Carl Sandburg called the President "a not perfect man and yet more precious than fine gold."[71]

Despite his failure, Willkie had made lasting contributions. He had speeded the process of making New Deal reforms part of a common heritage which would not be affected by change of the party in power. He had helped educate the businessman to a new sense of social responsibility. Most important, despite his frenzied rhetoric in the last weeks of the campaign, he had declined to challenge the assumptions of Roosevelt's internationalist foreign policy. Yet this last commitment had one unfortunate consequence. It convinced a number of Americans that they had been refused a genuine choice in the decision to support the Allies, with its risk of involvement in war, since they had been given two internationalist candidates from whom to choose. At an America First meeting in Kansas City on June 19, 1941, Senator Nye declared: "I shall be surprised if history does not show that beginning at the Republican convention at Philadelphia a conspiracy was carried out to deny the American people a chance to express themselves. . . ."[72] The conviction that they had never been defeated fairly

[69] Samuel Eldersveld, "The Influence of Metropolitan Pluralities in Presidential Elections Since 1920," *American Political Science Review*, XLIII (1949), 1197; Louis Bean, Frederick Mosteller and Frederick Williams, "Nationalities and 1944," *Public Opinion Quarterly*, VIII (1944), 368–375; Samuel Lubell, "Post-Mortem: Who Elected Roosevelt?" *Saturday Evening Post*, CCXIII (January 25, 1941), pp. 91–92, 94. Cf. Irving Bernstein, "John L. Lewis and the Voting Behavior of the C.I.O.," *Public Opinion Quarterly*, V (1941), 233–249; Paul Lazarsfeld, Bernard Berelson, and Hazel Gaudet, *The People's Choice* (New York, 1944).

[70] Gunnar Myrdal, *An American Dilemma* (New York, 1944), p. 268. Cf. Will Alexander, COHC, pp. 358 ff.; *Crisis*, XLVII (1940), 343.

[71] *Personal Letters*, II, 1086.

[72] Rex Stout (ed.), *The Illustrious Dunderheads* (New York, 1941), p. 167. See Thomas Gore to Mary Wyatt, January 22, 1941, Gore MSS.

and squarely, that they were led into war by an internationalist minority, left vestiges of bitterness to be exploited in the years after the war.

Roosevelt, returned to an unprecedented third term, now sought to fashion a global foreign policy to contain the fascists. In Tokyo, Joseph Grew, who felt that his eight years of work for peace had "been swept away as if by a typhoon with little or nothing remaining to show for it," cautioned that Japan would not be deterred by bluff. "Only if they become certain that we mean to fight if called upon to do so," he wrote the President, "will our preliminary measures stand some chance of proving effective and of removing the necessity of war—the old story of Sir Edward Grey in 1914."[73] Within the Administration, advocates of a hard policy pressed for that final economic sanction, an embargo on oil. "We did not keep Japan out of Indo-China by continuing to ship scrap-iron, nor will we keep Japan out of the Dutch East Indies by selling it our oil," Harold Ickes remonstrated.[74]

But the main burden of advice Roosevelt received still cautioned restraint toward Japan. "Any strength that we might send to the Far East would, by just so much, reduce the force of our blows against Germany and Italy," Admiral Stark pointed out.[75] Admiral James Otto Richardson, in command of the fleet at Hawaii, even wished to return naval forces to the Pacific Coast. Roosevelt believed that the Japanese should be met with firmness—he ordered Richardson to hold the fleet at Pearl Harbor to deter Japan—but he agreed that a showdown with Tokyo should be eschewed. Despite the entreaties of men like Ickes, he continued to permit oil to flow to the island empire.[76]

Since resistance to Hitler had priority, the President directed the

[73] Grew, *Turbulent Era*, II, 1256–1257.

[74] Ickes to F.D.R., October 17, 1940, FDRL PSF 17. "Now we've stopped scrap iron, what about oil?" Eleanor Roosevelt asked. Eleanor Roosevelt to F.D.R., Nov. 12, 1940, FDRL PSF 35.

[75] Stark to F.D.R., November 12, 1940, FDRL Tully Safe File. "If Germany should win in Europe, Japan will run amuck in the Far East," observed Francis Sayre, High Commissioner in Manila, "whereas if England wins, I doubt if Japan will thereafter present much of a problem in this part of the world. . . . Under no circumstances," Sayre advised, should the United States "play into Hitler's hands by being sucked into a war with Japan." Sayre to F.D.R., November 13, 1940, FDRL PSF 17.

[76] U.S. Congress, *Pearl Harbor Attack*, Hearings before the Joint Committee on the Investigation of the Pearl Harbor Attack, 79th Cong., 1st Sess., pursuant to S. Con. Res. 27, Part 1, November 15–21, 1945 (Washington, 1946), pp. 264–266; F.D.R. to Eleanor Roosevelt, November 13, 1940, FDRL PSF 35.

energies of the government to converting the United States into a "great arsenal of democracy." He not only announced that half the country's future war production would be allotted to Britain but, in response to an urgent plea from Winston Churchill, unveiled the startling proposal to lend arms directly to Britain on the understanding that they would be returned or replaced when the war ended.[77] The nation still hoped that such subvention would be enough; that without any commitment of American military or naval forces, the Allies would be able to overcome the Axis and thus spare the United States the need to fight. When the William Allen White Committee in November, 1940, suggested it might be necessary for the U.S. Navy to convoy merchant ships to Britain, White, after first approving the statement, backed down in a public letter in which he called the convoy scheme "a silly thing, for convoys, unless you shoot, are confetti and it's not time to shoot now or ever. . . . If I was making a motto for the Committee to Defend America by Aiding the Allies," White added, "it would be 'The Yanks Are Not Coming.' " Early in January, after Fiorello La Guardia accused him of "doing a typical Laval," White resigned as committee chairman.[78]

White's irresolution represented the feeling of the average American at the end of 1940. He wanted to help Britain, even at the risk of war, yet he wished too to remain at peace. But it was by no means clear that economic aid alone would suffice. At a meeting in October, 1940, Secretary Knox burst out: "I can't escape saying that the English are not going to win this war without our help, I mean our military help." In December, Secretary Stimson reflected that it would be impossible to expand production to the levels needed "until we got into the war ourselves." By January, even White was writing: "The dictators are greedy for our wealth and have scorn for our liberty. Sooner or later we shall have to meet them with arms, how and when I don't know."[79] Increasingly, Washington discussed not "whether" but

[77] *Public Papers*, IX, 563, 605–613, 643; Churchill, *Their Finest Hour*, pp. 557–569.

[78] Walter Johnson, *William Allen White's America* (New York, 1947), pp. 544, 547; F.D.R. to White, June 16, 1941, White MSS., Box 324.

[79] Charles Tansill, *Back Door to War* (Chicago, 1952), p. 602; Theodore Carlson to Stephen Chadwick, December 21, 1940, Chadwick MSS.; Langer and Gleason, *Undeclared War*, p. 187; Stimson MS. Diary, December 13, 29, 1940; White to Rev. Allen Keedy, January 3, 1941, Walter Johnson (ed.), *Selected Letters of William Allen White* (New York, 1947), p. 422.

"when" we would enter the war. In the Pacific, the decision to impose sanctions on oil would not long be postponed. Even if such measures halted the Japanese advance toward the South Seas, war might come over American insistence on the liberation of China.[80] In the Atlantic, it seemed absurd to ship arms to Britain and then permit U-boats to send them to the bottom of the ocean. Inevitably, lend-lease would lead to convoying, and if the Nazi wolfpack menaced the convoy, would not U.S. naval vessels have to shoot? As the new Atlantic Fleet, organized on February 1, 1941, extended its patrol from the blue Caribbean to the icy waters of Greenland, quasi-belligerency shaded into war.

[80] Schroeder, *Axis Alliance*, pp. 203–204.

CHAPTER 14

The Roosevelt Reconstruction: Retrospect

IN EIGHT YEARS, Roosevelt and the New Dealers had almost revolutionized the agenda of American politics. "Mr. Roosevelt may have given the wrong answers to many of his problems," concluded the editors of *The Economist*. "But he is at least the first President of modern America who has asked the right questions." In 1932, men of acumen were absorbed to an astonishing degree with such questions as prohibition, war debts, and law enforcement. By 1936, they were debating social security, the Wagner Act, valley authorities, and public housing. The thirties witnessed a rebirth of issues politics, and parties split more sharply on ideological lines than they had in many years past. "I incline to think that for years up to the present juncture thinking Democrats and thinking Republicans had been divided by an imaginary line," reflected a Massachusetts congressman in 1934. "Now for the first time since the period before the Civil War we find vital principles at stake." Much of this change resulted simply from the depression trauma, but much too came from the force of Roosevelt's personality and his use of his office as both pulpit and lectern. "Of course you have fallen into some errors—that is human," former Supreme Court Justice John Clarke wrote the President, "but you have put a new face upon the social and political life of our country."[1]

[1] The Editors of the Economist, *The New Deal* (New York, 1937), p. 149; Representative Robert Luce to Herbert Claiborne Pell, November 14, 1934, Pell MSS., Box 7; Elliott Roosevelt (ed.), *F.D.R.: His Personal Letters, 1928–1945* (2 vols., New York, 1950), I, 723.

Franklin Roosevelt re-created the modern Presidency. He took an office which had lost much of its prestige and power in the previous twelve years and gave it an importance which went well beyond what even Theodore Roosevelt and Woodrow Wilson had done. Clinton Rossiter has observed: "Only Washington, who made the office, and Jackson, who remade it, did more than [Roosevelt] to raise it to its present condition of strength, dignity, and independence."[2] Under Roosevelt, the White House became the focus of all government—the fountainhead of ideas, the initiator of action, the representative of the national interest.

Roosevelt greatly expanded the President's legislative functions. In the nineteenth century, Congress had been jealous of its prerogatives as the lawmaking body, and resented any encroachment on its domain by the Chief Executive. Woodrow Wilson and Theodore Roosevelt had broken new ground in sending actual drafts of bills to Congress and in using devices like the caucus to win enactment of measures they favored. Franklin Roosevelt made such constant use of these tools that he came to assume a legislative role not unlike that of a prime minister. He sent special messages to Congress, accompanied them with drafts of legislation prepared by his assistants, wrote letters to committee chairmen or members of Congress to urge passage of the proposals, and authorized men like Corcoran to lobby as presidential spokesmen on the Hill. By the end of Roosevelt's tenure in the White House, Congress looked automatically to the Executive for guidance; it expected the administration to have a "program" to present for consideration.[3]

Roosevelt's most important formal contribution was his creation of the Executive Office of the President on September 8, 1939. Executive Order 8248, a "nearly unnoticed but none the less epoch-making event in the history of American institutions," set up an Executive Office staffed with six administrative assistants with a "passion for anonymity."[4] In 1939, the President not only placed obvious agencies like the

[2] Clinton Rossiter, *The American Presidency* (Signet edition, New York, 1956), p. 114.

[3] *Ibid.*, pp. 81–84; Edward S. Corwin, *The President: Office and Powers 1787–1957* (New York, 1957), pp. 274–275. Yet despite the growth of the Presidency, this was a period in which Congress had great influence. Much of the specific New Deal legislation was the consequence of the work of a Robert Wagner or a Robert La Follette, Jr. The expansion of the Presidency resulted in a reinvigoration of the whole political system.

[4] Luther Gulick, cited in Rossiter, *American Presidency*, p. 96.

White House Office in the Executive Office but made the crucial decision to shift the Bureau of the Budget from the Treasury and put it under his wing. In later years, such pivotal agencies as the Council of Economic Advisers, the National Security Council, and the Central Intelligence Agency would be moved into the Executive Office of the President. Roosevelt's decision, Rossiter has concluded, "converts the Presidency into an instrument of twentieth-century government; it gives the incumbent a sporting chance to stand the strain and fulfill his constitutional mandate as a one-man branch of our three-part government; it deflates even the most forceful arguments, which are still raised occasionally, for a plural executive; it assures us that the Presidency will survive the advent of the positive state. Executive Order 8248 may yet be judged to have saved the Presidency from paralysis and the Constitution from radical amendment."[5]

Roosevelt's friends have been too quick to concede that he was a poor administrator. To be sure, he found it difficult to discharge incompetent aides, he procrastinated about decisions, and he ignored all the canons of sound administration by giving men overlapping assignments and creating a myriad of agencies which had no clear relation to the regular departments of government.[6] But if the test of good administration is not an impeccable organizational chart but creativity, then Roosevelt must be set down not merely as a good administrator but as a resourceful innovator. The new agencies he set up gave a spirit of excitement to Washington that the routinized old-line departments could never have achieved. The President's refusal to proceed through channels, however vexing at times to his subordinates, resulted in a competition not only among men but among ideas, and encouraged men to feel that their own beliefs might win the day. "You would be surprised, Colonel, the remarkable ideas that have been turned loose just because men have felt that they can get a hearing," one senator confided.[7] The President's "procrastination" was his own way

---

[5] Rossiter, *American Presidency*, p. 100. Cf. Emile Giraud, *La Crise de la démocratie et le renforcement du pouvoir exécutif* (Paris, 1938).

[6] "At times Roosevelt acted as if a new agency were almost a new solution. His addiction to new organizations became a kind of nervous tic which disturbed even avid New Dealers." Arthur Schlesinger, Jr., *The Coming of the New Deal* (Boston, 1959), p. 535. Schlesinger has an excellent discussion of Roosevelt's administrative talent.

[7] Elbert Thomas to Colonel E. LeRoy Bourne, January 6, 1934, Elbert Thomas MSS., Box 23.

both of arriving at a sense of national consensus and of reaching a decision by observing a trial by combat among rival theories. Periods of indecision—as in the spring of 1935 or the beginning of 1938—were inevitably followed by a fresh outburst of new proposals.[8]

Most of all, Roosevelt was a successful administrator because he attracted to Washington thousands of devoted and highly skilled men. Men who had been fighting for years for lost causes were given a chance: John Collier, whom the President courageously named Indian Commissioner; Arthur Powell Davis, who had been ousted as chief engineer of the Department of the Interior at the demand of power interests; old conservationists like Harry Slattery, who had fought the naval oil interests in the Harding era. When Harold Ickes took office as Secretary of the Interior, he looked up Louis Glavis—he did not even know whether the "martyr" of the Ballinger-Pinchot affair was still alive—and appointed him to his staff.[9]

The New Dealers displayed striking ingenuity in meeting problems of governing. They coaxed salmon to climb ladders at Bonneville; they sponsored a Young Choreographers Laboratory in the WPA's Dance Theatre; they gave the pioneer documentary film maker Pare Lorentz the opportunity to create his classic films *The Plow That Broke the Plains* and *The River.* At the Composers Forum-Laboratory of the Federal Music Project, William Schuman received his first serious hearing. In Arizona, Father Berard Haile of St. Michael's Mission taught written Navajo to the Indians.[10] Roosevelt, in the face of derision from professional foresters and prairie states' governors, persisted in a bold scheme to plant a mammoth "shelterbelt" of parallel rows of trees from the Dakotas to the Panhandle. In all, more than two hundred million trees were planted—cottonwood and willow, hackberry and cedar, Russian olive and Osage orange; within six years, the President's visionary windbreak had won over his former critics.[11] The spirit

---

[8] Richard Neustadt, *Presidential Power* (New York, 1960), pp. 156–158.

[9] In Roosevelt's first year in office, he signed an order restoring Glavis to the civil service status he had lost when President Taft fired him. Ironically, Ickes found Glavis as intolerable a subordinate as Taft had, and concluded that he had "been very unjust to Ballinger all of these years." *The Secret Diary of Harold Ickes* (3 vols., New York, 1954), III, 111.

[10] John Collier to Louis Brandeis, April 5, 1937, Brandeis MSS., SC 19.

[11] H. H. Chapman, "Digest of Opinions Received on the Shelterbelt Project," *Journal of Forestry,* XXXII (1934), 952–957; Bristow Adams, "Some Fence!" *Cornell Countryman,* XXXII (1934), 4; *Science News Letter,* CXXXIV (1938), 409; "Prairie Tree Banks," *American Forester,* CXLVII (1941), 177.

behind such innovations generated a new excitement about the potentialities of government. "Once again," Roosevelt told a group of young
Democrats in April, 1936, "the very air of America is exhilarating."[12]

Roosevelt dominated the front pages of the newspapers as no other
President before or since has done. "Frank Roosevelt and the NRA
have taken the place of love nests," commented Joe Patterson, publisher of the tabloid New York *Daily News*. At his very first press conference, Roosevelt abolished the written question and told reporters
they could interrogate him without warning. Skeptics predicted the
free and easy exchange would soon be abandoned, but twice a week,
year in and year out, he threw open the White House doors to as many
as two hundred reporters, most of them representing hostile publishers,
who would crowd right up to the President's desk to fire their questions. The President joshed them, traded wisecracks with them, called
them by their first names; he charmed them by his good-humored ease
and impressed them with his knowledge of detail.[13] To a degree,
Roosevelt's press conference introduced, as some observers claimed, a
new institution like Britain's parliamentary questioning; more to the
point, it was a device the President manipulated, disarmingly and
adroitly, to win support for his program.[14] It served too as a classroom
to instruct the country in the new economics and the new politics.

Roosevelt was the first president to master the technique of reaching
people directly over the radio. In his fireside chats, he talked like a
father discussing public affairs with his family in the living room. As
he spoke, he seemed unconscious of the fact that he was addressing
millions. "His head would nod and his hands would move in simple,
natural, comfortable gestures," Frances Perkins recalled. "His face
would smile and light up as though he were actually sitting on the
front porch or in the parlor with them." Eleanor Roosevelt later ob-

[12] Samuel Rosenman (ed.), *The Public Papers and Addresses of Franklin D.
Roosevelt* (13 vols., New York, 1938–50), V, 165.

[13] Elmer Cornwell, Jr., "Presidential News: The Expanding Public Image,"
Journalism Quarterly, XXXVI (1959), 275–283; "The Chicago Tribune,"
Fortune, IX (May, 1934), 108; *Editor and Publisher*, March 4, 1933; Thomas
Stokes, *Chip Off My Shoulder* (Princeton, 1940), p. 367.

[14] Erwin Canham, "Democracy's Fifth Wheel," *Literary Digest*, CXIX (January 5, 1935), 6; Douglass Cater, *The Fourth Branch of Government* (Boston,
1959), pp. 13–14, 142–155; James Pollard, *The Presidents and the Press* (New
York, 1947), pp. 773–845.

served that after the President's death people would stop her on the street to say "they missed the way the President used to talk to them. They'd say 'He used to talk to me about my government.' There was a real dialogue between Franklin and the people," she reflected. "That dialogue seems to have disappeared from the government since he died."[15]

For the first time for many Americans, the federal government became an institution that was directly experienced. More than state and local governments, it came to be *the* government, an agency directly concerned with their welfare. It was the source of their relief payments; it taxed them directly for old age pensions; it even gave their children hot lunches in school. As the role of the state changed from that of neutral arbiter to a "powerful promoter of society's welfare," people felt an interest in affairs in Washington they had never had before.[16]

Franklin Roosevelt personified the state as protector. It became commonplace to say that people felt toward the President the kind of trust they would normally express for a warm and understanding father who comforted them in their grief or safeguarded them from harm. An insurance man reported: "My mother looks upon the President as someone so immediately concerned with her problems and difficulties that she would not be greatly surprised were he to come to her house some evening and stay to dinner." From his first hours in office, Roosevelt gave people the feeling that they could confide in him directly. As late as the Presidency of Herbert Hoover, one man, Ira Smith, had sufficed to take care of all the mail the White House received. Under Roosevelt, Smith had to acquire a staff of fifty people to handle the thousands of letters written to the President each week. Roosevelt gave people a sense of membership in the national community. Justice Douglas has written: "He was in a very special sense the people's President, because he made them feel that with him in the White House they shared the Presidency. The sense of sharing the

[15] Frances Perkins, *The Roosevelt I Knew* (New York, 1946), p. 72; Bernard Asbell, *When F.D.R. Died* (New York, 1961), p. 161.

[16] Felix Frankfurter, "The Young Men Go to Washington," *Fortune,* XIII (1936), 61; E. W. Bakke, *Citizens Without Work* (New Haven, 1940), pp. 52–53.

Presidency gave even the most humble citizen a lively sense of belonging."[17]

When Roosevelt took office, the country, to a very large degree, responded to the will of a single element: the white, Anglo-Saxon, Protestant property-holding class. Under the New Deal, new groups took their place in the sun. It was not merely that they received benefits they had not had before but that they were "recognized" as having a place in the commonwealth. At the beginning of the Roosevelt era, charity organizations ignored labor when seeking "community" representation; at the end of the period, no fund-raising committee was complete without a union representative. While Theodore Roosevelt had founded a lily-white Progressive party in the South and Woodrow Wilson had introduced segregation into the federal government, Franklin Roosevelt had quietly brought the Negro into the New Deal coalition. When the distinguished Negro contralto Marian Anderson was denied a concert hall in Washington, Secretary Ickes arranged for her to perform from the steps of Lincoln Memorial. Equal representation for religious groups became so well accepted that, as one priest wryly complained, one never saw a picture of a priest in a newspaper unless he was flanked on either side by a minister and a rabbi.

The devotion Roosevelt aroused owed much to the fact that the New Deal assumed the responsibility for guaranteeing every American a minimum standard of subsistence. Its relief programs represented an advance over the barbaric predepression practices that constituted a difference not in degree but in kind. One analyst wrote: "During the ten years between 1929 and 1939 more progress was made in public welfare and relief than in the three hundred years after this country was first settled." The Roosevelt administration gave such assistance not as a matter of charity but of right. This system of social rights was written into the Social Security Act. Other New Deal legislation abolished child labor in interstate commerce and, by putting a floor under wages and a ceiling on hours, all but wiped out the sweatshop.[18]

Roosevelt and his aides fashioned a government which consciously

[17] Richard Neuberger, "They Love Roosevelt," *Forum and Century*, CI (1939), 15; Corwin, *The President*, p. 471; William O. Douglas, *Being an American* (New York, 1948), p. 88.

[18] Josephine Chapin Brown, *Public Relief 1929–1939* (New York, 1940), p. ix; Thomas Paul Jenkin, *Reactions of Major Groups to Positive Government*

sought to make the industrial system more humane and to protect workers and their families from exploitation. In his acceptance speech in June, 1936, the President stated: "Governments can err, Presidents do make mistakes, but the immortal Dante tells us that divine justice weighs the sins of the cold-blooded and the sins of the warm-hearted in different scales.

"Better the occasional faults of a Government that lives in a spirit of charity than the constant omission of a Government frozen in the ice of its own indifference." Nearly everyone in the Roosevelt government was caught up to some degree by a sense of participation in something larger than themselves. A few days after he took office, one of the more conservative New Deal administrators wrote in his diary: "This should be a Gov't of humanity."[19]

The federal government expanded enormously in the Roosevelt years. The crisis of the depression dissipated the distrust of the state inherited from the eighteenth century and reinforced in diverse ways by the Jeffersonians and the Spencerians. Roosevelt himself believed that liberty in America was imperiled more by the agglomerations of private business than by the state. The New Dealers were convinced that the depression was the result not simply of an economic break-down but of a political collapse; hence, they sought new political instrumentalities. The reformers of the 1930's accepted almost unquestioningly the use of coercion by the state to achieve reforms.[20] Even Republicans who protested that Roosevelt's policies were snuffing out liberty voted overwhelmingly in favor of coercive measures.[21]

This elephantine growth of the federal government owed much to the fact that local and state governments had been tried in the crisis and found wanting. When one magazine wired state governors to ask their views, only one of the thirty-seven who replied announced that

---

in the United States, 1930–1940 (University of California Publications in Political Science [Berkeley and Los Angeles, 1945]), p. 284.

[19] *Public Papers,* V, 235; J. F. T. O'Connor MS. Diary, June 25, 1933.

[20] Paul Carter has noted the change in the social gospel. The editors of *The Baptist,* he has written, "recognized that the transfer of social privilege involves the use of social coercion, a fact which the Right and Center of the old Social Gospel had not always faced up to." Carter, "The Decline and Revival of the Social Gospel," (unpublished Ph.D. dissertation, Columbia University, 1954).

[21] On the compulsory Potato Act, only six Republicans (and just nine Democrats) voted in opposition.

he was willing to have the states resume responsibility for relief.[22] Every time there was a rumored cutback of federal spending for relief, Washington was besieged by delegations of mayors protesting that city governments did not have the resources to meet the needs of the unemployed.

Even more dramatic was the impotence of local governments in dealing with crime, a subject that captured the national imagination in a decade of kidnapings and bank holdups. In September, 1933, the notorious bank robber John Dillinger was arrested in Ohio. Three weeks later, his confederates released him from jail and killed the Lima, Ohio, sheriff. In January, 1934, after bank holdups at Racine, Wisconsin, and East Chicago, Indiana, Dillinger was apprehended in Tucson, Arizona, and returned to the "escape-proof" jail of Crown Point, Indiana, reputedly the strongest county prison in the country. A month later he broke out and drove off in the sheriff's car. While five thousand law officers pursued him, he stopped for a haircut in a barber shop, bought cars, and had a home-cooked Sunday dinner with his family in his home town. When he needed more arms, he raided the police station at Warsaw, Indiana.

Dillinger's exploits touched off a national outcry for federal action. State and local authorities could not cope with gangs which crossed and recrossed jurisdictional lines, which were equipped with Thompson submachine guns and high-powered cars, and which had a regional network of informers and fences in the Mississippi Valley. Detection and punishment of crime had always been a local function; now there seemed no choice but to call in the federal operatives. In July, 1934, federal agents shot down Dillinger outside a Chicago theater. In October, FBI men killed Pretty Boy Floyd near East Liverpool, Ohio; in November, they shot Baby Face Nelson, Public Enemy No. 1, near Niles Center, Illinois. By the end of 1934, the nation had a new kind of hero: the G-man Melvin Purvis and the chief of the Division of Investigation of the Department of Justice, J. Edgar Hoover. By the end of that year, too, Congress had stipulated that a long list of crimes would henceforth be regarded as federal offenses, including holding up a bank insured by the Federal Deposit Insurance Corporation. The

[22] *Today*, III (January 12, 1935), 4.

family of a kidnaped victim could call in the federal police simply by phoning National 7117 in Washington.[23]

Under the New Deal, the federal government greatly extended its power over the economy. By the end of the Roosevelt years, few questioned the right of the government to pay the farmer millions in subsidies not to grow crops, to enter plants to conduct union elections, to regulate business enterprises from utility companies to air lines, or even to compete directly with business by generating and distributing hydroelectric power. All of these powers had been ratified by the Supreme Court, which had even held that a man growing grain solely for his own use was affecting interstate commerce and hence subject to federal penalties.[24] The President, too, was well on his way to becoming "the chief economic engineer," although this was not finally established until the Full Employment Act of 1946. In 1931, Hoover had hooted that some people thought "that by some legerdemain we can legislate ourselves out of a world-wide depression." In the Roosevelt era, the conviction that government both should and could act to forestall future breakdowns gained general acceptance. The New Deal left a large legacy of antidepression controls—securities regulation, banking reforms, unemployment compensation—even if it could not guarantee that a subsequent administration would use them.[25]

In the 1930's, the financial center of the nation shifted from Wall Street to Washington. In May, 1934, a writer reported: "Financial news no longer originates in Wall Street." That same month, *Fortune* commented on a revolution in the credit system which was "one of the major historical events of the generation." "Mr. Roosevelt," it noted, "seized the Federal Reserve without firing a shot." The federal government had not only broken down the old separation of bank and state in the Reserve system but had gone into the credit business itself in a wholesale fashion under the aegis of the RFC, the Farm Credit

[23] "The Marines Are Coming," *Fortune,* X (August, 1934), 56 ff.; *Literary Digest,* CXVII (May 5, 1934), 9; CXVIII (July 28, 1934), 6; CXVIII (December 8, 1934), 7; *Time,* XXIII (March 12, 1934), 14; *Public Papers,* III, 242 ff.

[24] Wickard *v.* Filburn, 317 U.S. 111 (1942).

[25] Sidney Hyman, *The American President* (New York, 1954), pp. 263–264; Carl Degler, *Out of Our Past* (New York, 1959), pp. 391–393. In a few pages, Degler has written the best analysis of the permanent significance of the New Deal.

Administration, and the housing agencies. Legislation in 1933 and 1934 had established federal regulation of Wall Street for the first time. No longer could the New York Stock Exchange operate as a private club free of national supervision. In 1935, Congress leveled the mammoth holding-company pyramids and centralized yet more authority over the banking system in the federal government. After a tour of the United States in 1935, Sir Josiah Stamp wrote: "Just as in 1929 the whole country was 'Wall Street-conscious' now it is 'Washington-conscious.' "[26]

Despite this encroachment of government on traditional business prerogatives, the New Deal could advance impressive claims to being regarded as a "savior of capitalism." Roosevelt's sense of the land, of family, and of the community marked him as a man with deeply ingrained conservative traits. In the New Deal years, the government sought deliberately, in Roosevelt's words, "to energize private enterprise." The RFC financed business, housing agencies underwrote home financing, and public works spending aimed to revive the construction industry. Moreover, some of the New Deal reforms were Janus-faced. The NYA, in aiding jobless youth, also served as a safety valve to keep young people out of the labor market. A New Deal congressman, in pushing for public power projects, argued that the country should take advantage of the sea of "cheap labor" on the relief rolls. Even the Wagner Act and the movement for industrial unionism were motivated in part by the desire to contain "unbalanced and radical" labor groups. Yet such considerations should not obscure the more important point: that the New Deal, however conservative it was in some respects and however much it owed to the past, marked a radically new departure. As Carl Degler writes: "The conclusion seems inescapable that, traditional as the words may have been in which the New Deal expressed itself, in actuality it was a revolutionary response to a revolutionary situation."[27]

[26] Ferdinand Lundberg, "Wall Street Dances to Washington's Tune," *Literary Digest*, CXVII (May 12, 1934), 46; "Federal Reserve," *Fortune*, IX (May, 1934), 65–66, 125; Sir Josiah Stamp, "Six Weeks in America," *The Times* (London), July 4, 1935.

[27] *Public Papers*, IX, 11; Walter Pierce to Bureau of Publicity, Democratic National Committee, January 4, 1940, Pierce MSS., File 7.1; Robert Wagner to Harold McCollom, April 24, 1935, Wagner MSS.; Sidney Lens, *Left, Right &*

Not all of the changes that were wrought were the result of Roosevelt's own actions or of those of his government. Much of the force for change came from progressives in Congress, or from nongovernmental groups like the C.I.O., or simply from the impersonal agency of the depression itself. Yet, however much significance one assigns the "objective situation," it is difficult to gainsay the importance of Roosevelt. If, in Miami in February, 1933, an assassin's bullet had been true to its mark and John Garner rather than Roosevelt had entered the White House the next month, or if the Roosevelt lines had cracked at the Democratic convention in 1932 and Newton Baker had been the compromise choice, the history of America in the thirties would have been markedly different.

At a time when democracy was under attack elsewhere in the world, the achievements of the New Deal were especially significant. At the end of 1933, in an open letter to President Roosevelt, John Maynard Keynes had written: "You have made yourself the trustee for those in every country who seek to mend the evils of our condition by reasoned experiment within the framework of the existing social system. If you fail, rational change will be gravely prejudiced throughout the world, leaving orthodoxy and revolution to fight it out." In the next few years, teams of foreigners toured the TVA, Russians and Arabs came to study the shelterbelt, French writers taxed Léon Blum with importing "Rooseveltism" to France, and analysts characterized Paul Van Zeeland's program in Belgium as a "New Deal." Under Roosevelt, observed a Montevideo newspaper, the United States had become "as it was in the eighteenth century, the victorious emblem around which may rally the multitudes thirsting for social justice and human fraternity."[28]

In their approach to reform, the New Dealers reflected the tough-

---

Center (Hinsdale, Ill., 1949), pp. 286 ff.; Degler, *Out of Our Past,* p. 416. Not only did the New Deal borrow many ideas and institutions from the Progressive era, but the New Dealers and the progressives shared more postulates and values than is commonly supposed. Nonetheless, the spirit of the 1930's seems to me to be quite different from that of the Progressive era.

[28] *The New York Times,* December 31, 1933; Ludovic Naudeau, "Le Rooseveltisme ou la troisième solution," *L'Illustration,* XCCV (November 28, 1936), 375; Otto Veit, "Franklin Roosevelts Experiment," *Die Neue Rundschau,* XLV (1934), 718–734; Nicholas Halasz, *Roosevelt Through Foreign Eyes* (Princeton, 1961); Donald Dozer, *Are We Good Neighbors?* (Gainesville, 1959), p. 30.

minded, hard-boiled attitude that permeated much of America in the thirties. In 1931, the gangster film *Public Enemy* had given the country a new kind of hero in James Cagney: the aggressive, unsentimental tough guy who deliberately assaulted the romantic tradition. It was a type whose role in society could easily be manipulated; gangster hero Cagney of the early thirties was transformed into G-man hero Cagney of the later thirties. Even more representative was Humphrey Bogart, creator of the "private eye" hero, the man of action who masks his feelings in a calculated emotional neutrality.[29] Bogart, who began as the cold desperado Duke Mantee of *Petrified Forest* and the frightening Black Legionnaire, soon turned up on the right side of anti-Fascist causes, although he never surrendered the pose of noninvolvement. This fear of open emotional commitment and this admiration of toughness ran through the vogue of the "Dead End Kids," films like *Nothing Sacred,* the popularity of the St. Louis Cardinals' spike-flying Gas House Gang, and the "hardboiled" fiction of writers like James Cain and Dashiell Hammett.

Unlike the earlier Progressive, the New Dealer shied away from being thought of as sentimental.[30] Instead of justifying relief as a humanitarian measure, the New Dealers often insisted it was necessary to stimulate purchasing power or to stabilize the economy or to "conserve manpower." The justification for a better distribution of income was neither "social justice" nor a "healthier national life," wrote Adolf Berle. "It remained for the hard-boiled student to work out the simple equation that unless the national income was pretty widely diffused there were not enough customers to keep the plants going."[31] The reformers of the thirties abandoned—or claimed they had abandoned—the old Emersonian hope of reforming man and sought only to change institutions.[32] This meant that they did not seek to "uplift"

[29] Lewis Jacobs, *The Rise of the American Film* (New York, 1939), pp. 509–512; Lincoln Kirstein, "James Cagney and the American Hero," *Hound and Horn,* V (1932), 465–467; Alistair Cooke, *A Generation on Trial* (New York, 1950), p. 11.

[30] A New Deal social worker, obviously moved by something she had seen, would preface her report apologetically: "At the risk of seeming slobbery . . ." Martha Gellhorn to Harry Hopkins, April 25, 1935, Hopkins MSS.

[31] A. A. Berle, Jr., "The Social Economics of the New Deal," *The New York Times Magazine,* October 29, 1933, pp. 4–5.

[32] Edgar Kemler, *The Deflation of American Ideals* (Washington, 1941), p. 69.

the people they were helping but only to improve their economic position. "In other words," Tugwell stated bluntly, "the New Deal is attempting to do nothing to *people,* and does not seek at all to alter their way of life, their wants and desires."[33]

Reform in the 1930's meant *economic* reform; it departed from the Methodist-parsonage morality of many of the earlier Progressives, in part because much of the New Deal support, and many of its leaders, derived from urban immigrant groups hostile to the old Sabbatarianism. While the progressive grieved over the fate of the prostitute, the New Dealer would have placed Mrs. Warren's profession under a code authority. If the archetypical progressive was Jane Addams singing "Onward, Christian Soldiers," the representative New Dealer was Harry Hopkins betting on the horses at Laurel Race Track. When directing FERA in late 1933, Hopkins announced: "I would like to provide orchestras for beer gardens to encourage people to sit around drinking their beer and enjoying themselves. It would be a great unemployment relief measure." "I feel no call to remedy evils," Raymond Moley declared. "I have not the slightest urge to be a reformer. Social workers make me very weary. They have no sense of humor."[34]

Despite Moley's disclaimer, many of the early New Dealers like himself and Adolf Berle did, in fact, hope to achieve reform through regeneration: the regeneration of the businessman. By the end of 1935, the New Dealers were pursuing a quite different course. Instead of attempting to evangelize the Right, they mobilized massive political power against the power of the corporation. They relied not on converting industrial sinners but in using sufficient coercion. New Dealers like Thurman Arnold sought to ignore "moral" considerations altogether; Arnold wished not to punish wrongdoers but to achieve price flexibility. His "faith" lay in the expectation that "fanatical alignments between opposing political principles may disappear and a competent,

[33] Tugwell, *The Battle for Democracy* (New York, 1935), p. 319. "The excuse for us being in this thing," Aubrey Williams explained to NYA leaders, "is that we are trying to reform the structure of things rather than try to reform the people." National Advisory Committee, NYA, Minutes, August 15, 1937, Charles Taussig MSS., Box 6. In a speech to TVA employees, David Lilienthal derided "uplift." George Fort Milton to Lilienthal, July 10, 1936, Milton MSS., Box 20.

[34] *Time,* XXIII (January 1, 1934), 10; XXI (May 8, 1933), 10. Cf. Richard Hofstadter, *The Age of Reform* (New York, 1955), pp. 300–322.

practical, opportunistic governing class may rise to power."[35] With such expectations, the New Dealers frequently had little patience with legal restraints that impeded action. "I want to assure you," Hopkins told the NYA Advisory Committee, "that we are not afraid of exploring anything within the law, and we have a lawyer who will declare anything you want to do legal."[36]

In the thirties, nineteenth-century individualism gave ground to a new emphasis on social security and collective action.[37] In the twenties, America hailed Lindbergh as the Lone Eagle; in the thirties, when word arrived that Amelia Earhart was lost at sea, the *New Republic* asked the government to prohibit citizens from engaging in such "useless" exploits. The NRA sought to drive newsboys off the streets and took a Blue Eagle away from a company in Huck Finn's old town of Hannibal, Missouri, because a fifteen-year-old was found driving a truck for his father's business. Josef Hofmann urged that fewer musicians become soloists, Hollywood stars like Joan Crawford joined the Screen Actors Guild, and Leopold Stokowski canceled a performance in Pittsburgh because theater proprietors were violating a union contract.[38] In New York in 1933, after a series of meetings in Heywood Broun's penthouse apartment, newspapermen organized the American Newspaper Guild in rebellion against the dispiriting romanticism of Richard Harding Davis.[39] "We no longer care to develop the individual as a unique contributor to a democratic form," wrote the mordant Edgar Kemler. "In this movement each individual sub-man is important, not for his uniqueness, but for his ability to lose himself in the mass, through his fidelity to the trade union, or cooperative organization, or political party."[40]

The liberals of the thirties admired intellectual activity which had

[35] Arnold, *Symbols of Government*, pp. 270–271. Cf. Sidney Hook, *Reason, Social Myths and Democracy* (New York, 1940), pp. 41–61.

[36] National Advisory Committee, NYA, Minutes, August 15, 1935, Charles Taussig MSS., Box 6.

[37] In Ernest Hemingway's *To Have and Have Not*, Harry Morgan says: "No matter how a man alone ain't got no bloody . . . chance." Hemingway, *To Have and Have Not* (New York, 1937), p. 225.

[38] *New Republic*, XCI (1937), 262; *Time*, XXIII (February 19, 1934), 14; Herbert Harris, *American Labor* (New Haven, 1938), p. 175.

[39] Cf. *Editor and Publisher*, LXVI (December 23, 1933), 7, 28.

[40] Kemler, *Deflation of Ideals*, pp. 109–110. Kemler, it hardly need be said, grossly overstated his argument.

a direct relation to concrete reality. Stuart Chase wrote of one government report: "This book is live stuff—wheelbarrow, cement mixer,
steam dredge, generator, combine, power-line stuff; library dust does
not gather here."[41] If the poet did not wish to risk the suspicion that
his loyalties were not to the historic necessities of his generation, wrote
Archibald MacLeish, he must "soak himself not in books" but in the
physical reality of "by what organization of men and railroads and
trucks and belts and book-entries the materials of a single automobile
are assembled."[42] The New Dealers were fascinated by "the total man
days per year for timber stand improvement," and Tugwell rejoiced
in the "practical success" of the Resettlement Administration demonstrated by "these healthy collection figures." Under the Special Skills
Division of the RA, Greenbelt was presented with inspirational paintings like *Constructing Sewers, Concrete Mixer,* and *Shovel at Work.*
On one occasion, in attempting to mediate a literary controversy, the
critic Edmund Wilson wrote: "It should be possible to convince Marx-
ist critics of the importance of a work like 'Ulysses' by telling them that
it is a great piece of engineering—as it is."[43] In this activist world of
the New Dealers, the aesthete and the man who pursued a life of
contemplation, especially the man whose interests centered in the past,
were viewed with scorn. In Robert Sherwood's *The Petrified Forest,*
Alan Squier, the ineffectual aesthete, meets his death in the desert and
is buried in the petrified forest where the living turn to stone. He is an
archaic type for whom the world has no place.

The new activism explicitly recognized its debt to Dewey's dictum
of "learning by doing" and, like other of Dewey's ideas, was subject to
exaggeration and perversion. The New Deal, which gave unprecedented authority to intellectuals in government, was, in certain important respects, anti-intellectual. Without the activist faith, perhaps not
nearly so much would have been achieved. It was Lilienthal's convic-

[41] Stuart Chase, "Old Man River," *New Republic,* LXXXII (1935), 175.
"I speak in dispraise of dusty learning, and in disparagement of the historical
technique," declared Tugwell. "Are our plans wrong? Who knows? Can we tell
from reading history? Hardly." Tugwell, *Battle for Democracy,* pp. 70–71.

[42] MacLeish, *A Time to Speak* (Boston, 1941), p. 45. Cf. MacLeish, "The
Social Cant," *New Republic,* LXXIII (1932), 156–158.

[43] *The New York Times,* October 29, 1936; Paul Conkin, *Tomorrow a New
World* (Ithaca, 1959), p. 196; Edmund Wilson, "The Literary Class War: I,"
New Republic, LXX (1932), 323.

tion that "there is almost nothing, however fantastic, that (given competent organization) a team of engineers, scientists, and administrators cannot do today" that helped make possible the successes of TVA.[44] Yet the liberal activists grasped only a part of the truth; they retreated from conceptions like "tragedy," "sin," "God," often had small patience with the force of tradition, and showed little understanding of what moved men to seek meanings outside of political experience. As sensitive a critic as the poet Horace Gregory could write, in a review of the works of D. H. Lawrence: "The world is moving away from Lawrence's need for personal salvation; his 'dark religion' is not a substitute for economic planning."[45] This was not the mood of all men in the thirties—not of a William Faulkner, an Ellen Glasgow—and many of the New Dealers recognized that life was more complex than some of their statements would suggest. Yet the liberals, in their desire to free themselves from the tyranny of precedent and in their ardor for social achievement, sometimes walked the precipice of superficiality and philistinism.

The concentration of the New Dealers on public concerns made a deep mark on the sensibility of the 1930's. Private experience seemed self-indulgent compared to the demands of public life. "Indeed the public world with us has *become* the private world, and the private world has become the public," wrote Archibald MacLeish. "We live, that is to say, in a revolutionary time in which the public life has washed in over the dikes of private existence as sea water breaks over into the fresh pools in the spring tides till everything is salt."[46] In the thirties, the Edna St. Vincent Millay whose candle had burned at both ends wrote the polemical *Conversation at Midnight* and the bitter "Epitaph for the Race of Man" in *Wine From These Grapes*.

The emphasis on the public world implied a specific rejection of the

[44] David Lilienthal, *TVA: Democracy on the March* (New York, 1944), p. 3.

[45] Waldo Frank, "Our Guilt in Fascism," *New Republic,* CII (1940), 603–608; Murray Kempton, *Part of Our Time* (New York, 1955); p. 2; Cushing Strout, "The Twentieth-Century Enlightenment," *American Political Science Review,* XLIX (1955), 321–339; Morton White, *Social Thought in America* (New York, 1949), p. 241; *New Republic,* LXXIII (1932), 133.

[46] MacLeish mocked "the nineteenth-century poet, the private speaker, the whisperer to the heart, the unworldly romantic, the quaint Bohemian, the understander of women, the young man with the girl's eyes." MacLeish, *A Time to Speak,* pp. 62, 88.

values of the 1920's. Roosevelt dismissed the twenties as "a decade of debauch," Tugwell scored those years as "a decade of empty progress, devoid of contribution to a genuinely better future," Morris Cooke deplored the "gilded-chariot days" of 1929, and Alben Barkley saw the twenties as a "carnival" marred by "the putrid pestilence of financial debauchery."[47] The depression was experienced as the punishment of a wrathful God visited on a nation that had strayed from the paths of righteousness.[48] The fire that followed the Park Avenue party in Thomas Wolfe's *You Can't Go Home Again*, like the suicide of Eveline at the end of John Dos Passos' *The Big Money*, symbolized the holocaust that brought to an end a decade of hedonism.[49] In an era of reconstruction, the attitudes of the twenties seemed alien, frivolous, or —the most cutting word the thirties could visit upon a man or institution—"escapist." When Morrie Ryskind and George Kaufman, authors of the popular *Of Thee I Sing*, lampooned the government again in *Let 'em Eat Cake* in the fall of 1933, the country was not amused. The New York *Post* applauded the decision of George Jean Nathan and his associates to discontinue the *American Spectator:* "Nihilism, dadaism, smartsetism—they are all gone, and this, too, is progress."[50] One of H. L. Mencken's biographers has noted: "Many were at pains to write him at his new home, telling him he was a sophomore, and those writing in magazines attacked him with a fury that was suspect because of its very violence."[51]

[47] *Public Papers*, V, 179; Tugwell, *Battle for Democracy*, p. 54; Morris Cooke to Louis Howe, July 3, 1933, Cooke MSS., Box 51; *Literary Digest*, CXXII (July 4, 1936), 27.

[48] "We were all miserable sinners," announced the Harvard economist Oliver Sprague. *The New York Times*, December 10, 1933. Cf. "Special Week of Penitence and Prayer, October 2–8," *Federal Council Bulletin*, XV (September, 1932), 14–15; Milton Garber, Radio Address, 1932, Garber MSS.

[49] Melvin Landsberg, "A Study of the Political Development of John Dos Passos from 1912 to 1936" (unpublished Ph.D. dissertation, Columbia University, 1959).

[50] Hiram Motherwell, "Political Satire Scores on the Stage," *Today*, II (July 28, 1934), 24; *The New York Times*, November 12, 1933; New York *Post, Press Time* (New York, 1936), pp. 317 ff.

[51] William Manchester, *Disturber of the Peace* (New York, 1951), p. 258. "Empty stomachs became more important than hurt sensibilities," commented one critic. "The vexations and hurt feelings of a Carol Kennecott or the spiritual frustration of an Amory Blaine in *This Side of Paradise*, or the frantic dash for personal freedom of a Janet Marsh seemed trifling themes when the

Commentators on the New Deal have frequently characterized it by that much-abused term "pragmatic." If one means by this that the New Dealers carefully tested the consequences of ideas, the term is clearly a misnomer. If one means that Roosevelt was exceptionally anti-ideological in his approach to politics, one may question whether he was, in fact, any more "pragmatic" in this sense than Van Buren or Polk or even "reform" Presidents like Jackson and Theodore Roosevelt. The "pragmatism" of the New Deal seemed remarkable only in a decade tortured by ideology, only in contrast to the rigidity of Hoover and of the Left.

The New Deal was pragmatic mainly in its skepticism about utopias and final solutions, its openness to experimentation, and its suspicion of the dogmas of the Establishment. Since the advice of economists had so often been wrong, the New Dealers distrusted the claims of orthodox theory—"All this is perfectly terrible because it is all pure theory, when you come down to it," the President said on one occasion—and they felt free to try new approaches.[52] Roosevelt refused to be awed by the warnings of economists and financial experts that government interference with the "laws" of the economy was blasphemous. "We must lay hold of the fact that economic laws are not made by nature," the President stated. "They are made by human beings."[53] The New Dealers denied that depressions were inevitable events that had to be borne stoically, most of the stoicism to be displayed by the most impoverished, and they were willing to explore novel ways to make the social order more stable and more humane. "I am for experimenting . . . in various parts of the country, trying out schemes which are supported by reasonable people and see if they work," Hopkins told a conference of social

dominant feature of the national scene was twelve million unemployed." Halford Luccock, *American Mirror* (New York, 1940), p. 36. Cf. Margaret Mitchell to Mrs. Julian Harris, April 28, 1936, Mitchell MSS.

[52] *Public Papers*, II, 269. "In the months and years following the stock market crash," Professor Galbraith has concluded, "the burden of reputable economic advice was invariably on the side of measures that would make things worse." John Kenneth Galbraith, *The Great Crash* (Boston, 1955), pp. 187–188. For a typical example, see N. S. B. Gras to Edward Costigan, July 22, 1932, Costigan MSS., V.F. 1.

[53] *Public Papers*, I, 657. The Boston *Transcript* commented: "Two more glaring misstatements of the truth could hardly have been packed into so little space." J. Joseph Huthmacher, *Massachusetts People and Politics, 1919–1933* (Cambridge, 1959), p. 244. Cf. Eccles, *Beckoning Frontiers*, p. 73.

workers. "If they do not work, the world will not come to an end."[54]

Hardheaded, "anti-utopian," the New Dealers nonetheless had their Heavenly City: the greenbelt town, clean, green, and white, with children playing in light, airy, spacious schools; the government project at Longview, Washington, with small houses, each of different design, colored roofs, and gardens of flowers and vegetables; the Mormon villages of Utah that M. L. Wilson kept in his mind's eye—immaculate farmsteads on broad, rectangular streets; most of all, the Tennessee Valley, with its model town of Norris, the tall transmission towers, the white dams, the glistening wire strands, the valley where "a vision of villages and clean small factories has been growing into the minds of thoughtful men."[55] Scandinavia was their model abroad, not only because it summoned up images of the countryside of Denmark, the beauties of Stockholm, not only for its experience with labor relations and social insurance and currency reform, but because it represented the "middle way" of happy accommodation of public and private institutions the New Deal sought to achieve. "Why," inquired Brandeis, "should anyone want to go to Russia when one can go to Denmark?"[56]

Yet the New Deal added up to more than all of this—more than an experimental approach, more than the sum of its legislative achievements, more than an antiseptic utopia. It is true that there was a certain erosion of values in the thirties, as well as a narrowing of horizons, but the New Dealers inwardly recognized that what they were doing had a deeply moral significance however much they eschewed ethical pretensions. Heirs of the Enlightenment, they felt themselves part of a broadly humanistic movement to make man's life on earth more tolerable, a movement that might someday even achieve a co-operative

[54] *Public Papers*, II, 302; V, 497; Josephine Chapin Brown, *Public Relief*, p. 152. Cf. Clarke Chambers, "FDR, Pragmatist-Idealist," *Pacific Northwest Quarterly*, LII (1961), 50–55; F. S. C. Northrop, *The Meeting of East and West* (New York, 1947), p. 152; Jacob Cohen, "Schlesinger and the New Deal," *Dissent*, VIII (1961), 466–468.

[55] Tugwell, *Battle for Democracy*, p. 22. This is a vision caught, in different ways, in the paintings of Paul Sample, Charles Sheeler, Grant Wood, and Joe Jones.

[56] Marquis Childs, *Sweden: The Middle Way* (New Haven, 1936); David Lilienthal to George Fort Milton, July 9, 1936; Milton to F.D.R., July 8, 1936, Milton MSS., Box 20; Irving Fisher to F.D R., September 28, 1934, Fisher MSS.; John Commons to Edward Costigan, July 25, 1932, Costigan MSS., V.F. 1; Arthur Schlesinger, Jr., *The Politics of Upheaval* (Boston, 1960), p. 221.

commonwealth. Social insurance, Frances Perkins declared, was "a fundamental part of another great forward step in that liberation of humanity which began with the Renaissance."[57]

Franklin Roosevelt did not always have this sense as keenly as some of the men around him, but his greatness as a President lies in the remarkable degree to which he shared the vision. "The new deal business to me is very much bigger than anyone yet has expressed it," observed Senator Elbert Thomas. Roosevelt "seems to really have caught the spirit of what one of the Hebrew prophets called the desire of the nations. If he were in India today they would probably decide that he had become Mahatma—that is, one in tune with the infinite."[58] Both foes and friends made much of Roosevelt's skill as a political manipulator, and there is no doubt that up to a point he delighted in schemes and stratagems. As Donald Richberg later observed: "There would be times when he seemed to be a Chevalier Bayard, *sans peur et sans reproche,* and times in which he would seem to be the apotheosis of a prince who had absorbed and practiced all the teachings of Machiavelli." Yet essentially he was a moralist who wanted to achieve certain humane reforms and instruct the nation in the principles of government. On one occasion, he remarked: "I want to be a *preaching President*—like my cousin."[59] His courtiers gleefully recounted his adroitness in trading and dealing for votes, his effectiveness on the stump, his wicked skill in cutting corners to win a point. But Roosevelt's importance lay not in his talents as a campaigner or a manipulator. It lay rather in his ability to arouse the country and, more specifically, the men who served under him, by his breezy encouragement of experimentation, by his hopefulness, and—a word that would have embarrassed some of his lieutenants—by his idealism.

The New Deal left many problems unsolved and even created some perplexing new ones. It never demonstrated that it could achieve prosperity in peacetime. As late as 1941, the unemployed still numbered six million, and not until the war year of 1943 did the army of the jobless

---

[57] Frances Perkins, "Basic Idea Behind Social Security Program," *The New York Times,* January 27, 1935.

[58] Thomas to Colonel E. LeRoy Bourne, January 6, 1934, Elbert Thomas MSS.

[59] Donald Richberg, *My Hero* (New York, 1954), p. 279; Schlesinger, *Coming of New Deal* (Boston, 1959), p. 558.

finally disappear. It enhanced the power of interest groups who claimed to speak for millions, but sometimes represented only a small minority.[60] It did not evolve a way to protect people who had no such spokesmen, nor an acceptable method for disciplining the interest groups. In 1946, President Truman would resort to a threat to draft railway workers into the Army to avert a strike. The New Deal achieved a more just society by recognizing groups which had been largely unrepresented—staple farmers, industrial workers, particular ethnic groups, and the new intellectual-administrative class. Yet this was still a halfway revolution; it swelled the ranks of the bourgeoisie but left many Americans—sharecroppers, slum dwellers, most Negroes—outside of the new equilibrium.

Some of these omissions were to be promptly remedied. Subsequent Congresses extended social security, authorized slum clearance projects, and raised minimum-wage standards to keep step with the rising price level. Other shortcomings are understandable. The havoc that had been done before Roosevelt took office was so great that even the unprecedented measures of the New Deal did not suffice to repair the damage. Moreover, much was still to be learned, and it was in the Roosevelt years that the country was schooled in how to avert another major depression. Although it was war which freed the government from the taboos of a balanced budget and revealed the potentialities of spending, it is conceivable that New Deal measures would have led the country into a new cycle of prosperity even if there had been no war. Marked gains had been made before the war spending had any appreciable effect. When recovery did come, it was much more soundly based because of the adoption of the New Deal program.

Roosevelt and the New Dealers understood, perhaps better than their critics, that they had come only part of the way. Henry Wallace remarked: "We are children of the transition—we have left Egypt but we have not yet arrived at the Promised Land." Only five years separated Roosevelt's inauguration in 1933 and the adoption of the last of the New Deal measures, the Fair Labor Standards Act, in 1938. The New Dealers perceived that they had done more in those years than had been done in any comparable period in American history, but they also saw that there was much still to be done, much, too, that continued to baffle them. "I believe in the things that have been done," Mrs.

[60] Henry Kariel, *The Decline of American Pluralism* (Stanford, 1961).

Roosevelt told the American Youth Congress in February, 1939. "They helped but they did not solve the fundamental problems. . . . I never believed the Federal government could solve the whole problem. It bought us time to think." She closed not with a solution but with a challenge: "Is it going to be worth while?"[61]

"This generation of Americans is living in a tremendous moment of history," President Roosevelt stated in his final national address of the 1940 campaign.

"The surge of events abroad has made some few doubters among us ask: Is this the end of a story that has been told? Is the book of democracy now to be closed and placed away upon the dusty shelves of time?

"My answer is this: All we have known of the glories of democracy— its freedom, its efficiency as a mode of living, its ability to meet the aspirations of the common man—all these are merely an introduction to the greater story of a more glorious future.

"We Americans of today—all of us—we are characters in the living book of democracy.

"But we are also its author. It falls upon us now to say whether the chapters that are to come will tell a story of retreat or a story of continued advance."[62]

[61] Henry Wallace, *The Christian Bases of World Order* (New York, 1943), p. 17; Dorothy Dunbar Bromley, "The Future of Eleanor Roosevelt," *Harper's,* CLXXX (1939), 136.
[62] *Public Papers,* IX, 545.

Bibliography

MANUSCRIPTS

ROOSEVELT AND THE NEW DEALERS

The most important source is the Franklin D. Roosevelt MSS. at the Franklin D. Roosevelt Library (FDRL) at Hyde Park, New York. The major subdivisions of the collection are Official File (OF), President's Personal File (PPF), and President's Secretary's File (PSF). Most of the papers of Roosevelt's first cabinet are still closed, but I was granted access to the Frances Perkins memoir at the Columbia Oral History Collection (COHC). The George Dern MSS. at the Library of Congress (LC) and the Claude Swanson MSS., Duke University, are unrevealing. For the Roosevelt circle and the New Deal administrators, I found most helpful the Thurman Arnold MSS., University of Wyoming; Wendell Berge MSS., LC; Holger Cahill's tape recording at the Archives of American Art, Detroit Institute of Arts; Holger Cahill, COHC; John Carmody MSS., FDRL; Morris Cooke MSS., FDRL; Democratic National Committee MSS., FDRL; Mary Dewson MSS., FDRL; Charles Fahy, COHC; Edward Flynn, COHC; Leon Henderson MSS., FDRL; Harry Hopkins MSS., FDRL; Louis Howe MSS., FDRL; Jesse Jones, LC; Lowell Mellett MSS., FDRL; J. F. T. O'Connor MSS., University of California, Berkeley; George Peek MSS., University of Missouri; Georgia Lindsey Peek MSS., University of Missouri; Donald Richberg MSS., LC; James Roosevelt MSS., FDRL; J. D. Ross MSS., University of Washington, Seattle; Harry Slattery MSS., Duke University; Charles Taussig, MSS., FDRL; Rexford Tugwell MSS., FDRL; and the Aubrey Williams MSS., FDRL.

CONGRESS

Papers of congressmen vary greatly in quality, from huge, flatulent collections like the Tom Connally papers to invaluable collections like the Carter Glass manuscripts. Even the best of the collections contain an inordinate amount of correspondence over post-office patronage, but even the least rewarding include some items of interest. Among the papers of members of Congress which I consulted were the Josiah Bailey MSS., Duke University; Graham Barden MSS., Duke; Alben Barkley MSS., University of Kentucky; William Borah MSS., LC; Lyle Boren MSS., University of Oklahoma; Jonathan Bourne, Jr., MSS., University of Oregon; Joseph Bryson MSS., Duke; Arthur Capper MSS., Kansas State Historical Society, Topeka; Wilburn Cartwright MSS., Oklahoma; Tom Connally MSS., LC; Edward Costigan MSS., University of Colorado; James J. Davis MSS., LC; Robert Doughton MSS., University of North Carolina; John Erickson MSS., LC; Henry Prather Fletcher MSS., LC; Milton Garber MSS., Oklahoma; Joseph Gavagan, COHC; Carter Glass MSS., University of Virginia; Thomas Gore MSS., Oklahoma; Clifford Hope MSS., Kansas; Hiram Johnson MSS., University of California, Berkeley; Jed Johnson MSS., Oklahoma; Eugene Keogh, COHC; John Hosea Kerr MSS., North Carolina; Fiorello La Guardia MSS., Municipal Archives, New York City; James Hamilton Lewis MSS., LC; William McAdoo MSS., LC; Pat McCarran MSS., Nevada State Museum, Carson City; Charles McNary MSS., LC; Charles McNary MSS., Oregon; Samuel McReynolds MSS., North Carolina; James Murray MSS., Montana State University, Missoula; Jack Nichols MSS., Oklahoma; George Norris MSS., LC; Joseph O'Mahoney MSS., Wyoming; Robert Owen MSS., LC; Walter Pierce MSS., Oregon; Key Pittman MSS., LC; Henry Rainey MSS., LC; Lewis Schwellenbach MSS., LC; Lewis Schwellenbach MSS., University of Washington; James Scrugham MSS., Nevada State Historical Society, Reno; T. V. Smith MSS., University of Chicago; Frederick Steiwer MSS., Oregon; Robert Taft MSS., Yale University; Elbert Thomas MSS., FDRL; Elmer Thomas MSS., Oklahoma; Harry Truman MSS., Truman Library, Independence, Missouri; James Wadsworth, COHC; Robert Wagner MSS., Georgetown University; David I. Walsh MSS., Holy Cross Library, Worcester, Massachusetts; Thomas Walsh MSS., LC; Lindsay Warren MSS., North Carolina; and Wallace White, Jr., MSS., LC.

HOOVER AND THE CONSERVATIVE OPPOSITION

The Herbert Hoover MSS. are still closed, although it is anticipated that they will be opened shortly. Save for some of the congressional papers listed above, there are very few papers of Republican leaders of this period now available to scholars. The more significant papers are the Frank Knox MSS., LC; Alfred Landon MSS., Kansas; Agnes Meyer MS. Diary, Columbia University; Eugene Meyer MSS., Columbia; Ogden Mills MSS., LC; Gifford

Pinchot MSS., LC; Chester Rowell MSS., University of California, Berkeley; Henry Stimson MSS., Yale; Henry Stimson, COHC; and the Wendell Willkie MSS., Yale. Neither the Willkie nor Taft papers at Yale represent the main body of these collections, both of which are still closed. Conservative opposition to the New Deal outside the Republican party leadership may be studied in the James Truslow Adams MSS., Columbia; James Babb MSS., Yale; William Wilson Cumberland, COHC; and the Amos Pinchot MSS., LC.

For noncongressional conservative opposition to the New Deal within the Democratic party, the most fruitful collections are the Newton Baker MSS., LC; W. W. Ball MSS., Duke; Stephen Chadwick MSS., University of Washington; Bainbridge Colby MSS., LC; George Creel MSS., LC; John W. Davis, COHC; Thomas Lomax Hunter MSS., Virginia; David C. Reay MSS., privately held; and the Jouett Shouse MSS., University of Kentucky.

SUPREME COURT

Since Supreme Court justices rarely discuss cases in their correspondence, most of their papers are unrevealing. An exception is the Louis Brandeis collection, School of Law, University of Louisville, which does not have much of interest on litigation, but which contains the fascinating letters Felix Frankfurter wrote to Brandeis in an almost indecipherable scrawl on long sheets of yellow paper. Of marginal value are the Benjamin Cardozo MSS., Columbia (most of the Cardozo papers are reported to have been destroyed); Charles Evans Hughes MSS., LC; James Clark McReynolds MSS., Virginia; and the George Sutherland MSS., LC.

ECONOMIC POLICY

The best single collection is the George Leslie Harrison MSS., Columbia University; Harrison dictated frequent memos on the business of the day of the Federal Reserve Bank of New York. Other papers of varying interest are the Leonard Ayres MSS., LC; Irving Fisher MSS., Yale; Emanuel Goldenweiser MSS., LC; Charles Sumner Hamlin MSS., LC; Jackson Reynolds, COHC; James Harvey Rogers MSS., Yale; Gerard Swope, COHC; and the Gerard Swope MSS., Columbia.

AGRICULTURE

In addition to collections like the Peek MSS., listed above, the most rewarding papers were the W. W. Alexander memoir in the COHC; Louis Bean, COHC; Xenophon Caverno MSS., Missouri; George P. Clements MSS., University of California, Los Angeles; Chester Davis, COHC; Mordecai Ezekiel, COHC; Jerome Frank, COHC; Lyle Hague MSS., Oklahoma; William Hirth MSS., Missouri; John Hutson, COHC; H. L. Mitchell, COHC; John

Simpson MSS., Oklahoma; O. C. Stine, COHC; Louis Taber, COHC; Frank Tannenbaum MSS., privately held; and the M. L. Wilson memoir, COHC.

LABOR

The most interesting single collection was the Oral History interviews on automobile unionism at the Wayne State University Library, Detroit. Other helpful collections were the John Brophy memoir, COHC; Ewan Clague, COHC; Katherine Edson MSS., UCLA; Julius Emspak, COHC; John Frey MSS., LC; John Frey, COHC; Pelham Glassford MSS., UCLA; William Jett Lauck MSS., Virginia; Homer Martin MSS., Wayne; Wayne Morse MSS., Oregon; W. A. Pat Murphy collection, Oklahoma; W. B. Pine MSS., Oklahoma; and William Pollock, COHC.

PROGRESSIVE AND RADICAL SENTIMENT

Much of this can be traced in collections like the Simpson MSS. listed elsewhere in this essay. Of special interest are the Francis Heney MSS., University of California, Berkeley; Judson King MSS., LC; Robert Morss Lovett MSS., Chicago; Progressive National Committee MSS., LC; Socialist Party MSS., Duke; Lincoln Steffens MSS., Columbia; Norman Thomas MSS., New York Public Library; Townsend Plan Organization, 12th Regional District MSS., UCLA; Mary Van Kleeck MSS., Smith College; Lillian Wald MSS., New York Public Library; Frank Walsh MSS., New York Public Library.

FOREIGN AFFAIRS

The Henry Stimson MSS., Yale, have been noted above. The Cordell Hull MSS., LC, are closed for this period. Collections of value for foreign policy include the Hugh Byas MSS., Yale; Wilbur Carr MSS., LC; Josephus Daniels MSS., LC; Malcolm Davis, COHC; Norman Davis MSS., LC; William Dodd MSS., LC; W. Cameron Forbes MSS., Harvard University; Joseph Grew MSS., Harvard; Samuel Harper MSS., Chicago; Florence Jaffray (Mrs. J. Borden) Harriman MSS., LC; Edward House MSS., Yale; Nelson Trusler Johnson MSS., LC; Breckenridge Long MSS., LC; Jay Pierrepont Moffat MSS., Harvard; R. Walton Moore, MSS., FDRL; Herbert Claiborne Pell MSS., FDRL; William Phillips MSS., Harvard; William Phillips, COHC; and the James Warburg memoir, COHC.

For pacifist sentiment, the best single source is the Swarthmore College Peace Collection, which includes the Jane Addams MSS., American Peace Mobilization MSS., Emily Balch MSS., Henry Wadsworth Dana MSS., Hannah Clothier Hull and William I. Hull MSS., National Committee for the War Referendum MSS., National Council for the Prevention of War MSS., Sydney Strong MSS., and the Lydia Wentworth MSS. Also of value for the peace movement are the Carrie Chapman Catt MSS. and the Josephine Schain MSS. at Smith College, and the Salmon Levinson MSS. at the University of Chicago.

STATE POLITICS

Of particular interest for the "little New Deals" and state politics in the 1930's are the Wilbur Cross MSS., Yale; James Michael Curley Scrapbooks, Holy Cross; Richard Dillon MSS., University of New Mexico; Holm Bursum MSS., New Mexico; George Earle III MSS., Bryn Mawr, Pennsylvania (privately held); Charles Hay MSS., Missouri; Henry Johnston MSS., Oklahoma; Arthur Langlie MSS., University of Washington; and the Miguel Otero MSS., New Mexico. The Herbert Lehman MSS. are closed, but I had access to some of the material in interviewing the Senator for the Columbia Oral History Collection.

JOURNALISTS

The papers of journalists were, without exception, of high merit; their diaries were especially lively sources. The more important collections are the Frederick Lewis Allen MSS., LC; Raymond Clapper MSS., LC; Bernard DeVoto MSS., Stanford University; Henry Morrow Hyde MSS., Virginia; Arthur Krock, COHC; George Fort Milton MSS., LC; Fremont Older MSS., University of California, Berkeley; Robert Sawyer MSS., Oregon; J. David Stern, COHC; Dorothy Thompson MSS., Yale; Oswald Garrison Villard MSS., Harvard; William Allen White MSS., Columbia, and the major collection of the White MSS. at the Library of Congress.

MISCELLANY

Other papers of interest for this period include the George Biddle MSS., Philadelphia Museum of Art (on microfilm at the Archives of American Art, Detroit Institute of Arts); Bishop James Cannon, Jr., MSS., Duke; Edward Dickson MSS., UCLA; Burton Emmett MSS., North Carolina; Ellen Glasgow MSS., Smith; Franklin Hichborn MSS., UCLA; Emil Hurja MSS., FDRL; Benjamin Marsh MSS., LC; Langdon Marvin, COHC; Margaret Mitchell MSS., Smith; Homer Rainey MSS., Missouri; Henry Sheldon MSS., Oregon; Forbes Watson MSS., Archives of American Art, Detroit; and the Robert Woolley MSS., LC.

PUBLISHED WORKS

The following account is a highly selective discussion of major books and monographs, and makes no attempt to be exhaustive, nor to list all of the books previously cited in the footnotes.

GENERAL HISTORIES

Arthur M. Schlesinger, Jr.'s *The Age of Roosevelt* (Boston, 1957–60) is the major interpretation of the New Deal. The three volumes that have thus far appeared go down through 1936, but do not discuss foreign policy, save for the London Conference. The second volume, *The Coming of the New Deal*, is a particularly valuable contribution. The other volumes are entitled *The Crisis of the Old Order, 1919–1933*, and *The Politics of Upheaval*. The thirties are surveyed, in whole or in part, in Charles and Mary Beard, *America in Midpassage* (2 vols., New York, 1939); Carl Degler, *Out of Our Past* (New York, 1959); The Editors of the Economist, *The New Deal; An Analysis and Appraisal* (New York, 1937); Mario Einaudi, *The Roosevelt Revolution* (New York, 1959); Eric Goldman, *Rendezvous with Destiny* (New York, 1952); Walter Johnson, *1600 Pennsylvania Avenue* (Boston, 1960); Ernest K. Lindley, *Half Way with Roosevelt* (New York, 1936), and *The Roosevelt Revolution: First Phase* (New York, 1933); Dexter Perkins, *The New Age of Franklin Roosevelt 1932–45* (Chicago, 1957); Edgar E. Robinson, *The Roosevelt Leadership, 1933–1945* (Philadelphia, 1955); and Dixon Wecter, *The Age of the Great Depression 1929–1941* (New York, 1948).

The best account of the depression is Irving Bernstein, *The Lean Years* (Boston, 1960). Robert and Helen Lynd's *Middletown in Transition* (New York, 1937) is indispensable for society in the 1930's and there is important material, too, in a number of the Warner studies. Other accounts of value are Clarence Enzler, *Some Social Aspects of the Depression* (Washington, 1939); Federal Writers' Project, *These Are Our Lives* (Chapel Hill, 1939); William Ogburn (ed.), *Social Changes During Depression and Recovery* (Chicago, 1935); and Gilbert Seldes, *The Years of the Locust* (Boston, 1933).

ROOSEVELT AND HIS CIRCLE

Frank Freidel is writing the authoritative biography of Roosevelt. The three volumes that have thus far appeared—*Franklin D. Roosevelt: The Apprenticeship; Franklin D. Roosevelt: The Ordeal; Franklin D. Roosevelt: The Triumph* (Boston, 1952–56)—carry the story through the election of 1932. The two best single-volume studies of F.D.R. are James MacGregor Burns, *Roosevelt: The Lion and the Fox* (New York, 1956), a spirited critique, and Rexford Tugwell, *The Democratic Roosevelt* (Garden City, N.Y., 1957), a thoughtful appraisal which emphasizes both Roosevelt's greatness and his error in moving from a collectivistic to an atomistic policy. Tugwell's analysis also appeared in articles in the *Western Political Quarterly* (June, 1948), *Political Quarterly* (July–September, 1950), *Antioch Review* (December, 1953), and *Ethics* (1953–54). Of the personal memoirs about Roosevelt, Frances Perkins, *The Roosevelt I Knew* (New York, 1946), is already a classic. Roosevelt's own public papers and selective correspondence may be found in Samuel Rosenman (ed.), *The Public Papers and Addresses of Franklin D. Roosevelt* (13 vols., New York, 1938–50), and Elliott Roosevelt, *F.D.R.: His Personal Letters, 1928–1945* (2 vols., New York, 1950).

Of the scores of studies written about Roosevelt, the following are among the more important: Bernard Bellush, *Franklin D. Roosevelt as Governor of New York* (New York, 1955); Daniel Fusfeld, *The Economic Thought of Franklin D. Roosevelt and the Origins of the New Deal* (New York, 1956); Harold Gosnell, *Champion Campaigner: Franklin D. Roosevelt* (New York, 1952); Thomas Greer, *What Roosevelt Thought* (East Lansing, 1958); John Gunther, *Roosevelt in Retrospect* (New York, 1950); the essay on Roosevelt in Richard Hofstadter, *The American Political Tradition* (New York, 1948); Gerald Johnson, *Roosevelt: Dictator or Democrat?* (New York, 1941); Carroll Kilpatrick (ed.), *Roosevelt and Daniels* (Chapel Hill, 1952); James Roosevelt and Sidney Shalett, *Affectionately, F.D.R.* (New York, 1959); Samuel Rosenman, *Working with Roosevelt* (New York, 1952); and Grace Tully, *F.D.R. My Boss* (New York, 1949). Miss Tully, Roosevelt's secretary, 'has commented: "In reading some of the books my friends have written about Franklin Delano Roosevelt, I have been struck by the fact that in the second or third chapter the hero of the story often turns out to be the author, while Roosevelt vanishes from sight or is reduced to a small walk-on part."

Of the making of memoirs by New Dealers there is no end. Many have only ephemeral interest, but some are of lasting value. Raymond Moley, *After Seven Years* (New York, 1939), is much the best of the conservative critiques. Harold Ickes has revealed himself in *The Secret Diary of Harold Ickes* (3 vols., New York, 1954), *The Autobiography of a Curmudgeon* (New York, 1948), and in a series of eight articles in the *Saturday Evening Post* in 1948. John Morton Blum, *From the Morgenthau Diaries* (Boston, 1959), is a work of exceptional interest. Eleanor Roosevelt, *This I Remember* (New York, 1949); Marriner Eccles, *Beckoning Frontiers* (New York, 1951); and Louis Brownlow, *A Passion for Anonymity* (Chicago, 1958), are particularly discerning. The New Dealers are analyzed in Joseph Alsop and Robert Kintner, *Men Around the President* (New York, 1939); "Unofficial Observer" [John Franklin Carter], *The New Dealers* (New York, 1934); Robert E. Sherwood, *Roosevelt and Hopkins* (New York, 1948); Alfred Rollins, Jr., *Roosevelt and Howe* (New York, 1962); and Lela Stiles, *The Man Behind Roosevelt* (Cleveland, 1954).

HOOVER AND THE CONSERVATIVES

Hoover's views are set forth in his *Memoirs* (3 vols., New York, 1951–52); William Starr Myers (ed.), *The State Papers and Other Public Writings of*

Herbert Hoover (2 vols., New York, 1934); Ray Lyman Wilbur and Arthur Mastick Hyde, *The Hoover Policies* (New York, 1937); and in a succession of books written after Hoover's tenure in the White House. William Starr Myers and Walter Newton, *The Hoover Administration* (New York, 1936); Harris Warren, *Herbert Hoover and the Great Depression* (New York, 1959); Theodore Joslin, *Hoover Off the Record* (Garden City, N.Y., 1934); and Edgar Eugene Robinson and Paul Carroll Edwards (eds.), *The Memoirs of Ray Lyman Wilbur, 1875–1949* (Stanford, 1960), survey the Hoover Presidency. Important for conservatism in the 1930's are George Creel, *Rebel at Large* (New York, 1947); Morton Keller, *In Defense of Yesterday* (New York, 1958); George Peek, with Samuel Crowther, *Why Quit Our Own* (New York, 1936); Louis Wehle, *Hidden Threads of History* (New York, 1953); and George Wolfskill, *The Revolt of the Conservatives* (Boston, 1962). Walter Johnson, *William Allen White's America* (New York, 1947), and Johnson (ed.), *Selected Letters of William Allen White* (New York, 1947), display a salty Republican progressive. For the 1940 G.O.P. campaign, see Mary Earhart Dillon, *Wendell Willkie 1892–1944* (Philadelphia, 1952), and Donald Bruce Johnson, *The Republican Party and Wendell Willkie* ("Illinois Studies in the Social Sciences," XLVI [Urbana, 1960]).

POLITICS

The more important analyses of the politics of the 1930's are J. Joseph Huthmacher, *Massachusetts People and Politics, 1919–1933* (Cambridge, 1959); V. O. Key, Jr., *Southern Politics* (New York, 1949); Samuel Lubell, *The Future of American Politics* (New York, 1951); Roy Peel and Thomas Donnelly, *The 1932 Campaign* (New York, 1935); and Edgar Eugene Robinson, *They Voted for Roosevelt* (Stanford, 1947). Robert Burke, *Olson's New Deal for California* (Berkeley and Los Angeles, 1953), and Richard Keller, "Pennsylvania's Little New Deal" (unpublished Ph.D. dissertation, Columbia University, 1960), are valuable for state politics. Personal memoirs include Tom Connally, as told to Alfred Steinberg, *My Name Is Tom Connally* (New York, 1954); James A. Farley, *Behind the Ballots* (New York, 1938), and *Jim Farley's Story* (New York, 1948); Maury Maverick, *A Maverick American* (New York, 1937); Charles Michelson, *The Ghost Talks* (New York, 1944); and Jerry Voorhis, *Confessions of a Congressman* (Garden City, N.Y., 1948). Some of the leading figures are studied in Fred Israel, "Nevada's Key Pittman" (unpublished Ph.D. dissertation, Columbia University, 1959); Marian McKenna, *Borah* (Ann Arbor, 1961); and Bascom Timmons, *Garner of Texas* (New York, 1948).

SUPREME COURT

Apart from accounts in general works of constitutional history, the best books on the Court in the 1930's are Joseph Alsop and Turner Catledge, *The 168 Days* (Garden City, N.Y., 1938); John Frank, *Mr. Justice Black* (New York, 1949); Eugene Gerhart, *America's Advocate: Robert H. Jackson* (Indianapolis, 1958); Samuel Hendel, *Charles Evans Hughes and the Supreme Court* (New York, 1951); Robert Jackson, *The Struggle for Judicial Supremacy* (New York, 1941); Alpheus Mason, *Brandeis: A Free Man's Life* (New York, 1946), and *Harlan Fiske Stone* (New York, 1956); Dexter Perkins, *Charles Evans Hughes and American Democratic Statesmanship* (Boston, 1956); Harlan Phillips (ed.), *Felix Frankfurter Reminisces* (New York, 1960); and Merlo Pusey, *Charles Evans Hughes* (2 vols., New York, 1951). Few of these works, however, are as penetrating as the better articles in the law reviews, which are the best source for constitutional issues in this period.

THE ECONOMY

No decade of American history stimulated as many economic studies as the 1930's, and the historian has available to him innumerable first-rate studies turned out under the sponsorship either of the National Bureau of Economic Research or the Brookings Institution. A number of these works are cited in the footnotes. An especially lucid Brookings study is Lewis Kimmel, *Federal Budget and Fiscal Policy 1789–1958* (Washington, 1959). The economic journals, especially the *American Economic Review,* are indispensable. Among the syntheses and monographs on the economy in the thirties, some of the more helpful are the fifth volume of Joseph Dorfman, *The Economic Mind in American Civilization* (New York, 1959); John Kenneth Galbraith, *The Great Crash* (Boston, 1955); H. V. Hodson, *Slump and Recovery 1929–1937* (London, 1938); Broadus Mitchell, *Depression Decade* (New York, 1947); Lionel Robbins, *The Great Depression* (London, 1934); Kenneth Roose, *The Economics of Recession and Revival* (New Haven, 1954); Joseph Schumpeter, *Business Cycles* (2 vols., New York, 1939); Henry Villard, *Deficit Spending and the National Income* (New York, 1941); and Thomas Wilson, *Fluctuations in Income & Employment* (London, 1942).

ECONOMIC POLICY

Important analyses or expressions of the new economic viewpoints are Arthur Burns and Donald Watson, *Government Spending and Economic Expansion* (Washington, 1940); Seymour Harris (ed.), *The New Economics* (New York, 1947); Richard Gilbert *et al., An Economic Program for American Democracy* (New York, 1938); Alvin Hansen, *Full Recovery or Stagnation?* (New York, 1938), and *Fiscal Policy and Business Cycles* (New York, 1941); and Rudolph Weissman (ed.), *Economic Balance and a Balanced Budget* (New York, 1940). There is no good history of the NRA or, indeed, of most of the individual New Deal operations. For the NRA, see the early Douglass V. Brown *et al., The Economics of the Recovery Program* (New York, 1934); Leverett Lyon *et al., The National Recovery Administration* (Washington, 1935); Hugh Johnson, *The Blue Eagle from Egg to Earth* (Garden City, N.Y., 1935); and Donald Richberg, *The Rainbow* (New York, 1936). The most significant studies of the money question are Arthur Whipple Crawford, *Monetary Management under the New Deal* (Washington, 1940); Allan Everest, *Morgenthau, the New Deal and Silver* (New York, 1950); G. Griffith Johnson, Jr., *The Treasury and Monetary Policy, 1933–1938* (Cambridge, 1939); James Daniel Paris, *Monetary Policies of the United States 1932–1938* (New York, 1938); Joseph Reeve, *Monetary Reform Movements* (Washington, 1943); and Ray Westerfield, *Our Silver Debacle* (New York, 1936). Monographs on particular topics include Alfred Bernheim and Margaret Grant Schneider (eds.), *The Security Markets* (New York, 1935); Robert Moore Fisher, *20 Years of Public Housing* (New York, 1959); J. K. Galbraith and G. G. Johnson, Jr., *The Economic Effects of the Federal Public Works Expenditures* (Washington, 1940); C. Lowell Harriss, *History and Policies of the Home Owners' Loan Corporation* (New York, 1951); David Lynch, *The Concentration of Economic Power* (New York, 1946); and Timothy McDonnell, S.J., *The Wagner Housing Act* (Chicago, 1957). Invaluable for insights into the economic policies of the New Deal is "Proceedings of the National Emergency Council December 19, 1933—April 18, 1936" (Washington, 1955), on microfilm at FDRL.

RELIEF AND SECURITY

Among the numerous accounts of the experience of unemployment, the best are E. Wight Bakke, *Citizens Without Work* (New Haven, 1940), and *The Unemployed Worker* (New Haven, 1940); Eli Ginzberg, *The Unemployed* (New York, 1943); and U.S. Works Progress Administration, *The Personal Side,* by Jessie Bloodworth and Elizabeth Greenwood, mimeographed (Washington, 1939). For the special problems of youth, the most helpful studies are Kenneth Holland and Frank Ernest Hill, *Youth in the CCC* (Washington, 1942); Betty and Ernest K. Lindley, *A New Deal for Youth* (New York, 1939); and George Rawick, "The New Deal and Youth" (unpublished Ph.D. dissertation, University of Wisconsin, 1957). Paul Conkin, *Tomorrow a New World* (Ithaca, 1959), is a first-rate study of the New Deal communities. Other significant works on social welfare include Edith Abbott, *Public Assistance* (Chicago, 1940); Grace Abbott, *From Relief to Social Security* (Chicago, 1941); Grace Adams, *Workers on Relief* (New Haven, 1939); Louise Armstrong, *We Too Are the People* (Boston, 1938); Josephine Chapin Brown, *Public Relief 1929–1939* (New York, 1940); Searle Charles,

"Harry L. Hopkins: New Deal Administrator, 1933–38" (unpublished Ph.D dissertation, University of Illinois, 1953); Paul Douglas, *Social Security in the United States* (New York, 1936); Hallie Flanagan, *Arena* (New York, 1940); Corrington Gill, *Wasted Manpower* (New York, 1939); Harry Hopkins, *Spending to Save* (New York, 1936); Donald Howard, *The WPA and Federal Relief Policy* (New York, 1943); Harold Ickes, *Back to Work* (New York, 1935); Mirra Komarovsky, *The Unemployed Man and His Family* (New York, 1940); Lewis Meriam, *Relief and Social Security* (Washington, 1946); Edward Ainsworth Williams, *Federal Aid for Relief* (New York, 1939); and Edwin E. Witte, *Development of the Social Security Act* (Madison, Wis., 1962).

AGRICULTURE

The Brookings Institution sponsored a number of studies of New Deal farm policies; the most important is Edwin Nourse *et al., Three Years of the Agricultural Adjustment Act* (Washington, 1937). Gilbert Fite, *George N. Peek and the Fight for Farm Parity* (Norman, Okla., 1954), is an excellent study. Richard Kirkendall, "The New Deal Professors and the Politics of Agriculture" (unpublished Ph.D. dissertation, University of Wisconsin, 1958), is illuminating. Other important studies are Dean Albertson, *Roosevelt's Farmer* (New York, 1961); Murray Benedict, *Can We Solve the Farm Problem?* (New York, 1955), and *Farm Policies of the United States 1790–1950* (New York, 1953); Benedict and Oscar Stine, *The Agricultural Commodity Programs* (New York, 1956); Russell Lord, *The Wallaces of Iowa* (Boston, 1947); Grant McConnell, *The Decline of Agrarian Democracy* (Berkeley and Los Angeles, 1953); Arthur Raper, *Preface to Peasantry* (Chapel Hill, 1936); and Carl Schmidt, *American Farmers in the World Crisis* (New York, 1941).

CONSERVATION

The major collection of sources is Edgar Nixon (ed.), *Franklin D. Roosevelt and Conservation, 1911–1945* (2 vols., Hyde Park, 1957). Other significant contributions are David Lilienthal, *TVA: Democracy on the March* (New York, 1944); Roscoe Martin (ed.), *TVA: The First Twenty Years* (University, Ala., and Knoxville, Tenn., 1956); C. Herman Pritchett, *The Tennessee Valley Authority* (Chapel Hill, 1943); Joseph Ransmeier, *The Tennessee Valley Authority* (Nashville, 1942); Anna Lou Riesch, "Conservation under Franklin D. Roosevelt" (unpublished Ph.D. dissertation, University of Wisconsin, 1952); Philip Selznick, *TVA and the Grass Roots* (Berkeley and Los Angeles, 1953); Kenneth Trombley, *The Life and Times of a Happy Liberal: A Biography of Morris Llewellyn Cooke* (New York, 1954); and Norman Wengert, *Valley of Tomorrow* (Knoxville, 1952)

LABOR

Milton Derber and Edwin Young (eds.), *Labor and the New Deal* (Madison, Wis., 1957), is an important collection of essays. Irving Bernstein, *The New Deal Collective Bargaining Policy* (Berkeley and Los Angeles, 1950). is a model monograph. The fruits of Sidney Fine's research have appeared in a number of articles, notably in the *Mississippi Valley Historical Review* (June, 1958), pp. 23–50; the *Journal of Economic History* (September, 1958), pp. 249–282; and the *Michigan Alumnus Quarterly Review* (1960), pp. 249–259. Other studies of value are Robert R. R. Brooks, *As Steel Goes . . .* (New Haven, 1940), and *Unions of Their Own Choosing* (New Haven, 1939); Robert Christie, *Empire in Wood* (Ithaca, 1956); Walter Galenson, *The CIO Challenge to the AFL* (Cambridge, 1960); Herbert Harris, *American Labor* (New Haven, 1938); Vernon Jensen, *Nonferrous Metals Industry Unionism 1932–1954* (Ithaca, 1954); Alfred Winslow Jones, *Life, Liberty, and Property* (Philadelphia, 1941); Matthew Josephson, *Sidney Hillman* (Garden City, N.Y., 1952); Henry Kraus, *The Many and the Few* (Los Angeles, 1947); Sidney Lens, *Left, Right & Center* (Hinsdale, Ill., 1949); Edward Levinson, *Labor on the March* (New York, 1938); Ruth McKenney, *Industrial Valley* (New York, 1939); James Morris, *Conflict within the AFL* (Ithaca, 1958); Philip Taft, *The A.F. of L. from the Death of Gompers to the Merger* (New York, 1959); Mary Heaton Vorse, *Labor's New Millions* (New York, 1938); and James Wechsler, *Labor Baron* (New York, 1944).

POLITICAL FERMENT

For Huey Long, see Forrest Davis, *Huey Long* (New York, 1935); Harnett Kane, *Louisiana Hayride* (New York, 1941); Allan Sindler, *Huey Long's Louisiana* (Baltimore, 1956); and T. Harry Williams, "The Gentleman from Louisiana: Demagogue or Democrat," *Journal of Southern History,* XXVI (1960), 3–21. The best study of Townsend is Abraham Holtzman, "The Townsend Movement: A Study in Old Age Pressure Politics" (unpublished Ph.D. dissertation, Harvard University, 1952). There is no adequate study of Coughlin, but James Shenton, "The Coughlin Movement and the New Deal," *Political Science Quarterly,* LXXIII (1958), 352–373; Alfred McClung Lee and Elizabeth Briant Lee, *The Fine Art of Propaganda* (New York, 1939); and Lowell Dyson, "The Quest for Power: Father Coughlin and the Election of 1936" (unpublished M.A. essay, Columbia University, 1957), are helpful. For various shades of progressivism, see Alfred Bingham and Selden Rodman (eds.), *Challenge to the New Deal* (New York, 1934); Paul Carter, *The Decline and Revival of the Social Gospel* (Ithaca, 1956); Bruce Bonner Mason, "American Political Protest, 1932–1936" (unpublished Ph.D. dissertation, University of Texas, 1953); Donald McCoy, *Angry Voices* (Lawrence, Kan., 1958); Russel Nye, *Midwestern Progressive Politics* (Lansing, 1951); and Theodore Saloutos and John Hicks, *Agricultural Discontent in the Middle West 1900–1939* (Madison, Wis., 1951). George Mayer, *The Political Career of Floyd B. Olson* (Minneapolis, 1951), is a skillful account. Harper & Brothers published a series of important studies on the Negro, of which the most important is the classic Gunnar Myrdal, *An American Dilemma* (New York, 1944). For radicalism, see Daniel Aaron, *Writers on the Left* (New York, 1961); Irving Howe and Lewis Coser, *The American Communist Party* (Boston, 1957); Murray Kempton, *Part of Our Time* (New York, 1955); August Raymond Ogden, *The Dies Committee* (Washington, 1945); David Shannon, *The Socialist Party of America* (New York, 1955); and James Wechsler, *The Age of Suspicion* (New York, 1953). On the domestic fascists, especially good is Leo Lowenthal and Norbert Guterman, *Prophets of Deceit* (New York, 1949).

INTELLECTUAL CURRENTS

The intellectual history of the 1930's has yet to be written. Among the standard accounts are Maxwell Geismar, *Writers in Crisis* (Boston, 1942); Leo Gurko, *The Angry Decade* (New York, 1947); and Halford Luccock, *American Mirror* (New York, 1940). Works that illuminate the intellectual life of the decade are Harold Clurman, *The Fervent Years* (New York, 1957); Alistair Cooke, *A Generation on Trial* (New York, 1950); Edgar Kemler, *The Deflation of American Ideals* (Washington, 1941); Max Lerner, *Ideas for the Ice Age* (New York, 1941); Cushing Strout, *The Pragmatic Revolt in American History* (New Haven, 1958); and Edmund Wilson, *The Shores of Light* (New Haven, 1952). Representative expressions of New Deal thinking are Thurman Arnold, *The Folklore of Capitalism* (New Haven, 1937), and *The Symbols of Government* (New Haven, 1935); Rexford Tugwell, *The Battle for Democracy* (New York, 1935); and Henry Wallace, *New Frontiers* (New York, 1934).

FOREIGN AFFAIRS

The major sources on foreign affairs are the publications of the Department of State and the materials yielded by the German and Japanese war trials. The annual volumes of *The United States in World Affairs,* edited by Whitney Shepardson and William Scroggs for the Council on Foreign Relations, are particularly helpful. The best accounts of Roosevelt's foreign policy are Robert Divine, *The Illusion of Neutrality* (Chicago, 1962); Donald Drummond, *The Passing of American Neutrality 1937–1941* (Ann Arbor, 1955); and the authoritative volumes by William Langer and S. Everett Gleason, *The Challenge to Isolation 1937–1940* and *The Undeclared War,*

1940–1941 (New York, 1952–53). Other useful studies are Joseph Alsop and Robert Kintner, *American White Paper* (New York, 1940); Allan Nevins, *The New Deal and World Affairs* (New Haven, 1950); and Pierre Monniot, *Les États-Unis et la neutralité de 1939 à 1941* (Paris, 1946). Charles Tansill, *Back Door to War* (Chicago, 1952), and Charles A. Beard, *American Foreign Policy in the Making 1932–1940* (New Haven, 1946), are critical of Roosevelt's policies. Basil Rauch, *Roosevelt from Munich to Pearl Harbor* (New York, 1950), replies to Beard.

The ideas of policy makers are set forth in Herbert Feis, *Seen from E.A.* (New York, 1947); Cordell Hull, *The Memoirs of Cordell Hull* (2 vols., New York, 1948); Nancy Harvison Hooker (ed.), *The Moffat Papers* (Cambridge, 1956); William Phillips, *Ventures in Diplomacy* (Boston, 1952); Francis Sayre, *Glad Adventure* (New York, 1957); and Sumner Welles, *The Time for Decision* (New York, 1944). William Dodd, Jr., and Martha Dodd (eds.), *Ambassador Dodd's Diary, 1933–1938* (New York, 1941), must be used with caution. An astute commentary is Franklin Ford, "Three Observers in Berlin: Rumbold, Dodd, and François-Poncet," in Gordon Craig and Felix Gilbert (eds.), *The Diplomats, 1919–1939* (Princeton, 1953). For Hoover's foreign policy, see Richard Current, *Secretary Stimson* (New Brunswick, 1954); Robert Ferrell, *American Diplomacy in the Great Depression* (New Haven, 1957); Elting Morison, *Turmoil and Tradition* (Boston, 1960); William Myers, *The Foreign Policies of Herbert Hoover* (New York, 1940); and Henry Stimson and McGeorge Bundy, *On Active Service in Peace and War* (New York, 1948).

Volumes which treat special phases of Roosevelt's foreign policy are Robert Paul Browder, *The Origins of Soviet-American Diplomacy* (Princeton, 1953); F. Jay Taylor, *The United States and the Spanish Civil War* (New York, 1956); H. L. Trefousse, *Germany and American Neutrality 1939–1941* (New York, 1951); and Mark Skinner Watson, *Chief of Staff: Prewar Plans and Preparations* (Washington, 1950). For the Good Neighbor policy, see Francisco Cuevas Cancino, *Roosevelt y la Buena Vecindad* (Mexico City, 1954); Alexander De Conde, *Herbert Hoover's Latin-American Policy* (Stanford, 1951); Donald Dozer, *Are We Good Neighbors?* (Gainesville, 1959); Edward Guerrant, *Roosevelt's Good Neighbor Policy* (Albuquerque, 1950); Robert Smith, *The United States and Cuba* (New York, 1960); and Bryce Wood, *The Making of the Good Neighbor Policy* (New York, 1961). E. David Cronon, *Josephus Daniels in Mexico* (Madison, Wis., 1960), is an admirable analysis.

Different aspects of isolationism and pacifism are studied in Selig Adler, *The Isolationist Impulse* (London, 1957); Elton Atwater, *American Regulation of Arms Exports* (Washington, 1941); Edwin Borchard and William

Lage, *Neutrality for the United States* (New Haven, 1940); Wayne Cole, *America First* (Madison, Wis., 1953); Dorothy Detzer, *Appointment on the Hill* (New York, 1948); and Walter Johnson, *The Battle Against Isolation* (Chicago, 1944).

The growing crisis in the Far East is examined in the volumes on Stimson cited above and in T. A. Bisson, *American Policy in the Far East 1931–1941* (New York, 1941); Herbert Feis, *The Road to Pearl Harbor* (Princeton, 1950); Joseph Grew, *Ten Years in Japan* (New York, 1944), and *Turbulent Era,* edited by Walter Johnson (2 vols., Boston, 1952); and F. C. Jones, *Japan's New Order in East Asia* (London, 1954). Paul Schroeder, *The Axis Alliance and Japanese-American Relations 1941* (Ithaca, 1958), is a well-reasoned interpretation.

Index

ing of coal and iron police in Pennsylvania, 187; *see also* Miners, United Mine Workers

Coburn, Charles, 126

Cohen, Benjamin, and Wall Street regulatory legislation, 148–149; and legislation on breakup of holding companies, 155; on recession, 247; and investigation of economic power, 257

Cole, G. D. H., 55

Collective bargaining, part of New Deal, 12; guarantee for labor, 57, 58; cotton code and, 65; under 7 (a) of NIRA, 107; and NLRA, 151

Collective Security, and Hull, 203; and arms embargo, 220; and neutrality legislation, 225; advocates of, and Japan, 290

Collier, John, New Deal and Indians, 86, 329

Collins, William, 109

Committee to Defend America by Aiding the Allies, 310, 324

Committee on Economic Security, to develop social insurance system, 130

Committee for Industrial Organization within AFL, 111

Commodity Credit Corporation, RFC subsidiary, 71; and loans to farmers, 74

Commons, John, 36

Communists, American, and Bonus Army, 13; and unemployed demonstrations, 25; and 1932 elections, 27; and strikes of 1934, 111–114; and Dies, 280–281; strength of, 281–283; and nonintervention, 311

Compton, Arthur, repeal of embargo, 223

Comstock, William, and bank holiday in Michigan, 38

Congress of Industrial Organizations, 111; supports Roosevelt in 1936, 188–189; bid for power, 239; backs Olin Johnston, 268; and Communists, 281, 282; and change, 337

Congress of United States, and sales tax fight in 1932, 4; and "lame duck" session of, 1932–33, 27–28; special session of March 9, 1933,

42–44, 45–46, 61; passes farm bill, 51–52; passes public works and relief bill, 52–53; passes HOLA, 53; passes TVA, 55; passes FDIC, 60; and relief appropriations, 120, 125; abolition of Federal Theatre and Federal Writers' projects, 127; and social security legislation, 131–132; and establishment of Farm Security Administration, 141; legislative program of summer, 1935, 150–163; and "little NRA" legislation, 161–162; and Bonus Bill, 171; and conservation, 173; and civil rights legislation, 186; and Democratic victory in 1936, 196; and tariffs, 204; and neutrality legislation, 222, 224–225; and *Panay* attack, 229; and Roosevelt, 250–251; appropriation for public works, 257; and TNEC, 257; and wages and hours law, 261–263; and New Deal in 1939, 272–273; and reorganization bill, 278; and repeal of embargo, 294–295; appropriation for defense, 300; and legislative functions of President, 327

Connally, Tom, Senator, and reputation of Wilson, 218; and court packing bill, 234

Connally Act of 1935, 162

Conservation, in New York state, 4; of soil, and TVA, 55; and Taylor Act, 86–87; and Roosevelt program, 93; under FERA, 123; Soil Conservation Act, 172–173; activities for, 173–174; and AAA statute of 1938, 255

Conservatives, in 1932 elections, 27; and NIRA, 58; on spending, 84–85; in Democratic party, 92; and work relief, 124; attitudes toward social security legislation, 131; and decisions of Supreme Court, 144; oppose Eccles bill, 159–160; opposition to court packing bill, 234; and strikes, 242–243; and recession, 254, 273; and TNEC, 258; on wages and hours law, 261–262; and Democratic party purge of, 266–270; op-